The Authors

Dr Mary Burgis is a freshwater biologist w
particular interest in planktonic animals. S
studying Lake George in Uganda as part of
International Biological Programme, she ha
a continuing interest in the lakes of Africa. S
helped to organise an international workshop on African Inland Waters, held at Nairobi in 1979, and has recently been involved in a project assembling information on African Wetlands and Shallow Water Bodies. She has been a senior lecturer in Biological Sciences at the City of London Polytechnic since 1973, where she teaches ecology, conservation and freshwater biology, and has extended her research interests to the biology of gravel-pit lakes.

Dr Pat Morris is a lecturer in zoology in the University of London, Council member of the Mammal Society and prominent member of several conservation bodies. He has written many scientific papers for zoological journals and has written or edited a number of popular natural history books, including *The Natural History of the British Isles* (Country Life) and *The Living Countryside* partwork (Eaglemoss). He frequently contributes to BBC and local radio programmes such as *Wildlife* and *The Living World*, and was consultant for the very popular TV film *The Great Hedgehog Mystery.* He has travelled widely with his wife, the co-author, to study and photograph the wildlife of five continents.

Front cover photograph
A shallow lake in southern Alaska: a legacy of the Ice Age

Back cover photographs
Top left: Giant water lily, *Victoria amazonica*
Top right: Bird diversity at Lake Nakuru, Kenya
Bottom left: Alligator
Bottom right: Pike, *Esox lucius*

Designed by Hobson Street Studio Limited

Printed in Great Britain

The Natural History of
LAKES

MARY J. BURGIS
and
PAT MORRIS

with illustrations by
GUY TROUGHTON

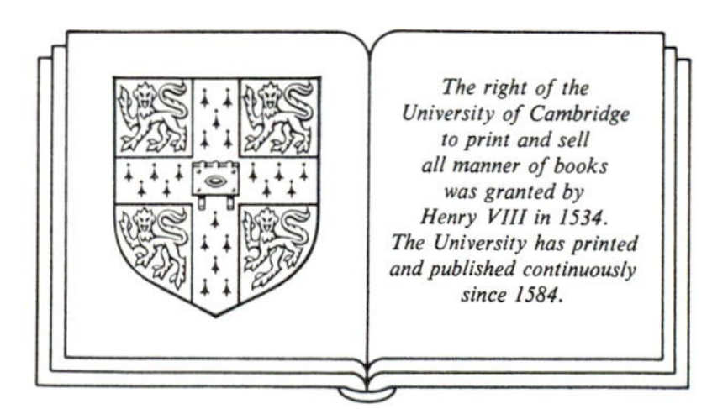

CAMBRIDGE UNIVERSITY PRESS
Cambridge
New York New Rochelle
Melbourne Sydney

Published by the Press Syndicate of the University of Cambridge
The Pitt Building, Trumpington Street, Cambridge CB2 1RP
32 East 57th Street, New York, NY 10022, USA
10 Stamford Road, Oakleigh, Melbourne 3166, Australia

First published 1987

Printed in Great Britain at the University Press, Cambridge

British Library cataloguing in publication data

Burgis, Mary J.
The natural history of lakes.
1. Lakes
I. Title II. Morris, Pat
551.48′2 GB1603

Library of Congress cataloguing in publication data

Burgis, Mary.
The natural history of lakes.
1. Limnology. 2. Lakes. I. Morris, Pat. II. Title.
QH96.15.B87 1987 551.48′2 86-18776

ISBN 0 521 30793 7 hard covers

DS

CONTENTS

PREFACE

Archaeologists, ornithologists, geologists – most people know what these specialists study but who knows what a Limnologist does? Despite its impeccable etymology (*limnos* is Greek for lake), the name for people who study lakes is not widely recognised, even among scientists, and this creates an unfortunate communication blockage. Many limnologists also trade under alternative names which may be more comprehensible – chemist, hydrologist, entomologist, algologist – but these are not substitutes for the word limnologist because they do not encompass the study of the whole lake system.

Naturalists, bird watchers and conservationists sometimes are impatient of the specialist's apparent obsession with detail at the expense of bigger and seemingly more important things. This too is a pity because it is only with a foundation of detailed studies that it is possible to arrive at an informed overview, which is what we have tried to provide in this book. For their part, specialists are sometimes tempted to belittle the subject of Natural History, perhaps because it implies a broad but superficial knowledge which does not accord with their own more detailed but narrow approach. But of course, scientific details are best appreciated in the context of the whole picture.

This book is not written for limnologists but any who do glance at it will recognise many of our principal sources of information. Some of these are listed at the back of the book, but it is appropriate here to say how indebted we are to the numerous colleagues who have painstakingly summarised the work on particular lakes. A most important stimulus for the study of lakes as whole ecosystems was the International Biological Programme which officially lasted from 1966 to 1972, but whose network of limnologists has continued to keep in touch and undertake co-operative projects aimed at synthesising and extending the work done during the Programme itself. MJB was privileged to work on an IBP project in Uganda and to participate in some of the subsequent activities. This experience gave her an enduring fascination with lakes and a desire that their interest and hidden complexity should be understood more widely.

In this book we have tried above all to make science accessible and readable. For this reason we have decided to omit detailed references from the text, but sources of data used for the tables and figures are listed at the end. We hope that our book will be enjoyed by many who are interested in fishing, bird watching, sailing, geology, geography, and engineering, and who want to know more about the lake ecosystems which support these activities, as well as by those who enjoy reading about familiar and far-away places. In trying to provide for them all we are aware of many omissions but make no apology (except if the omission has led to error) since in the end the choice of content is personal. This is our choice of interesting lake ecosystems with which we hope readers will share our fascination.

We are acutely aware that our sins of omission are numerous, because this is a big subject necessarily reduced to a small book. However, we hope that our committed sins are fewer and are grateful to Rosalind Pontin and Marion Jowett for bringing at least some of them to our notice. We are also grateful to Guy Troughton whose illustrations add informed elegance to our book.

1 LAKES in the landscape

Beauty is in the eye of the beholder. In the case of landscape appreciation, some prefer mountains, others the wide open spaces of a prairie; many find beauty in snowbound wastes, while others would choose the scorched sand and rocks of a desert landscape, or the dense complexity of a forest. Whatever your own preference among these extremes, it is clear that for many people lakes enhance the scenic quality of landscapes. This was certainly evident during the days of the great landscape gardeners of eighteenth-century England: streams were dammed and the land moulded to form lakes which would become the centrepieces of grand designs for 'natural' parkland to surround the stately homes of the rich and titled gentry. The lake in the park at Blenheim Palace (where Winston Churchill was born) is one of the most famous; it was created by Capability Brown as part of the palace and park given to the Duke of Marlborough by a grateful nation for his military successes.

Although we shall consider man-made lakes in Chapter 9, we are primarily concerned with natural lakes which not only enhance the landscape but also play a vital role in many people's lives, as well as containing fascinating communities of plants and animals whose lives also depend on the well-being of the lake. Indeed, a lake sometimes seems to behave like an animal itself – critically dependent upon the health and proper functioning of its parts. The central message of this book is that lakes do not end at their shores and cannot be isolated from the land around them. They are profoundly affected by their locality and by changes taking place on land, even at great distances from the lake itself.

A lake is a water-filled hollow in the earth's surface, inland from the ocean. In most countries there are many different words to describe lakes of different sizes or different types, or the lakes in different regions. We shall use the word 'lake' for all of these, but we shall exclude ponds. The distinction between a pond and a lake is not at all clear but most people know intuitively what they would regard as a pond rather than a lake. Lakes are generally bigger and deeper. Most of the lakes we shall be considering are relatively large and almost all have a high proportion of their total area of water free from emergent plants. Some people try to distinguish ponds from lakes by calling a pond any body of water sufficiently shallow to allow light to penetrate to the bottom, and thus allow water weeds to grow all over its area, but this would exclude Lake Okeechobee, the largest lake in the southern United States which is only about 4 m deep and full of water weeds. If we must attempt a definition, let us say that a 'pond' is something you could reasonably expect to wade across in a couple of minutes without getting completely wet (though probably pretty filthy!); a lake is something you would not attempt to cross in this manner.

Seventy percent of the Earth's surface is covered with water. Most of it is in the oceans; fresh water covers only 2%, which is an area of about 2.5 million km^2. Moreover, many lakes do not contain what is generally thought of as 'fresh' water since they often contain many chemicals, referred to collectively as salts. It has been estimated that the volume of inland saline waters roughly equals that of the true freshwater lakes: 104 000 km^3 saline water (i.e. containing more than 3 g of salts per litre), as against 125 000 km^3 fresh water. About one fifth of the world's fresh water is contained in a single lake, Baikal, in Siberia. It has a volume of 23 000 km^3, a maximum depth of 1620 m and a mean depth of 740 m. These three 'vital statistics' tell us a lot about a lake and we can calculate the area by dividing the mean depth

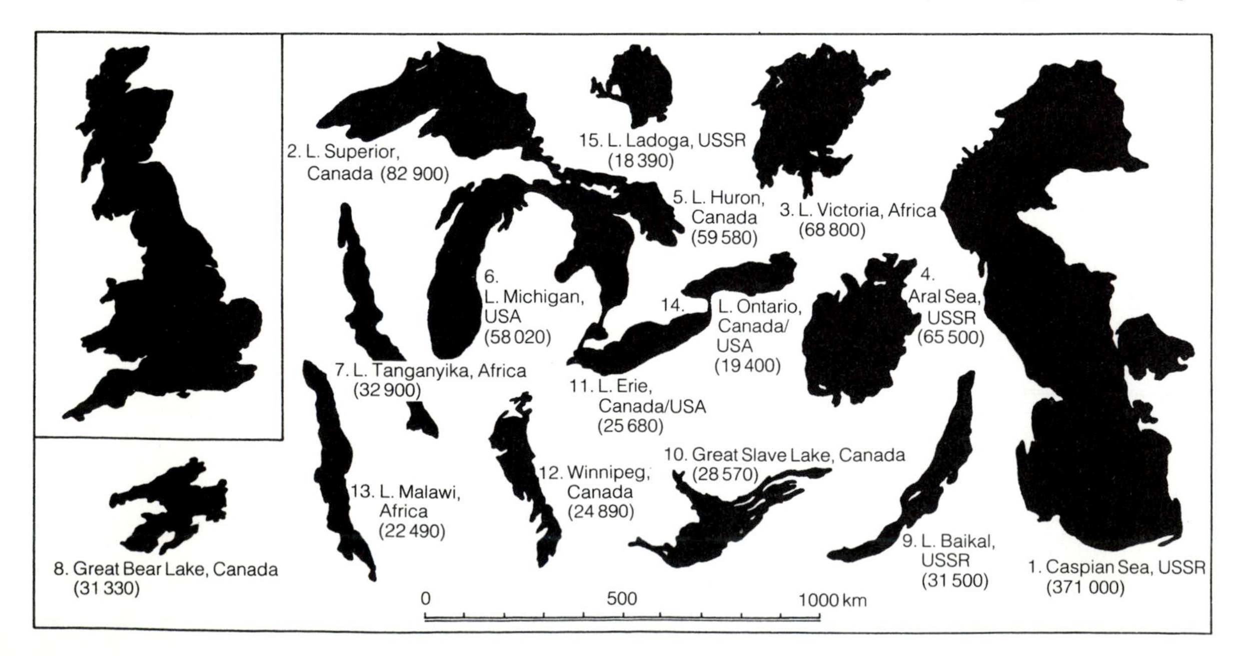

The fifteen largest lakes in the world (and the outline of Great Britain) all drawn to the same scale. The numbers indicate rank and the figures in brackets the surface areas in square kilometres. (After Ruttner, 1963.)

into the volume. Although it is the deepest lake in the world, Lake Baikal is by no means the largest in area. That distinction belongs to the Caspian Sea which is, as its name suggests, more of an inland sea than a lake. It and the Aral Sea both contain saline water and are thought to be remnants of a former marine system which have become isolated from the oceans as a result of geological events. The largest single area of fresh water is Lake Superior, one of the North American Great Lakes. Together, these Great Lakes form the biggest continuous mass of fresh water on Earth, with a total area of 245 240 km^2 (larger than the whole of the Federal Republic of Germany) and combined volume of 24 620 km^3.

Geographical aspects

Lakes are not distributed evenly over the surface of the world. Africa has more than twenty times the volume of lakes and reservoirs found in South America, although the latter receives more than twice as much rainfall per unit area. South America is drained by the Amazon and many other huge rivers but there are no great lake basins like those of Lake Victoria or Lake Tanganyika in Africa. Even within a continent the distribution of lakes is uneven; for example, southern Africa has very few lakes of any size compared with eastern Africa. Both Europe and North America are well endowed with lakes, and Canada contains a disproportionate number of the world's largest lakes as well as many thousands of smaller ones. Great Britain has 5505 lakes, reservoirs and other water bodies more than 4 ha in area (1 ha = 0.01 km^2). Including those less than 4 ha would bring the total to something in excess of 60 000, with an absolute total area of 2451 km^2, which is just over 1% of the land area. There are many more lakes in the north-west than in other regions of Britain and this correlates with predominantly high ground and the areas of highest rainfall. Roughly one half of the 5505 waters counted are located in the Highlands and Islands of Scotland. Loch Ness alone contains more water than all the lakes and reservoirs in England and Wales put together.

Although the area of a lake is its most obvious characteristic from looking at a map or the landscape itself, its other vital statistics are also a part of the landscape in which it lies. In fact, it is essential to realise that although a lake seems to be a clearly defined, self-contained entity, both on a map and on the ground, it is actually an integral part of the whole landscape and geographical area and should always be considered in that context.

The hollow in which water collects to form a lake, the lowest area of the catchment, is the **lake basin**; this is the physical boundary, the sides and bottom, of the lake itself.

The region from which water drains and finally arrives in the lake basin is known as the **catchment area** or **drainage basin** of that lake. The water may drain straight into the lake or first into a collecting system of streams and rivers; it may even have passed through other lakes on the way, but all these are included in the catchment area. The lake itself may be within the catchment area of another lake

Table 1.1 *Number of lakes in various size classes in Great Britain (from Maitland, 1979).*

Lake area	Number
<4 ha	56 176
4–250 ha	4 416
0.25–1 km^2	848
1–4 km^2	177
4–16 km^2	51
16–32 km^2	9
>32 km^2	4

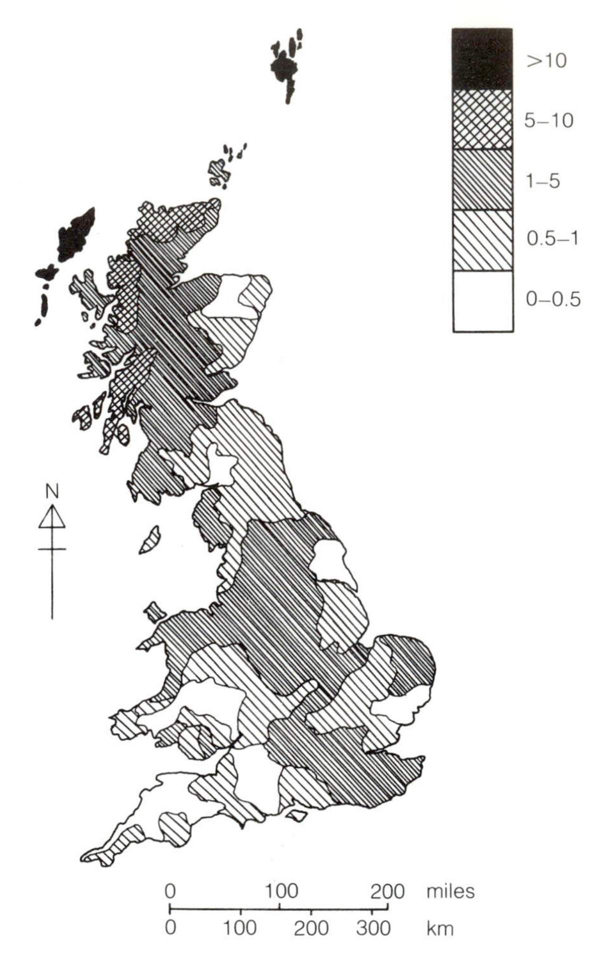

Map of Great Britain showing the density of standing waters per 100 km^2. (After Smith and Lyle, 1979.)

Scotland's largest lochs. *Three of the four largest British water bodies are natural lakes and are the largest lakes (lochs) in Scotland (the other is a reservoir). If the largest lochs are ranked in terms of their lengths, areas, maximum and mean depths, and volumes, and the top twelve taken in each case, then a list of sixteen waters is obtained which, taken overall, are the largest lakes in Scotland. (From Maitland, 1981.)*

Loch	Length (km)	Area (km^2)	Max. depth (m)	Mean depth (m)	Volume (million m^3)
1. Lomond	36	71	190	37	2628
2. Awe	41	39	94	32	1230
3. Ness	39	57	230	132	7452
4. Morar	19	27	310	87	2307
5. Shiel	28	20	128	41	793
6. Lochy	16	15	162	70	1132
7. Treig	8	6	133	63	417
8. Tay	23	19	155	58	1697
9. Shin	28	23	49	16	371
10. Ericht	23	19	156	58	1141
11. Katrine	13	12	151	61	818
12. Maree	22	29	112	38	1156
13. Rannoch	16	19	134	51	1032
14. Arkaig	19	16	110	47	797
15. Glass	7	5	111	49	248
16. Earn	10	10	88	42	433

downstream, which is itself outside the catchment of the lake under consideration. The extent of the catchment and the amount of rain falling on it ultimately determine the volume of water entering a lake. The condition of the inflowing water and the pattern of supply depend very much on the nature of the catchment area, particularly its geology and vegetation, as well as on human activities which can significantly alter the nature and volume of the lake's input as well as the lake itself.

The lake's regional geology is perhaps the most important influence on its life and style. Old rocks tend to be hard and dense (the softer ones having been worn away long ago), are resistant to erosion and so tend to form angular landscapes of steep slopes and jagged peaks. Areas of bare rock occur, from which water runs off quickly. As hard rocks do not wear away readily, so the run-off contains little sediment or dissolved minerals. What goes into a lake in such a region is thus little different from what fell from the sky – just rain water.

By contrast, geologically new rocks include many softer varieties that are readily eroded and washed into lakes as silt and dissolved salts. Moreover, the erosion of such materials leads to the formation of good soils, often used for farming, with further consequences for catchments and the lakes they feed.

The steep slopes formed by hard rocks often continue below the surface of a lake, resulting in a deep, steep-sided lake basin. Softer

The area of land from which a lake receives water is known as its catchment and the nature of the lake depends very much on the type of rocks, vegetation and human activity within its catchment area as well as the shape and location of the lake basin itself.

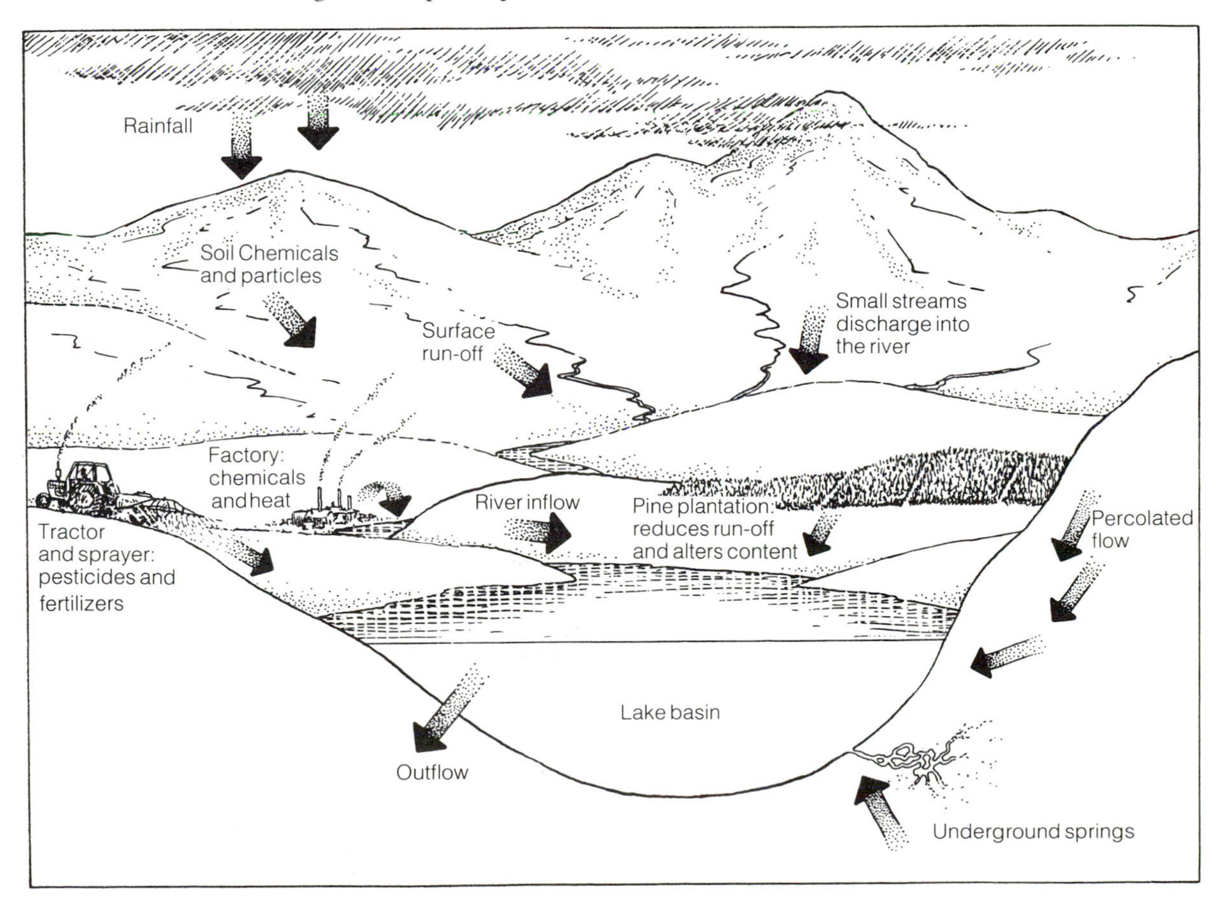

rocks result in gentle slopes and shallower lakes which silt up comparatively quickly, making them shallower still. The depth of a lake is possibly the single most crucial characteristic, as will become clear in later chapters. Thus geology influences not only the landscape but both the appearance and biology of the lakes within it.

In areas of dense impermeable rock, run-off following rain is immediate and considerable. In areas of porous rocks, the ground itself soaks up the water and releases it more slowly.

Not all the water that falls on a catchment ends up in the lake. Plants, especially trees, soak up a lot of it and transpire moisture back into the atmosphere. Forest vegetation and thick soil cover also trap the water from rainstorms, releasing it to the lake as a slow and relatively continuous supply. On bare mountains and moorlands, by contrast, heavy rain runs off more or less directly into the lake and its feeder streams, producing surges of inflow corresponding to the weather conditions. Thus in open landscapes there is a much greater difference between the maximum and minimum daily inputs than in forested areas where the plant communities help to retard inflow and buffer the lake from the effects of sudden downpours of rain.

The underwater landscape

Underwater the 'landscape' of the lake is, to a considerable extent, a mirror of its surroundings. Although the shape of the lake basin is hidden by the water it contains, one can visualise the angle at which the land continues down beneath the water, based on the type of shoreline. Gently sloping shores are likely to lead to gentle slopes underwater in a lake, unless the water level is unusually high and there is a sudden increase in depth not far offshore. By the same logic, where steep mountains or cliffs fall directly to the water's edge it is likely that they continue down below its surface. But only with the aid of detailed soundings can we actually discover the shape of the bottom of a lake basin. In earlier days this would have been done by dropping a weight on a line to the bottom at numerous points on transects across the lake; nowadays more detailed data can be obtained far more quickly by using echo-sounders which not only tell us the depth of the lake but frequently the thickness of sediments on the bottom as well.

Lake evolution

On geological time scales, lakes are transient features of the landscape because, theoretically, even the deepest lakes are gradually being filled in by the accumulation of sediments. In the final stages the lake turns to swamp and then to dry land. This process is known as **succession** and one can see where this has happened in many places. In other cases there seems to be a sort of equilibrium, at least in the short term, where organic components of the sediment are decomposed as fast as they accumulate, or where sediments are washed out of the lake at roughly the same rate as they are brought in. In most lakes the sediments are indeed slowly accumulating and

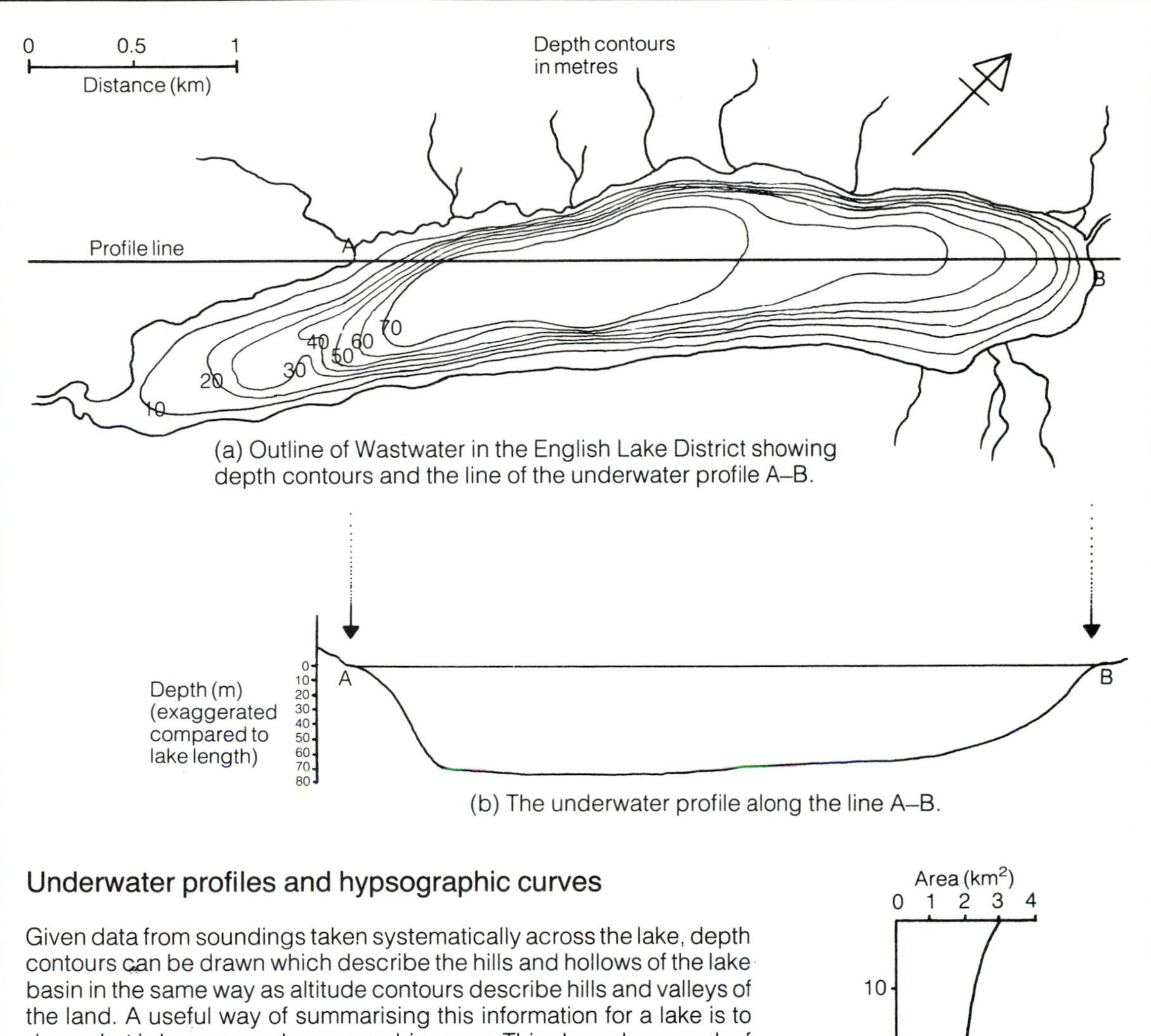

(a) Outline of Wastwater in the English Lake District showing depth contours and the line of the underwater profile A–B.

(b) The underwater profile along the line A–B.

Underwater profiles and hypsographic curves

Given data from soundings taken systematically across the lake, depth contours can be drawn which describe the hills and hollows of the lake basin in the same way as altitude contours describe hills and valleys of the land. A useful way of summarising this information for a lake is to draw what is known as a hypsographic curve. This shows how much of the area of the lake there is at each depth. From this can be calculated a similar curve showing how much of the volume of the lake is in each depth category. These are useful for comparing lakes: for example, in the English Lake District a deep, steep-sided lake such as Wastwater has about 50% of its area over 20 m deep while a shallower lake such as Bassenthwaite has none of its area as deep and only about 20% of its area deeper than 10 m.

From the depth contours we can also see whether the underwater landscape is simple, with only one area of deep water, as in Wastwater, or whether the continuous flat surface of the water hides more than one basin. In Windermere for instance it is easy to guess that there are two main basins because the lake has a narrow 'waist' across the middle, but it is not at all obvious that the shallow (mean depth 3.9 m) Loch Leven contains two much deeper areas known as 'kettle holes' which go down to 23 and 25 m. These holes were formed by the melting of large blocks of ice left behind after the glacier which covered that part of Scotland had retreated. They do not contain a large proportion of the lake's volume nor are they big enough to make the loch function as a 20 m deep lake, but their existence can only be established when the underwater landscape has been mapped in detail.

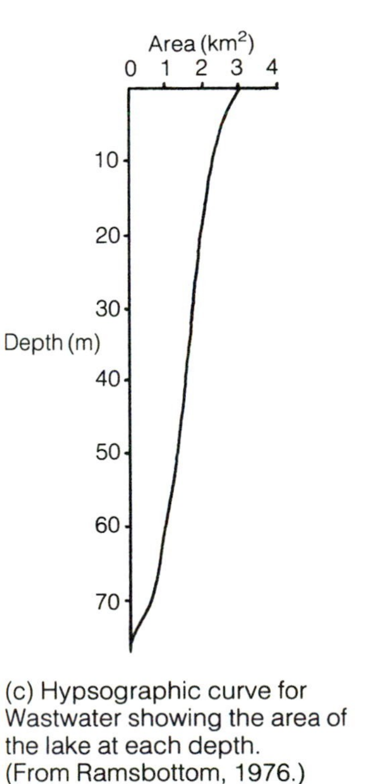

(c) Hypsographic curve for Wastwater showing the area of the lake at each depth. (From Ramsbottom, 1976.)

examination of their sequential layers reveals the history of the lake community and chemistry over the whole period of their existence. The study and interpretation of lake sediments is called palaeolimnology and can reveal fascinating details of the history of a lake against which the effects of contemporary human use can be compared. It is often possible to see not only the history of the lakes themselves, but also that of the vegetation on the surrounding hills, reflected in the pollen which has been preserved in the lake sediments.

Thus the landscape that we see all round a lake not only continues beneath the surface of the water but has a profound influence on the lake itself and, moreover, leaves a record of its history in the sediments accumulating on the bottom of the lake basin.

How lakes are formed

Lakes are formed as the results of a number of different processes, both natural and artificial. The causes which bring a lake into being are often interlinked and frequently result in lakes with a similar origin having similar physical characteristics, even though they are on different continents. There are sometimes biological similarities too, so it is helpful to survey lake types and group them according to their causation.

The formation of a kettle lake; one type of lake formed by glaciers.

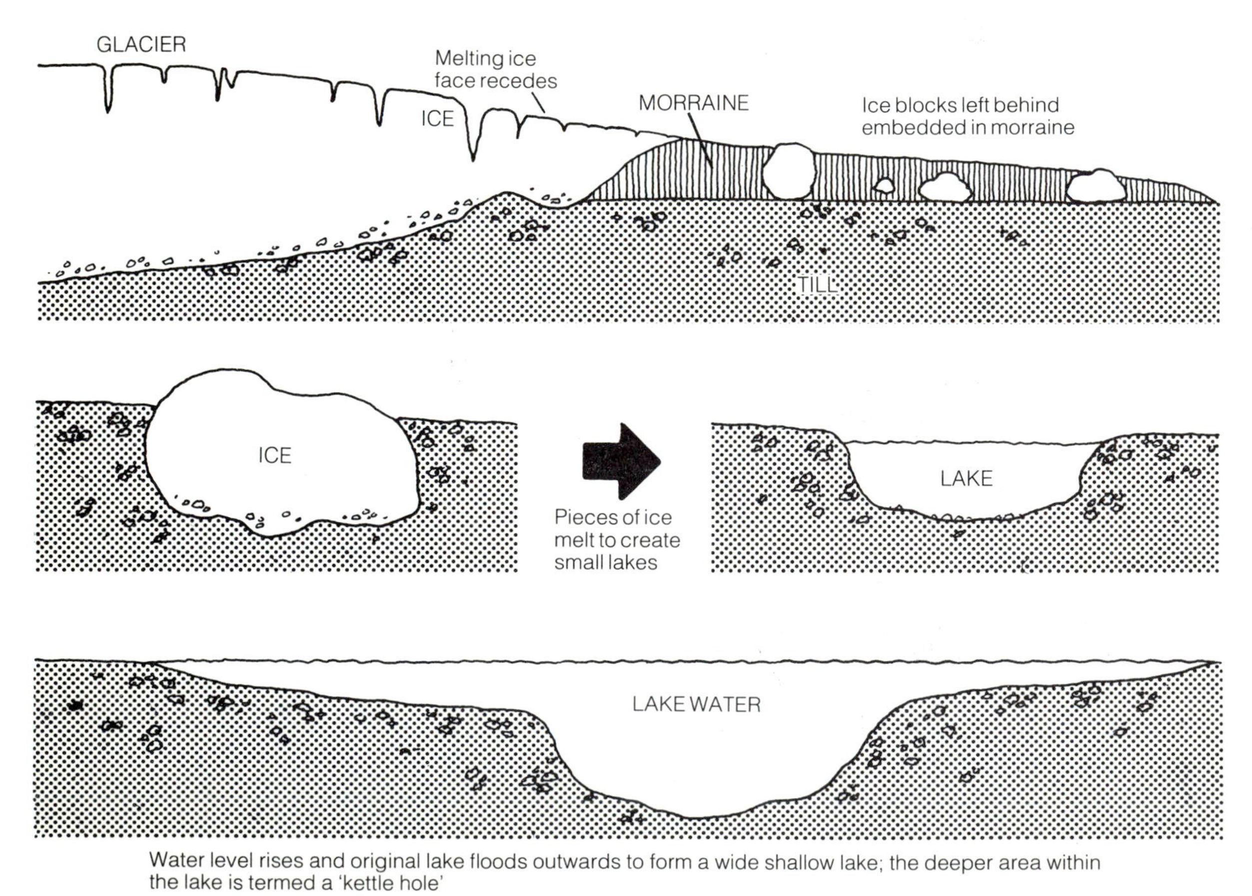

Water level rises and original lake floods outwards to form a wide shallow lake; the deeper area within the lake is termed a 'kettle hole'

Lakes formed by glaciers

The majority of lakes in the North Temperate Zone were formed by the gouging and scraping action of glaciers during the Pleistocene Ice Ages which ended about 10 000 years ago. There is much less land in the South Temperate Zone so the number of lakes formed by glaciers is less in the Southern Hemisphere but there are plenty of examples in New Zealand and in southern South America. Where glaciers followed existing valleys they deepened and widened them; when they retreated they left piles of moraine debris blocking valleys, damming streams and rivers. The formation of such lakes depended on the relationship between movement of the glaciers and the nature of the underlying rock, so they tend to be concentrated in particular areas and have given rise to many famous lake districts in different parts of the world. The English Lake District is one of the best known and most distinctive.

Ice-scour lakes

Where ice sheets moved over relatively flat surfaces of hard but jointed or fractured rock, basins were hollowed out and subsequently filled with water. The lakes of the Canadian Shield were mostly formed in this way. In Britain the small lakes of Sutherland and Wester Ross in Scotland are good examples of this, but the best example in Europe is the landscape of Finland. Here, the action of ice on very ancient rocks created a gently undulating landscape in which almost every dip is filled with water. There are no very high mountains and few deep lakes but an endless mosaic of land and water. Lakes cover a total of 10% of the country and the proportion is higher in the south where the greatest area of lakes is concentrated. Here it is possible to travel by boat for more than 100 km through winding channels and past innumerable islands. Sometimes the lakes are separated by only very narrow ridges of land, the longest of which may stretch for 6–7 km at a width of less than 100 m.

Lakes formed by movements of the earth's crust

Movements such as unwarping or faulting in the earth's crust itself caused the formation of basins sometimes known as tectonic lake basins. Different types of crustal movements form different kinds of lakes, so lakes formed in this way have rather less in common than those formed by glacial action.

When a portion of the earth's surface subsides in relation to its surroundings or, conversely, the sides are uplifted, a lake basin may be formed. The area and depth of water will depend on the amount of drainage from the surrounding land which will, in turn, depend on the rainfall and the size of the area from which water drains towards the lake rather than away from it into another basin. The great lakes of eastern Africa are probably the most spectacular example of lakes formed as a consequence of major movements in the earth's crust.

The system of faults which run from north to south through Africa forms a series of splits in the earth's crust which have filled with water in places to form some of the world's largest and deepest lakes. The Eastern Rift extends as far north as Israel, where it contains Lake Tiberias and the Dead Sea; Further south it is filled by the Red Sea. In Ethiopia parts of it are occupied by Lakes Zwai, Abiata, Shala, Abaya, Chamo and several others (the Ethiopian Rift Lakes). Lake Turkana in Kenya is in one of the lowest portions of the Rift Valley which continues southwards right through Kenya and Tanzania, where there are shallow saline lakes in basins with no outflows ('endorheic' basins).

To the west lies another split in the African continent, the Western Rift. On the floor of this valley lies Lake Mobutu Sese Seko (formerly Lake Albert), then Lakes Edward and George, all of which drain North to the Nile. South of Lake Edward the Rift Valley is blocked by the Virunga volcanoes and their associated lakes. Further south lie Lake Kivu and Lake Tanganyika, the second deepest lake in the world. We will consider Tanganyika in more detail in Chapter 6

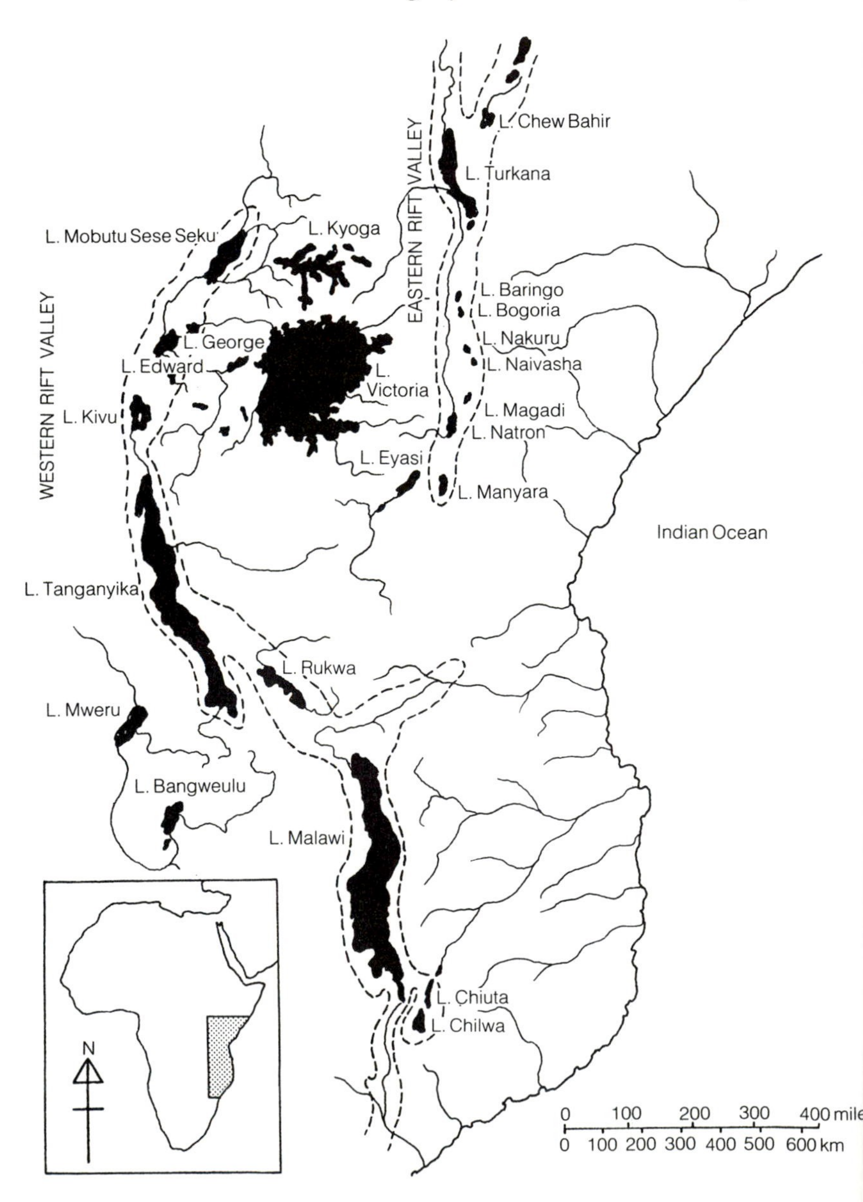

Eastern Africa, showing the Rift Valleys (dotted lines) and their lakes. Lake Victoria lies on the plateau between the two rifts.

but we note here that it completely fills the valley, which is 20–50 km wide at this point, and that the escarpments of the Rift rise straight from the water. They fall just as steeply down into the lake so that only at the northern and southern ends is there any shallow water. The floor of the lake is nearly 700 m below sea level and the lake has a maximum total depth of 1470 m.

At the junction of the Eastern and Western Rifts, the northern end of Lake Malawi lies in a separate Rift. It too is a very deep and steep-sided lake. Although not all Rift lakes are by any means the same, the nature of the crustal movements which caused their formation is very obvious in these cases. The Rift Valleys of Africa are relatively recent structures and their escarpments are not yet eroded to form gentle slopes. Indeed anyone who has travelled overland across East Africa from the coast towards the centre of the continent will be very aware of climbing first to the edge of the Eastern Rift, then travelling precipitously downwards before crossing the flat floor of the valley, and then toiling laboriously up the opposite escarpment on to the central plateau.

On this plateau is one of the largest areas of fresh water in the world, Lake Victoria. It is a huge, relatively shallow lake lying in a saucer-shaped basin and we shall discuss it in detail later in this chapter. The water of Lake Victoria flows out to the north, via Lake Kyoga, to the Nile. Lakes Kyoga and Kwania in central Uganda were formed by the ponding caused by upward movement of the Rift escarpments to east and west.

Where land has dropped between two parallel faults the resultant flat-bottomed valley is often called a '*graben*' and a lake formed between the two parallel faults is termed a '*graben* lake'. Lake Tahoe in the United States (see p. 204) is a good example; although 501 m deep it has a flat bottom and very steep sides. Two of the deepest lakes in Europe, Lakes Ohrid and Prespa are also *graben* lakes. The sides of Lake Ohrid are, in places, so steep that the water is 230 m deep only 300 m from the shore. Its maximum depth is 286 m. Indeed many of the world's deepest lakes are *graben* lakes, including Lake Baikal, the deepest of all. Other *graben* lakes include Lakes Balkhash, and Issyk-kul in Asia (the latter being 702 m deep), Lake Matana in Sulawesi (590 m deep) and Lake Torrens in Australia.

Lakes formed by volcanoes

Many lakes owe their origin to volcanic eruptions. Perhaps the easiest to visualise is a crater lake. While extinct volcanoes can be recognised by their distinctive shape, unless you climb to the top or fly over in an aeroplane, you cannot see that there is often a steep-sided lake basin inside. Crater Lake, in Oregon, is one of the most splendid examples of this type of lake. It is almost circular, about 10 km in diameter, and the deep blue water is more than 600 m deep. The original volcano, Mount Mazama, was much higher (about 4000 m) than the present mountain, which reaches about 2300 m. It erupted some time within the last 7000 years and spread lava over the surrounding countryside up to 55 km away. So great was the quantity of molten rock thrown out that a great void was left inside the mountain and the cone collapsed inwards (forming a **caldera**) which filled with water. In places the present rim still rises straight up to more than 500 m above the water. The lake is fed only by rainfall and run-off from the crater walls. This is almost exactly balanced by losses from evaporation and seepage, so the water level remains almost constant. The whole magnificent catchment is protected as a National Park.

Crater lakes are also found in areas where there have been a series of relatively minor volcanic explosions fairly close to each other, rather than one massive eruption. A good example of this is found in western Uganda in the southern foothills of the Ruwenzori Mountains. Here there are more than 80 small lakes of various depths and types of water, many in the bottom of classically conical craters. In some the rims are not high or have been eroded so that their volcanic origin is less obvious but they are all thought to have been formed during the same period of intense volcanic activity about 8000–10 000 years ago. Similar groups of lakes are found in the Eifel district of the Federal Republic of Germany and the Auvergne district of France. Most of the lakes in the North Island of New Zealand were formed by volcanic action (in contrast to the South Island where most of the lakes are of glacial origin) and include Lake Taupo, the largest. Lake Rotomahana occupies a series of confluent craters and this gives it a highly irregular outline. The present lake on this site was formed as recently as 1886, after a period of intense local volcanic activity.

In other cases the volcano was so big and exploded so long ago, that the floor of the enormous crater is relatively flat and water only accumulates in the lowest places, as it would in any other landscape. A good example of this is found in one of Africa's most famous wildlife areas, the Ngorongoro Crater in Tanzania, which contains several lakes, some fresh and some saline. The caldera itself is more than 14 km across and 600–700 m deep, one of the world's largest.

Volcanic activity also forms lakes when the ash or lava from an eruption blocks natural drainage down a valley or when the formation of the volcano itself does so. A fine example of this can be seen where the eruption of the Virunga volcanoes, which mark the borders between Uganda, Rwanda and Zaire, caused the formation of the Kigezi lakes in Uganda and the lakes of Northern Rwanda when volcanic ash blocked steep-sided valleys. Above the lakes the hills are now covered in terraces on which crops are grown and only relatively small areas of the original forest remain. This spectacularly scenic landscape owes its appearance to the material thrown out by the volcanoes whose seven cones stride across the horizon from east to west. Their crowns are still forested and strenuous efforts are being made to conserve what remains as the last stronghold of the mountain gorilla.

Lakes formed by rivers

When the flow of a river is suddenly slowed by a decrease in gradient it takes a more winding course than hitherto and meanders across the land at a more sedate pace. Sediments are deposited on the inside of bends and eroded from the outside so that the channel becomes progressively more sinuous. Sometimes the river breaks through the narrow isthmus between two succeeding curves. This short-cut then leaves a loop of the river off to one side of the new water course. The isolated portion of the channel may retain enough water to form an 'oxbow' lake. These are the lakes known as 'billabongs' in Australia. Other lakes may be formed from hollows in the valley floor which are filled by the river when it overflows its banks during times of flood. The flat floor of a wide valley which is periodically flooded by the river is known as the floodplain.

Most floodplain lakes are shallow (less than 4 m) and variable in extent, particularly if they are in communication with the river. For example, the Grand Lac in the Delta of the Mekong River has about 2500 km^2 of open water in the dry season but covers up to 11 000 km^2 at the height of the flood season. The water in the channel which connects it to the river flows first one way and then the other, as the relative water levels of river and lake alter.

The area of permanent water varies greatly between floodplains. Within Africa there is a range from the Kafue floodplain, where lagoons occupy only 7% of the total area of permanent water on the floodplain in the dry season (the rest is swamp), to the Inner Delta of the River Niger, where 77% of the 3877 km^2 of low-season water is lagoons and lakes. The area of water on this floodplain increases five-fold during the wet season.

Since floodplain soils have been deposited by the river over many centuries, they are usually very fertile areas and people have settled in such places since the beginning of civilisation. Indeed civilisation in the sense of permanent settlements supported by agriculture was first established in the so-called 'Fertile Crescent' floodplains of the Tigris and Euphrates Rivers of the Near East. Once the floodplain has been settled and the soil cultivated, floods become a threat and strenuous efforts are made to confine the river to its channel and to control the flow of water. In Europe the Danube is the only remaining major river whose floodplains are still naturally subject to extensive flooding. Most of the major seasonal floodplains now occur in tropical regions and they are increasingly modified by development programmes.

The River Amazon is still entirely uncontrolled, although dams are now being built on some of its tributaries. The floodplain, called the *várzea*, covers 60 000 km^2 and contains thousands of lakes of considerable variety. They range from circular to dendritic in shape and from less than 100 m to more than 50 km in length. Because it is

A diagrammatic cross-section of the Amazon Valley to show how *várzea* lakes are formed and how their depths and areas increase when the river is in flood (after Lowe-McConnell, 1975.)

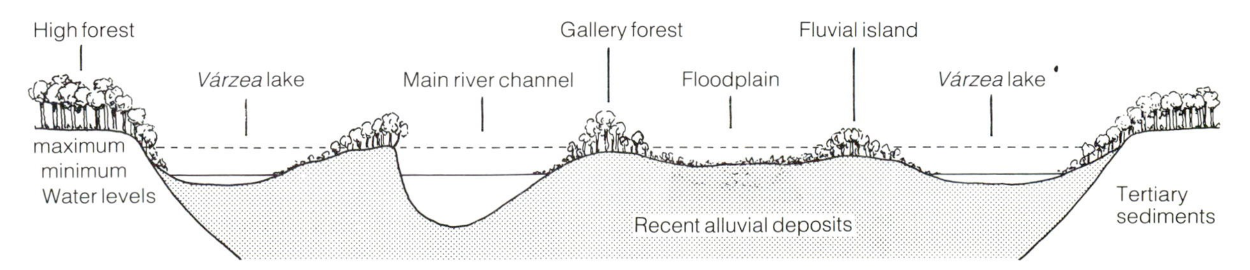

impossible to examine more than a few of these lakes at ground level, a survey was made using aerial photographs to establish their basic characteristics. Of more than 8000 discernable lakes, 10% were less than 250 m across and 10% more than 7 km and up to 60 km in length. The longest straight-line distance over open water (the 'fetch') gives an indication of how likely it is that the wind will be able to mix the water of the lake: the longer the fetch, the more likely is surface mixing by wind. This is particularly true in this region because the topographic relief of the floodplain is low and trees are frequently the highest structures around; so the differing lengths of these lakes also contribute to their biological diversity through the effects of wind-mixing of their waters. The size of these lakes also varies greatly with water level. For instance, the area of Lake Calado, near Manaus, changes from 2 to 8 km^2 seasonally.

Floodplain lakes vary seasonally not only in area but also in depth. At the low-water season they are mostly shallow lakes which are well mixed by the wind, but in the flood season they can become sufficiently deep to stratify (see Chapter 4) so that the water is much warmer at the top than at the bottom. The chemical composition of the water and the supplies of nutrients change as the flood rises and subsides. Large areas of semi-terrestrial vegetation become incorporated into the lake ecosystem during the flood, and fish (plus other animals) gain access to the lake that may be denied them for part of the year. Thus these lakes are typified by their wide range of seasonal variation, the extent of which depends on the size of the flood in any one year and their position relative to the river.

Coastal impoundments

In many parts of the world there are lakes which share the distinctive characteristic of being freshwater lagoons adjacent to the sea. They are often separated from the marine environment by a strip of sand or shingle only a few metres wide – an apparently precarious defence against becoming just another part of the ocean.

These freshwater lagoons are there because the passage of water from a river to the sea has been blocked by the sand or shingle barrier, a permeable dam through which water can seep but not flow. Provided the input of fresh water to the lagoon is sufficient, seepage outwards will prevent salt water getting in. A good example is Slapton Ley on the south coast of Devon (England) where the longshore drift (coastal current) has swept a shingle barrier across a wide bay, impounding the freshwater lagoon. Reed beds and freshwater fish survive only 50 m from the sea.

Bigger systems can develop behind barriers formed by sand dunes. Dunes develop from wind-born sand, picked up from sand banks exposed to the sun and wind at low tide and carried landwards by onshore winds. As the wind meets the rising beach, strandline debris or other obstructions, it is slowed and dumps its load of sand. In turn this slows the winds and traps more sand, gradually building a dune ridge. If the dunes form at a river mouth, the river can soon become blocked, especially in the dry season when its flow is so reduced that it cannot sweep away the sand. A coastal lake then forms.

Along the Indian Ocean coast of southern Africa, there is a series of lake systems separated from the sea by some of the highest dunes in the world. At Lake Sibaya, which is completely cut off from the sea, they reach 134 m in height, but further south at St Lucia they rise to 183 m. These dunes were probably formed when the sea level was considerably lower than it is now and they are covered with forest except where it has been cleared for sand mining. The Lake St Lucia system is still connected to the sea by a narrow channel up which salt water penetrates to mix with the fresh water flowing outwards. The boundary between the two fluctuates both with the tides and with the amount of water in the rivers flowing into the lakes.

Where lakes dominate the landscape

If we discount the Caspian Sea, which is truly an inland sea containing saline water and animals mostly derived from marine stock, the greatest areas of fresh water in the world are found in North America and East Africa. The basins of the Great Lakes of North America were formed by glacial action but they are so huge that the finger-like shape and smooth sided, U-shaped valleys so characteristic of smaller glacier lakes is less evident. Nevertheless, the ice gouged out deep, steep-sided basins which now contain the largest expanse of fresh water on earth. In East Africa the lakes were formed by movements of the earth's crust, most of them by faulting but the largest, Lake Victoria, by associated but different movements. The size of these lakes makes it difficult to see the overall picture except from a satellite far out in space. To earth-bound people there is little difference between the beaches of lakeshore and a sea-side except that the water is fresh. The water stretches beyond the far horizon; there is no other side visible.

The Great Lakes of North America

The basins of these five lakes were gouged by the successive advances and retreats of one huge ice sheet formed during the Wisconsin Glacial Period. At its greatest extent this glacier reached as far south as St Louis, about 18 000 years ago. Thereafter it retreated and the lakes gradually assumed their present shapes as the earth's crust, released from the enormous weight of more than 1.5 km thickness of ice, slowly rose. The lakes assumed their present form only about 10 000 years ago and are therefore contemporary with other glacial lakes at similar latitudes, like those of the English Lake District and the alpine lakes of Europe. The details of their shorelines are still changing as cliffs are eroded and sand spits lengthened; but these are cosmetic adjustments to the basic plan.

While waves roll up the beaches, ocean-going boats ply between huge ports that spill their effluent into the lake water. Some of the largest industrial cities in North America lie on the shores of the Great Lakes: Chicago on Lake Michigan and Detroit on the inflow to Lake Erie. Lakes Erie and Ontario, the lowest of the five lakes, are the most affected by urbanisation and this will be discussed further in

Chapter 10. Lake Superior is relatively untouched and most of the landscape along its shore is still in its natural state.

The Great Lakes receive water from a total area of nearly 750 000 km^2, equivalent to almost the whole of Great Britain and France put together. None of this area is particularly high or mountainous but it collects enough water to keep about 7000 m^3 flowing out of Lake Ontario every second, into the St Lawrence River. The land from which it drains is very varied. To the south of the lakes and on the tongue between Lakes Ontario, Erie and Huron it is covered with moraines left by the retreating glacier. This gives softly undulating country with deep soils and rounded hills. To the north and along the northern shore of Lake Superior lies the southern edge of the Canadian Shield which is composed of very ancient, hard rocks often exposed on the surface. It is on this land that the northern hardwood forests give way to the great boreal coniferous forests.

The lake shorelines reflect these geological differences and range from flat, swampy ground around Lake Erie, at whose western end it is difficult to tell where lake ends and land begins, to the sheer rock cliffs that rise in places more than 200 m above the northern shore of Lake Superior. In between there are extensive areas of sand dunes on the eastern shores of Lake Huron and at the eastern end of Lake

The Great Lakes of North America. Water flows from west to east, to end up in the St Lawrence River and, ultimately, in the Atlantic Ocean.

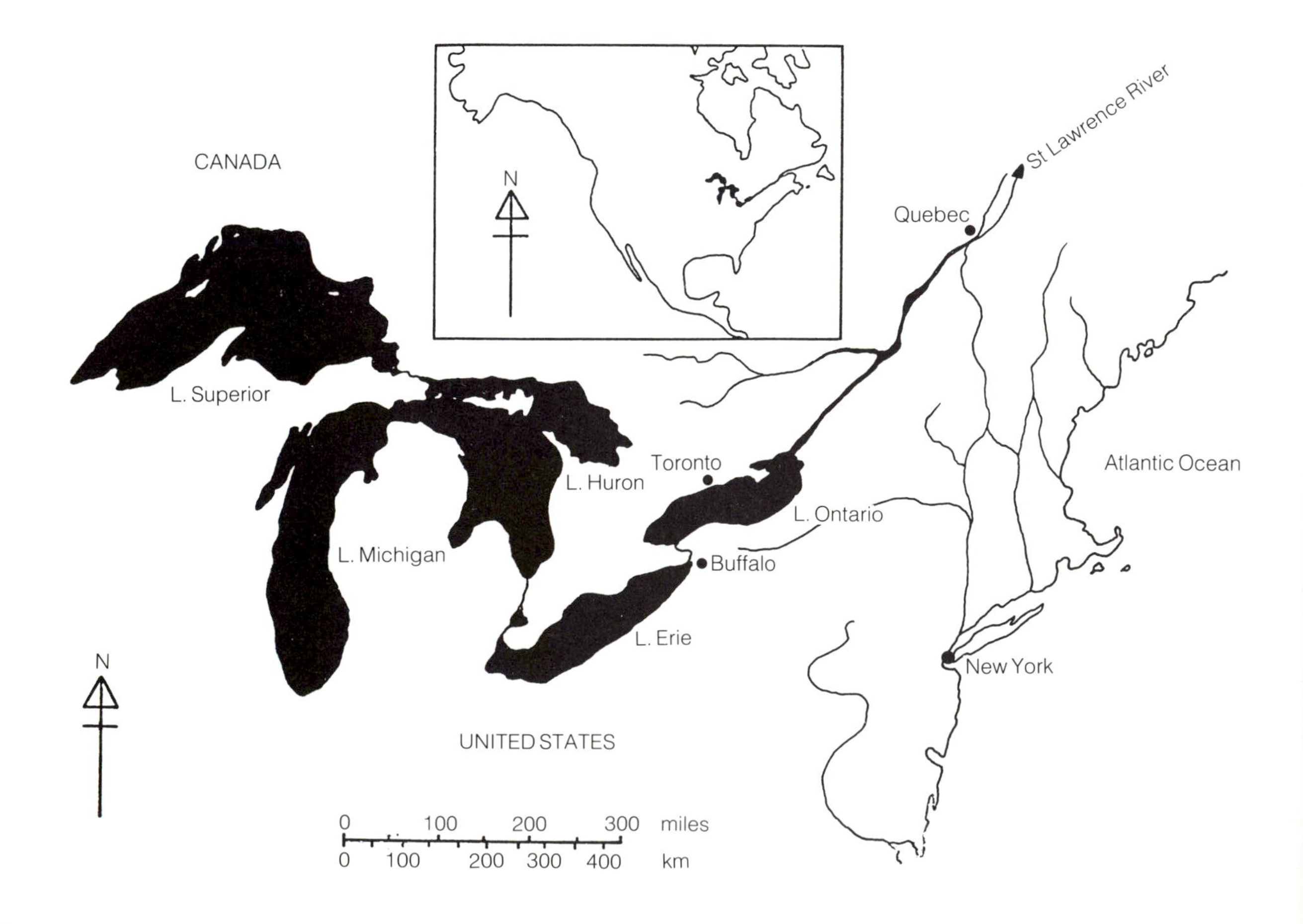

Ontario as well as smaller areas of dunes elsewhere. True tides (caused by the combined influence of the sun and moon) cause changes in water level of only about 3 cm at most; much larger changes in level are caused by winds and changes in barometric pressure. These result in regular 'slopping' of the lake water in its basin, a phenomenon known as a **seiche**. In Lake Erie seiche movement may be as much as 2.4 m up at one end and down at the other, within 14 hours. Storms cause even larger, sometimes catastrophic changes in level and even tidal waves. The lakes offer good conditions for the formation of waves; the region has strong winds, to which huge uninterrupted areas of deep water are exposed. Waves up to 6 m high have been recorded out on the open lake which is truly comparable with ocean conditions.

North and south of the Great Lakes, the landscape is covered with a myriad of smaller lakes gouged out by the force of the ice. Several of these, such as Lake Winnipeg, The Great Slave Lake and Great Bear Lake are among the world's largest lakes but most are very much smaller, particularly to the south in Wisconsin and Minnesota. The latter is truly the 'Land of 10 000 lakes' as its car licence plates proclaim. These lakes are not dotted at random over the landscape but are arranged in a series of arcs which mark each advance and retreat of the glacier. As it advanced it pushed a new pile of debris in front. This 'moraine' was left behind as the glacier retreated, damming the flow of meltwater and thus forming a new lake. If the next advance did not reach so far as the previous one, the lake was not obliterated but remained between one moraine ridge and the next.

Lake Victoria

As we have seen earlier, East Africa has many large lakes, particularly those lying in the Western Rift Valley. Lakes Tanganyika and Malawi are among the deepest lakes in the world and will be discussed elsewhere in this book, as will some of the much smaller and highly saline lakes which lie in the Eastern Rift Valley. But Lake Victoria, like the Great Lakes of North America, is so large that it is the dominant feature of its region. Moreover, it gives us an example of how not only the geographical features of the lake, but also the evolution of its fish community have been influenced by the processes which formed the lake.

Lake Victoria has an area of about 75 000 km^2. It has a long indented shoreline of varied structure, much of it swampy, and is surrounded by gently undulating landscape of low relief. The lake formed when the western edge of the Eastern Rift and the eastern edge of the Western Rift were gradually pushed upwards as the Rifts formed, and the land between them bent to form the basin in which the lake now lies. To the west this upwarping interrupted the passage of rivers which were previously flowing from east to west and draining into the Western Rift. They were now forced to reverse their flow and the eastward-flowing Katonga and Kafu Rivers, which now drain into Lake Victoria, both rise in swampy watersheds from which rivers drain in both directions. The Kagera River, now the biggest inflow to Lake Victoria, probably also reversed during the same period (probably about 25 000–30 000 years ago) and now receives large amounts of water from the Virunga volcanoes which block the Western Rift Valley in Burundi and Rwanda. There is evidence from raised beaches that the level of the lake was formerly much higher than at present. The lowest of these beaches, 3 m above the present lake level, has been dated by radio-carbon techniques to 3700 years before present. Although such large landscape features as Lake Victoria seem relatively stable within a human lifespan, changes are still occurring: in the early 1960s the level of the lake rose more than 2 m and flooded large areas of adjacent land. This was caused by increased rainfall on the lake itself, rather than further movement of the earth's crust; it has still not completely returned to the previous level. A lake as large as Victoria actually influences the local weather as well as being subject to the climatic regime of its catchment.

This immense saucer-shaped basin has a maximum depth of only 80 m and contains a complex community of which the fish are the best known components. They are zoologically interesting and also of major importance as food for the local people. More than 350 species of fish have been identified from Lake Victoria, the great majority of which belong to the family Cichlidae and are endemic to the lake (i.e. they are found nowhere else). Although the total number of fish species in the lake has still not been determined, they must all have evolved from ancestors that lived in the rivers which crossed the present basin. There are still thirteen species in the lake which are migratory: they go up the inflowing rivers to spawn and the young return to feed and grow in the lake. The other fish remain in the lake to breed.

None of the migratory species is endemic; they are found in rivers elsewhere and so are presumably much the same as the original species found in the rivers when the lake was formed. It is the cichlids that have evolved most since the formation of the lake basin, particularly species of the genus *Haplochromis*. They have adapted from the basic habit of eating midge larvae, which live in the bottom mud, to eating a wide variety of foods. They eat zooplankton, molluscs and other fish, and even steal the eggs out of the mouth of brooding parents and scrape the scales off the backs of other fish. Some species may be further separated by the actual prey species they consume or by where they hunt. For example, of two species which feed mainly on snails, *H. sauvagei* does so in shallower water than *H. prodromus* although there must be a good deal of overlap. Other species may be separated in the time at which they utilise the same resources in a particular location but nevertheless it seems certain that very similar species must be able to coexist. The proportion of species which eat other fish is remarkably high (about 40%) and, however much each predator specialises on particular prey species, there must be considerable overlap between feeding habits since it is unlikely to be a one predator to one prey relationship throughout the community.

There is evidence from a small lake close to Lake Victoria, Lake Nabugabo, of how rapidly cichlids can form new species. This lake was separated from Lake Victoria by a sand bar whose age was determined as 3700 years. Of the nine species of cichlid found in the lake, four are also found in Lake Victoria, but five (all *Haplochromis*), are found only in Lake Nabugabo although they do have close relatives, from which they have presumably evolved, in the main lake.

The evolution of this enormous variety of fish in Lake Victoria is thought to have been encouraged by fluctuations in water level during its formation. During dry periods there may have been several smaller lakes in the present basin within which isolated groups of fish evolved into new species. When the level rose they came together in the larger lake. Moreover, the great size, indented swampy shoreline and extensive littoral zone of the present lake also provide innumerable different habitats and locations to accommodate the complex community. Thus the evolution of a lake's community, as well as its shape and productivity are a result of its landscape history.

Much of this unique and fascinating community is now threatened with extinction because a large carnivorous fish, the Nile perch (*Lates niloticus*) was introduced to the lake just over 20 years ago. In that short time this predator has spread almost throughout this huge lake and in the process eaten many of the smaller *Haplochromis* species; some may have disappeared even before they had been described. This sad story will be discussed in more detail in Chapter 10.

Lands of many lakes

In many parts of the world there are regions where geological, or hydrological, events have left a legacy, not of huge lakes which dominate the landscape, but of many small lakes relatively close together. Such lake districts are particularly evident in areas affected by intense glacial activity and, where this has occurred in mountainous country, a landscape has formed which combines lake-filled valleys with hills and mountains to give scenery of quite spectacular beauty. This makes them particularly attractive to holiday-makers. The lakes of the European Alps have been popular tourist areas for centuries and the English Lake District, originally brought to public attention by the poems of Wordsworth and then made more accessible by the arrival of the railway, is now packed with visitors every year. The Southern Alps in New Zealand also have glacier-formed valleys now filled with lakes.

The English Lake District

Often the factors which result in the formation of lakes cause several to develop in the same geographical region. However, just because they all owe their origin to a common cause it does not mean that they are identical. Indeed, individual lakes in such a group may differ significantly in size, shape and ecology. A particularly good example is provided by the lakes of Cumbria which, with their spectacular mountainous surroundings, form a region known collectively as the English Lake District.

The core of the Lake District is a dome of old hard rocks from which a number of river valleys drain in a radial pattern like the spokes of a wheel. This pattern was already in existence before the

The English Lake District showing the radial pattern of lakes whose basins were formed by glaciers flowing outwards and downwards from the central dome. The table lists some of their physical characteristics.

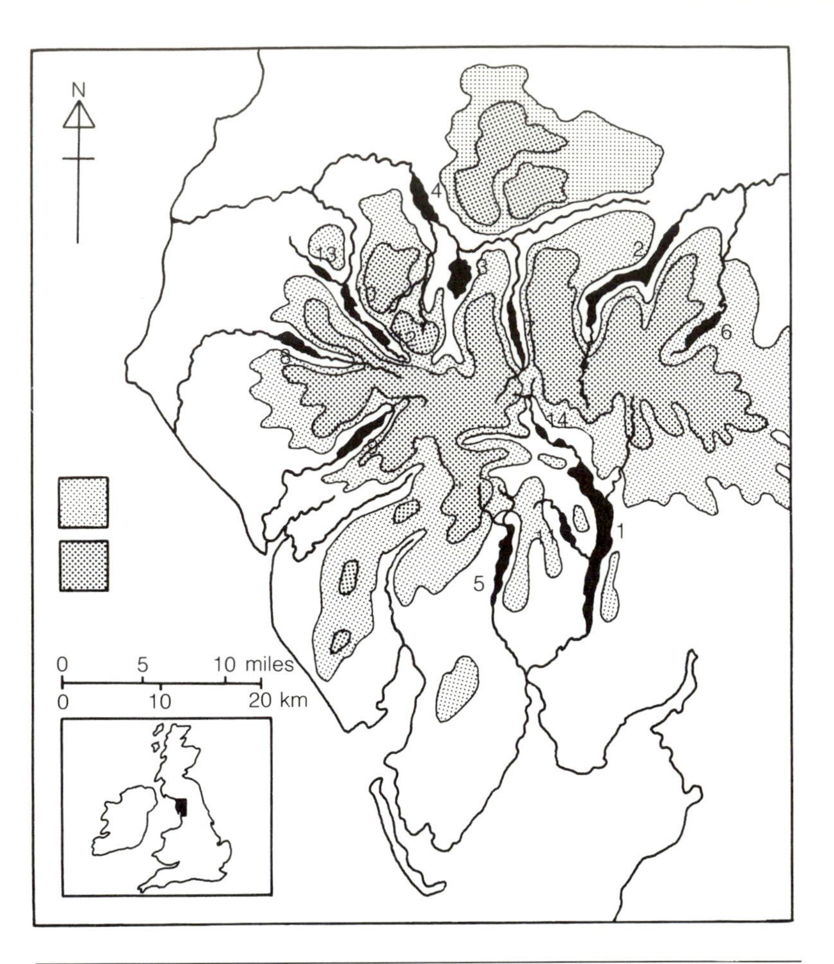

Data for some of the largest lakes in the English Lake District (from Ramsbottom, 1976)

	Area (km^2)	Length (km)	Max. depth (m)	Mean depth (m)	Volume (m^3) (10^6)	Area of drainage basin (km^2)
1. Windermere	14.8	17	64	21.3	314.5	230.5
2. Ullswater	8.9	11.8	62.5	25.3	223.0	145.5
3. Derwent Water	5.3	4.6	22	5.5	29	82.7
4. Bassenthwaite Lake	5.3	6.2	19	5.3	27.9	237.9
5. Coniston Water	4.9	8.7	56	24.1	113.3	60.7
6. Haweswater	3.9	6.9	57	23.4	76.6	26.6
7. Thirlmere	3.3	6.0	46	16.1	52.5	29.3
8. Ennerdale Water	3.0	3.8	42	17.8	53.2	44.1
9. Wastwater	2.9	4.8	76	39.7	115.6	48.5
10. Crummock Water	2.5	4.0	44	26.7	66.4	43.6
11. Esthwaite Water	1.0	2.5	15.5	6.4	6.4	17.1
12. Buttermere	0.9	2	28.6	16.6	15.2	16.9
13. Loweswater	0.6	1.8	16	8.4	5.4	8.9
14. Grasmere	0.6	1.6	21.5	7.7	4.9	27.9

onset of the Ice Ages. An ice cap formed on the centre of the dome and glaciers radiated outwards down its flanks; they gouged out the valleys, giving them U-shaped cross-sections, and frequently dumped moraines along the sides and across the ends as they retreated. Many of these valleys then filled with water to generate the familiar landscape of today: rounded hills, cleared of their forests by man and sheep, with long narrow lakes in the valleys. Sometimes the

Reading the history of a lake in its sediments

When lakes were first formed in front of retreating glaciers they were frozen each winter. During the summer thaw, melt water from the glacier brought in silt which was deposited on the bottom of the lake next winter when the water was still. This annual cycle is now evident in the layers of clay seen in cores taken down through the mud of the lake to the firm substrate in the bottom of the basin. Above this ancient layered mud lies the organic mud derived from living organisms. The lowest layers have a high mineral content but, working upwards from the bottom of the core, the content of organic matter gradually increases until the rich brown mud of the post-glacial period is reached. This mud contains diatom frustules (see Chapter 3), pollen and other plant parts from which the lake's, past ecology can be reconstructed.

There is a narrow band at the bottom in which most of the pollen comes from grasses, sedges and other plants typical of northern tundra areas today. Higher up the core, tree pollen increases and many more species are represented: birch, willow and pine pollens are abundant. At the top of this immediately post glacial deposit tree pollens decline and herbaceous pollens predominate which indicates that although the glaciers did not return, the climate had become cold again.

From here upwards (the top 4.8 m of sediment in the core from Esthwaite Water shown here) the more recent mud looks fairly uniform but shows some interesting changes in its pollen content. Firstly the proportion of pine tree pollen decreases and there are increasing amounts from deciduous trees, particularly alder; the proportion of grass pollen is quite low. The other major but more gradual change, is the decline of pollen from oak forest plants and the increasing proportion of grasses. This marks the period of forest clearances and establishment of the present open landscape of the Lake District. However, in some of the lakes, the most recent, superficial layers of the sediment contain increasing amounts of conifer pollen derived from the extensive modern forestry plantations on the fringes of the Lake District.

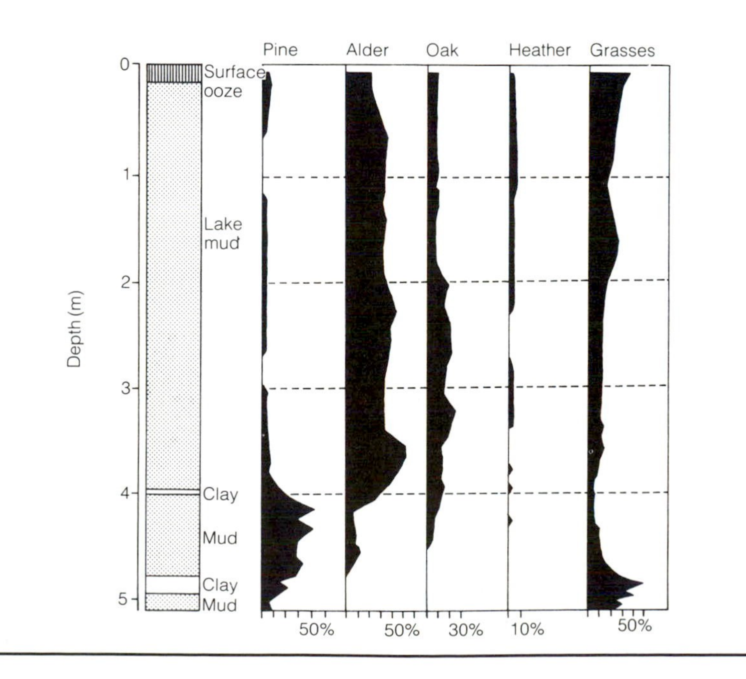

A diagram to show how the percentage of total pollen from different types of flowering plants changes with depth in a core of mud from the bottom of Esthwaite Water in the English Lake District. The bottom of the sequence was formed about 9000 years ago and the surface ooze is the present-day lake bottom. (Adapted from Franks and Pennington, 1961.)

lake occupied only the over-deepened rock basin but sometimes the moraine dam at the lower end further raised the water level to enlarge the lake. All of these lakes are about the same age, roughly 10 000 years (corresponding to the end of the last Ice Age), and their present differences stem from details of their basins, their surroundings and human influence. They range from Wastwater, which is cold and clear and where the hard bare rocks plunge down to the edge of the water, to Esthwaite Water which lies in a wide, green valley surrounded by fertile farmland and whose relatively shallow water becomes quite warm in summer and full of algae.

Professor Pearsall, who studied the Lake District lakes in the 1930s, recognised that the differences between the lakes were associated with the shapes and fertility of the valleys in which they lay. He arranged the lakes in a series from those with rocky shores and uncultivated catchments, such as Wastwater in which only 5% of the catchment area is suitable for cultivation and 73% of the lake shore is rocky, to those with soft shores and a good deal of cultivation in their catchments, such as Esthwaite in whose catchment 45% of the land is suitable for agriculture and only 12% of whose shoreline is rocky. Since then the differences have been increased because settlement, tourist development and agricultural improvement have proceeded apace, especially in the catchments with a higher proportion of flat, cultivable land.

In addition to the eleven large lakes, the Lake District contains innumerable small ones, called 'tarns'. These too owe their existence to the action of Pleistocene ice. Many of them lie in 'corries' (= *cwm* in Welsh; = *circque* in French) at the heads of the valleys. Here the ice has gouged out a deep bowl-shaped hollow below towering walls of rock and, in many cases, deepened the water it can hold by depositing a moraine on the open side. Sometimes the outflow stream has cut down through the moraine and either drained the tarn completely or left it with only the depth accommodated by the rock basin. Other tarns have formed in dips and hollows scoured by the ice in otherwise flatter areas of rock. Many of these are quite shallow and rich in aquatic plants.

The alpine lakes of Europe and New Zealand

The beautiful lakes of both the European and the Southern Alps were formed by glaciers which filled parallel valleys running down from a long ridge of mountains. In the South Island of New Zealand the Alps run north-east to south-west and large lakes fill many of their valleys. The European Alps also run roughly north-east to south-west with the lakes of Switzerland and Austria to the north and those of Italy to the south. In many cases a lake is deepened by a moraine dam at the lower end and in some cases morainic material forms a barrier across the centre of a valley, dividing a lake into two basins. A good example is provided by the Swiss Brienzer See and Thuner See, with the town of Interlaken in between. There was a tendency for the glaciers to fan out at lower elevations and this is mirrored in the lakes which now occupy their valleys: Lake Como is perhaps the best example, but the effect can also be seen in Lake Constance.

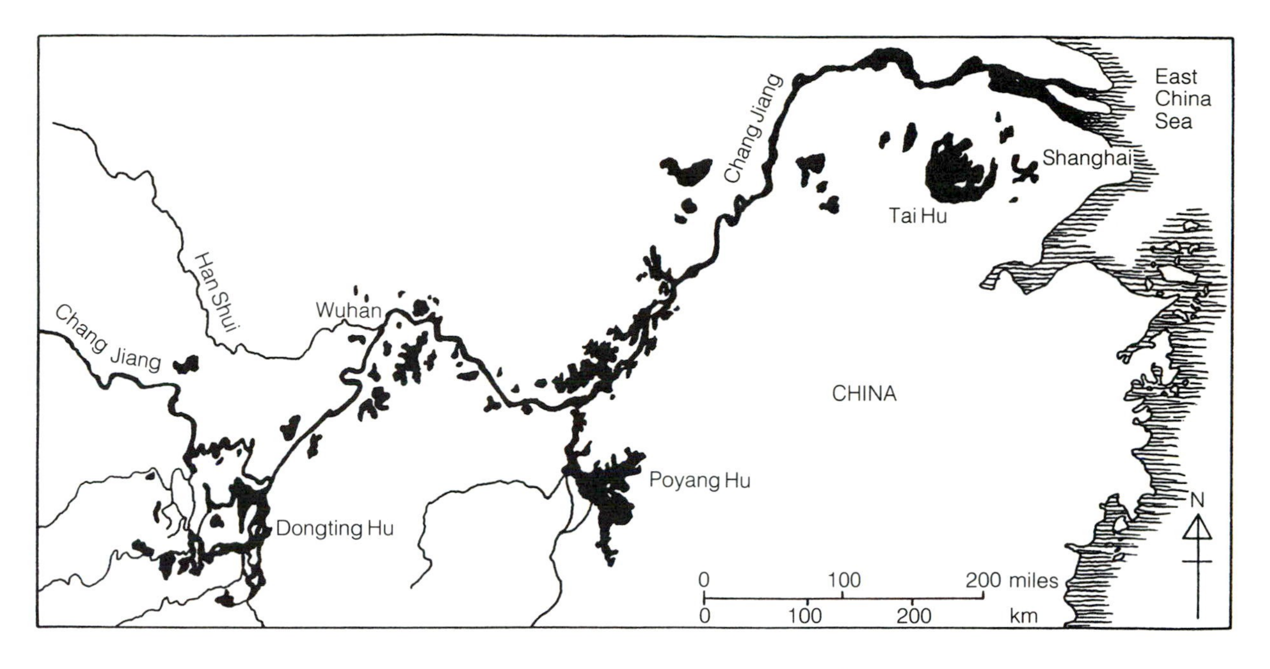

The lakes of the Middle and Lower Chang Jiang (Yangtze) River system. (Redrawn from Liu, 1984.)

Floodplain lakes of the Chang Jiang

River floodplains are other areas where many lakes occur together, but here they are found along the course of one river and its tributaries rather than in a number of different valleys. As we have seen earlier, the only really natural large floodplains now occur in some parts of Africa and in South America. Elsewhere they are highly regulated by man and in many cases the number of lakes which would occur naturally is much reduced. In Asia, however, fish culture is very important and many of the lakes are used as fish ponds.

Some of the earliest flood control programmes were implemented in China where not only is the floodplain soil put to good use by the human population but the productivity of the floodplain lakes is also extensively exploited for the benefit of the local people. The middle and lower reaches of the Chang Jiang (Yangtze River) were a nursery of Chinese civilization and, although the number of its floodplain lakes has diminished with increasing reclamation for arable use, there are still more than 20 000 km^2 of lakes associated with the 1560 river kilometres between Yichang and Zhenjiang. These 1760 lakes are mostly between 300 and 3000 ha in area, the largest are Donting Hu, Poyang Hu and Chao Hu. They rarely exceed 10 m depth and have a flat bottom of thick mud. They support very important commercial fisheries. However, the river is now kept within its channel, with the aid of dams, and so natural migrations of fish are impeded. Most of the commercially important species breed only in very large rivers and, during natural floods, their young would enter the lakes soon after hatching. Now, in the absence of natural flood cycles, fish fry are reared in hatcheries and then put into the lakes. Here they eat the natural food available in these very productive shallow waters and do not need to be fed artificially.

2

LAKE

water

Water is taken for granted by almost everyone except people who cannot get enough of it to drink, those suffering from floods, and perhaps those who work with it. Even people who live and work along the shores, or on the surface of, lakes are often not aware of the very special properties of the all-important substance we call water. Life first evolved in water, over 90% of living matter is water, and many plants and animals are confined to water. There are plenty of other liquids, but none of them has the same combination of characteristics as water.

Water is a compound of hydrogen and oxygen, both of which are gases at normal temperatures and pressures, and yet water is a liquid under most normal conditions. It is colourless, flows easily and is a liquid in which many other chemicals can dissolve. It is therefore an ideal medium for the chemical reactions of life's processes to take place.

If you stand at the edge of a lake the water may appear grey or blue because it reflects the colour of the sky. In some cases it may look green due to the millions of tiny plants suspended in it, or brown because it contains sediments or pigments from soil. If you look over the side of a boat while on a clear lake, go for a swim or dive with a face mask on, you will realise that whatever colour the lake seems to be, the water is more or less transparent. This enables light to penetrate the surface and pass down into the body of the lake to provide energy for green plants to make their own food by the process of photosynthesis. This is the basis of the lake ecosystem, the components of which we shall examine in Chapter 3.

Light and depth

When light passes down through a column of water the amount of light available at any particular depth decreases exponentially; most of the light is absorbed in the surface waters. Thus, if there are 100 units of light energy at the surface, they will decrease not in a simple linear manner such as 90 at 1 m, 80 at 2 m, 70 at 3 m, and so on, but so that at 1 m there may already be only 40 units, at 2 m only 20 units and at 3 m only 10 units. How far the light penetrates and the depth at which only 1% of the surface-incident light remains depend very much on the colour and turbidity of the water. Very clear infertile lakes may have a 1% depth greater than 50 m and the theoretical limit is set at about 200 m by the physical properties of pure water. In many lakes the 1% depth is less than 10 m and may be as little as 50 cm in lakes full of silt (such as those in front of glaciers) or those containing dense populations of phytoplankton suspended in the water.

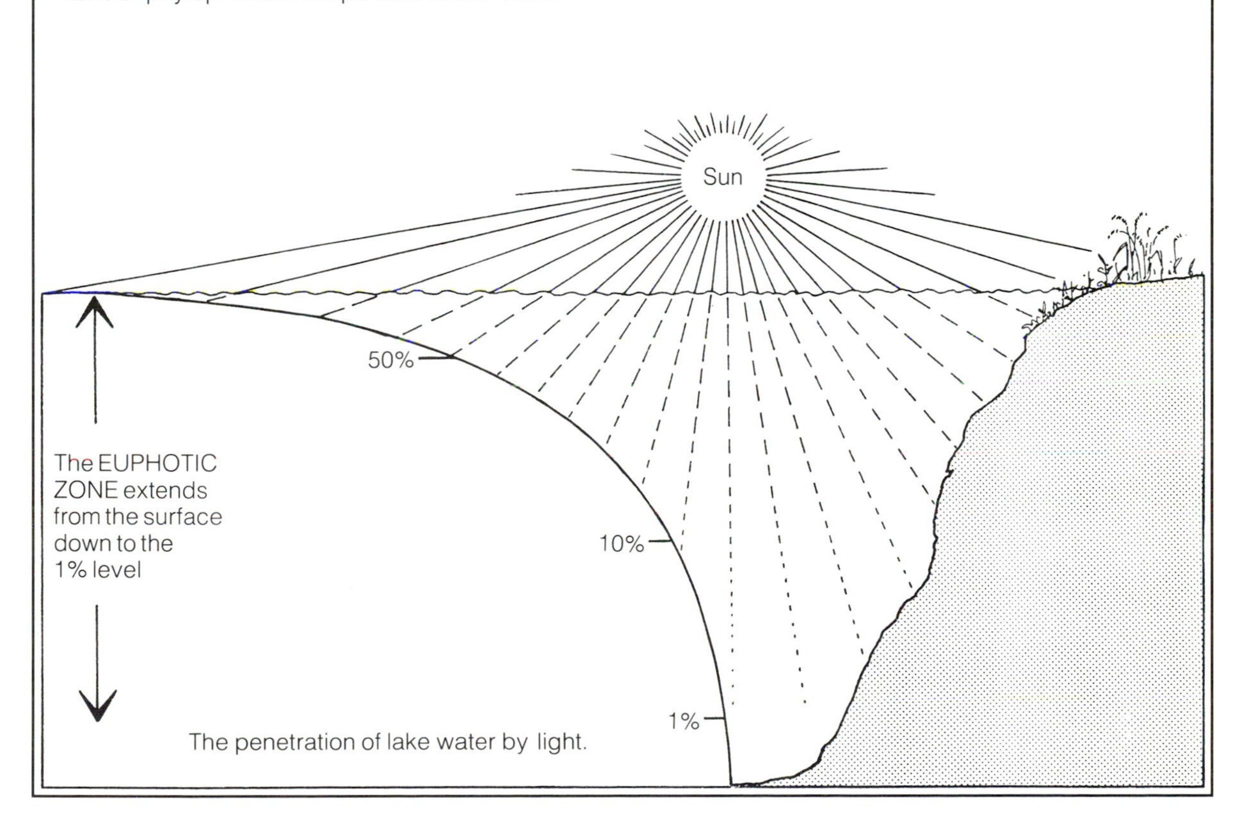

The penetration of lake water by light.

Physical characteristics

Absorption of light

Although pure water is colourless, as light passes down into a lake different parts of the light's spectrum are absorbed at different rates. The colours at the ends of the spectrum, red and violet, are absorbed most rapidly; blue-green light penetrates furthest. Eventually, if the water is deep enough, all the light is absorbed and darkness reigns. If the lake water is coloured by pigments, light is absorbed more rapidly. Where the lake water contains suspended particles, whether inert silt or plant cells, again light can not penetrate as far.

Plants can only synthesise food in the lighted, surface waters of a lake and this region is called the *euphotic zone*. The bottom of the euphotic zone is defined as the level by which 99% of the light has been filtered out, i.e. only 1% of the light at the surface remains. In a clear-water lake the euphotic zone may extend down to 20 m or more but in many lakes it is often about 3–5 m or even as little as 0.5 m deep. Below this zone, green plants cannot photosynthesise and so they cannot grow; if they sink into the unlighted zone they must possess a mechanism for floating or swimming back up through the water column, otherwise they will die and sink to the bottom of the lake. Animals which depend on these plants for food therefore either feed near the surface on living plants or wait for dead and dying plants to sink down to them in the darkness below.

There are parallels between this situation and that in a dense, tall forest. The sunlight falls on the tops of the trees as it does on the surface of the lake, and filters down through their leaves as it does through the algae in the lake water. At the forest floor, light is much reduced and only those plant species adapted to very low light conditions can grow, sometimes none at all. In the canopy of the forest the herbivores eat the leaves and fruits of the trees; on the forest floor are species which live off the fallen leaves and fruits of the trees and the waste products of those feeding in the canopy above. A similar situation exists in the oceans and also in the water of deep lakes.

Temperature

Sunlight warms the surface of the water of the lake and the extent to which the heat penetrates depends not only on the rate at which it is conducted downwards (rather a slow process) but also on the extent to which the surface water mixes with that below, mainly as a result of stirring by the wind. The importance of mixing for this and other aspects of the lake functioning will soon become evident, but first let us recall some general relations between water and heat. The temperature of water alters much less rapidly than that of air, so animals and plants which live in it are protected from sudden changes and only have to adjust gradually. There are no abrupt differences of temperature by night and day, as on land, and although the seasonal range of air temperature may be great, the lake temperature alters only slowly. The effects of seasonal changes will be discussed in Chapter 4.

Density

As the water temperature changes so does its density and this has enormous implications for the functioning of different lakes. It is a special peculiarity of water that its density is greatest at 4°C although it remains liquid and does not become solid (by turning to ice) until the temperature falls to 0°C. This means that ice is less dense than water just above zero and it therefore floats on the surface while slightly warmer, but denser, water remains below. This explains

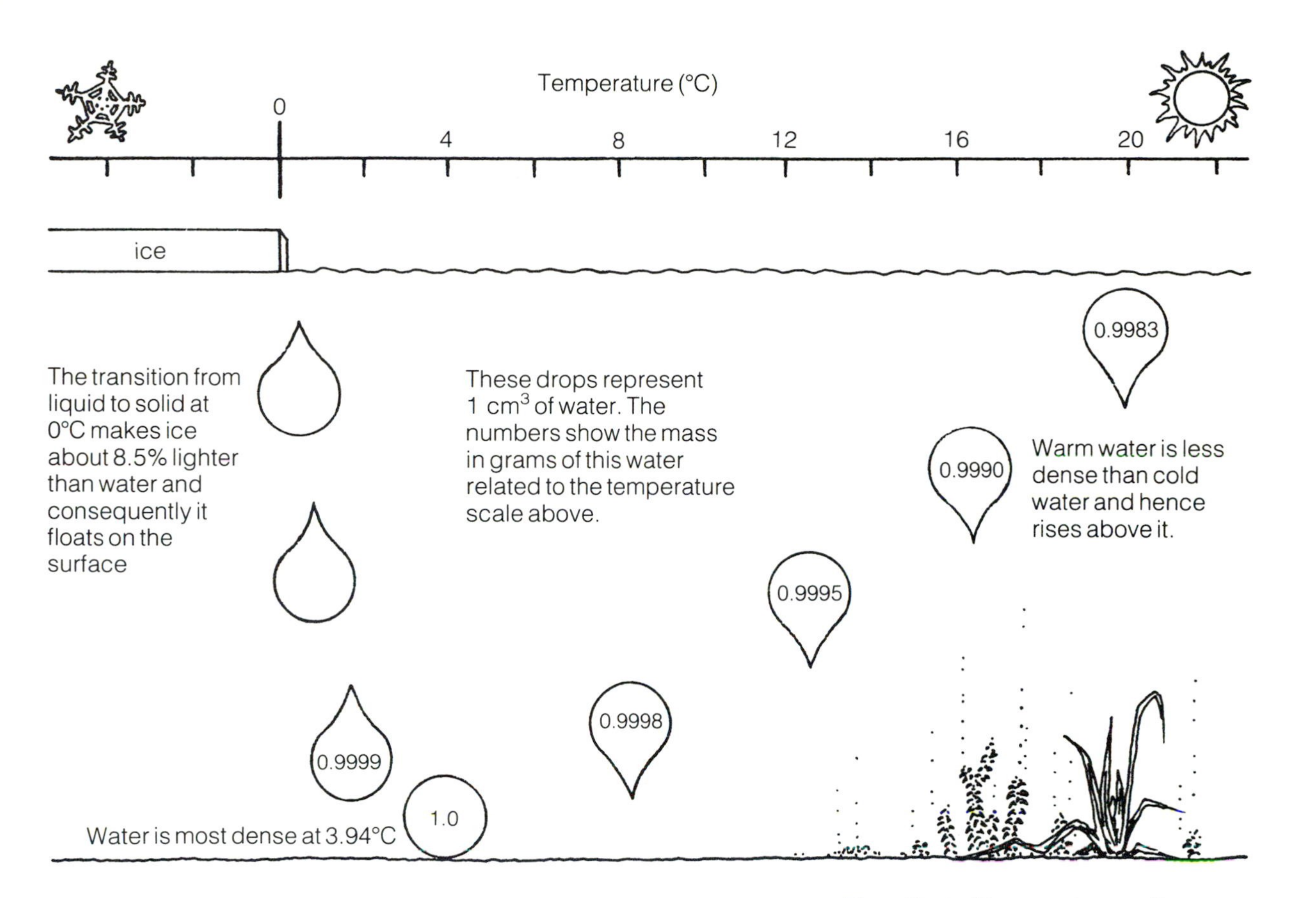

The effect of temperature on the density of water (pure water at a pressure equivalent to that at sea level.)

why ice forms first on the top of a lake and plants and animals are able to survive in the water underneath. The ice forms a barrier between colder air and slightly warmer water so the downwards growth of ice is slowed. Only very shallow lakes, in very cold places, freeze right to the bottom.

Stratification

When the surface waters of a lake begin to warm up, the heat takes a long time to penetrate down through the water, particularly if the weather is calm and the water is not stirred by strong winds. Warming and calm weather tend to go together. Gradually, a situation develops where there is a marked difference in temperature between the upper layers of the lake and the water lower down. This means that there is also a difference in density of the water: lighter water which can be stirred by gentle breezes floats on top of rather denser, cooler water. This does not mix with the upper water when light winds ruffle the surface. When a lake is divided like this into an upper, warmer, mixed layer and lower, colder, unmixed water, it is said to be *stratified*.

The two layers, or strata, are known as the *epilimnion* (the top layer) and the *hypolimnion* (below). Between them lies a relatively narrow zone sometimes called the metalimnion but more frequently

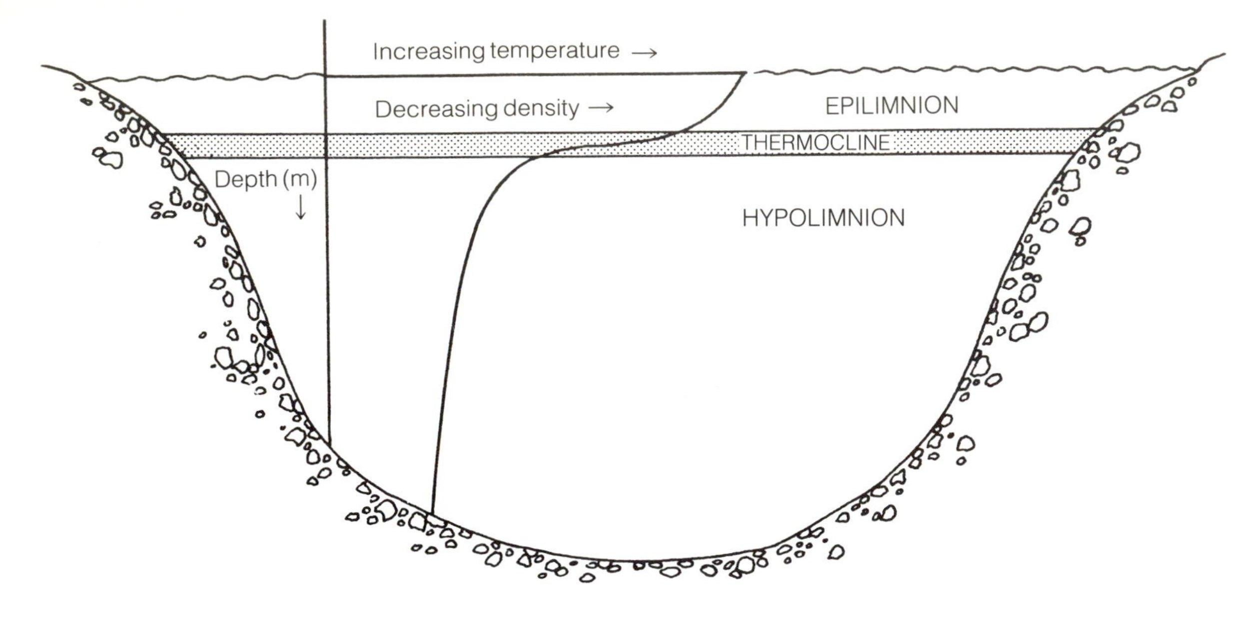

In a stratified lake the water near the surface is warmer and less dense than the water towards the bottom of the lake. The narrow transition zone between the two is the thermocline; above this lies the warmer less dense water of the epilimnion; the colder, denser water sinks and forms the hypolimnion. This only applies while the water is above 4°C (see p.27).

referred to as the *thermocline*. It is here that the temperature changes abruptly. It must be remembered that it is not the temperature itself which causes the stratification but rather the difference in the densities of the epilimnion and the hypolimnion. It is this difference which determines the amount of work the wind must do in order to mix together the layers of water. In sheltered places, or in lakes which are deep in relation to their surface area, there may be insufficient wind at the surface to achieve an effective stirring of the water; the lake may then remain stratified for weeks on end, with serious consequences for the life within it. By contrast, shallow lakes in exposed places may be so frequently stirred up that they do not get a chance to stratify for very long.

The nature of the relation between temperature and the density of water is such that the difference in density resulting from a difference of five Celsius degrees at cooler temperatures may be the same as that resulting from a difference of only two degrees at higher temperatures. This means that tropical lakes can be stratified as a result of very small differences (only one or two degrees) between the temperature of the epilimnion and the hypolimnion, while the cooler lakes of the temperate zones are not stratified until quite large differences in temperature are established.

There are a few lakes which are so deep and warm that they are permanently stratified and many shallow lakes that are never stratified for more than a few hours or days, but most lakes undergo a seasonal cycle of stratification, alternating with mixing, which is a major determinant of their ecology.

Water chemistry

It is probably true to say that no lake contains pure water. Apart from rain (or snow) falling on the surface of the lake, all the water in a lake has run over or through the rocks and soil of its catchment area. On the way it takes up chemicals into solution. Often these substances are referred to as *salts*. This is a general term used to describe chemicals which, when they are dissolved in water, separate into positively and negatively charged particles (called cations and anions, respectively). This creates confusion in the English language because the substance that everybody knows as 'salt' is only one specific example – sodium chloride (the cation is the sodium ion Na^+ and the anion is chloride Cl^-). Other common cations include calcium Ca^{2+}, potassium K^+ and magnesium Mg^{2+} ions; anions include not only chloride but also carbonate CO_3^-, sulphate SO_4^{2-}, nitrate NO_3^- and phosphate PO_4^-. They may come together in various combinations to form salts such as calcium carbonate $Ca(CO_3)_2$, calcium chloride $CaCl_2$, sodium sulphate Na_2SO_4, and potassium sulphate K_2SO_4. So lake water can contain a veritable cocktail of salts and yet have quite a different composition from sea water, whose chemical composition is dominated by sodium chloride (though it also contains a wide variety of other chemicals in lesser amounts).

Because the anions and cations carry electrical charges, water containing them can conduct electricity and its ability to do so gives a measure of the total quantity of charged particles (ions) dissolved in it. This is known as the **conductivity** of the water and will be used frequently in this book as a measure of the total amount of ions (and therefore salts) in different lake waters. Conductivity is measured in micro-Siemens per centimetre (usually written μS/cm). Water with a conductivity of less than 50 μS/cm does not contain many ions. The conductivity of average sea water is about 32 000 μS/cm; sea water is obviously salty, although not nearly as rich in chemicals as some of the lakes discussed in Chapter 8. The conductivities of some well-known freshwater lakes are shown in Table 2.2 and indicate that even the purest lake water is a little bit 'salty' and the term 'fresh water' is only relative.

In lakes which are not influenced by human activities the conductivity depends primarily on the nature of the rocks in the catchment area. Lakes whose water flows over very old, hard rocks that do not easily dissolve have a low conductivity. The Experimental Lakes of North-West Ontario in Canada have an average conductivity of only 19 μS/cm; the five largest Scottish lochs have mean annual conductivities ranging from 29 to 41 μS/cm. There is a small lake in Uganda, Lake Nabugabo, which receives water that has drained over very old hard rocks, from which almost nothing has dissolved. The water then filters through a bog dominated by *Sphagnum* moss (which extracts ions) before it enters the lake. The lake water has a conductivity of 25 μS/cm. These lakes contain water which is probably as near pure as any naturally occurring water bodies can be.

Rain is not pure water. It, too, contains ions in varying amounts, some of which may be picked up from sea spray, and others which are actually dissolved gases from the atmosphere through which the clouds and rain have passed. In coastal regions, rain can contain a high proportion of sodium and chloride ions. Clouds formed over urban and industrial regions may dissolve nitrate and sulphate ions

Table 2.1 *The effect of different types of catchment area on the chemical composition (mg/l) of stream water compared with that of rainfall in the same area (from Moss, 1980, and Gaudet & Melack, 1981).*

	New Hampshire, USA Igneous (insoluble) rocks Undisturbed forest		Norfolk, UK Chalk and glacial drift Lowland agriculture		Rift Valley, Kenya Thorn bush, and rangeland	
	Rainfall	Stream	Rainfall	Stream	Rainfall	Malewa R.
Na^+	0.12	0.87	1.2	32.5	0.54	9.0
K^+	0.07	0.23	0.74	3.1	0.31	4.3
Mg^{2+}	0.04	0.38	0.21	6.9	0.23	3.0
Ca^{2+}	0.16	1.65	3.7	100.0	0.19	8.0
Cl^-	0.47	0.55	<1.0	47.0	0.41	4.3
HCO_3^-	0.006	0.92	0	288.0	1.2	70.0
SO_4^{2-}					0.72	6.2
pH	4.14	4.92	3.5	7.7		

from the polluted atmosphere; these ions combine with water to form acids, leading to the phenomenon of 'acid rain' which often falls on places quite remote from its area of origin as clouds are blown along by the wind. The acidity of the rainfall is expressed as pH: the lower the pH value, the more the acidity. A neutral solution has a pH of 7 and above this the solution is alkaline; a value of 10 or more is strongly alkaline. Most rainfall is naturally on the acid side of neutrality with a pH between 4.5 and 5.7 – a result of dissolving carbon dioxide from the atmosphere to form a weak solution of carbonic acid. The addition of other gases (especially sulphur dioxide and the oxides of nitrogen) from the combustion of fossil fuels lowers the pH values of rainfall still further in many parts of the world and there is now great concern over the effects of this on life in lakes and rivers. Many animals are intolerant of a pH below 5.7, for reasons that are not fully understood.

Lakes which receive water that has flowed over or through rocks and soil rich in carbonate ions, have a high pH because the hydrogen ions have been neutralised. Water with a high concentration of calcium carbonate and calcium bicarbonate is called 'hard' water and often derives from chalk and limestone areas. The pH of 'hard' water is about 8–9. 'Soft' water comes from areas of dense, insoluble rocks and contains much less calcium; it has a lower pH value. We notice the difference through the reaction with soap, which is how the terminology originated – hard water needs more soap to form a lather. The amount of calcium carbonate and calcium bicarbonate in water is also very important to the plants and animals that live in it.

Carbonic acid in rainwater combines with calcium carbonate to form the relatively soluble compound calcium bicarbonate. In some cases the water becomes saturated with this; when the water comes out of the rocks into the air, it releases carbon dioxide, leaving calcium carbonate which, being much less soluble, comes out of solution as solid crystals. This is how stalactites and stalagmites are formed in caves, and also why spectacular terraces of travertine

pH – acidity, alkalinity and 'acid rain'

pH is a measure of the concentration of hydrogen ions in a solution. In acid water there is powerful hydrogen ion activity but, by a convention which is rather bewildering to the layman, this is denoted by the smaller figures on the pH scale. As the hydrogen ion activity decreases the solution becomes more alkaline and the pH figure increases. The scale runs from 1 to 14 and at pH 7 the solution is said to be neutral.

The highest natural values in alkaline waters probably occur in soda lakes such as Lake Nakuru, in East Africa, which has a pH of 12. By comparison, vinegar has a pH of about 1, stomach acid is pH 2 and saliva is a little bit alkaline at around pH 7.5. Lakes with a natural pH below 4 are rare and only occur in volcanic regions where comparatively strong acids drain into a lake. However, it has been observed that many lakes in areas of hard rocks have become increasingly acid due to industrial pollution and 'acid rain'. Although they still do not approach the most naturally acid lakes, it is the lowering of their pH below the levels which most animals can tolerate that has caused the disastrous changes recorded in these lakes.

Although it was damage to German forests which first made the general public aware of the phenomenon of 'acid rain', fisherman in southern Norway and Sweden had been concerned about it for much longer. Their lakes lie on very hard rocks and contain few calcium ions which could help to neutralise the increasing acidity of the rainwater. Salmon and trout disappeared, first from small upland lakes and then from larger lakes further downstream. By 1983 freshwater fish were nearly extinct across much of southern Norway. At least 1600 lakes had lost their brown trout and perch. The low pH causes the release of a poisonous form of aluminium from the soil into the water. It may be this which actually kills the fish, since in lakes where low acidity is due to naturally occurring humic acids, the fish do not die and in these lakes the aluminium is bound up with the humus and is not poisonous.

There is now very strong evidence that this increase in the acidity of lake waters is primarily due to industrial pollution of the air and Norway and Sweden are particularly unfortunate in that they not only have very hard rocks but they are also downwind of a high proportion of the polluted air from the rest of Europe. Britain is particularly guilty in this respect since her power stations pour out a lot of sulphur dioxide which blows towards Scandinavia and she is unwilling to reduce it, beyond the reductions already made from the even higher levels emitted previously. Research on the sediments of lakes in Galloway in Scotland, where the rocks are also very hard and the lakes naturally fairly acid, has shown that these lakes too have become more acid since the industrial revolution and that this process started before the hills around them were planted with trees. This counters the argument that changes in land-use rather than pollution might be responsible for the acidification of the lakes.

There is still controversy about what has caused the damage to forests in the Federal Republic of Germany, Scandinavia and elsewhere, which was originally ascribed to 'acid rain' and it seems likely that other forms of air pollution, particularly ozone, are also implicated.

(another form of calcium carbonate) are formed at places like Mammoth Springs in Wyoming, and why in many parts of the world mosses and other plants that grow around seepages of water from limestone rocks become covered in calcium carbonate and

Table 2.2 *The concentrations of the major ions in some famous lakes compared with sea water and the world average for fresh water.*

	Cations (mg/l)				Anions (mg/l)		
	Na^+	K^+	Mg^{2+}	Ca^{2+}	Cl^-	SO_4^{2-}	HCO_3^-
Loch Ness	3.8	0.27	0.8	1.9	7.53	n.d.	n.d.
Windermere	3.5	0.51	0.61	5.7	6.6	6.9	9.7
Lake Esrom	12.0	n.d.	5.6	42.0	22.0	8.2	140.0
Lake Kinneret	115.0	6.0	30.0	51.0	229.0	54.0	75.0
Lake Baikal			4.2	15.0	1.8	4.9	*c.*60.0
Lake Victoria	10.4	3.8	2.6	5.6	3.9	2.3	56.1
Lake Tanganyika	57.0	35.0	43.3	9.8	26.5	7.2	409.0
Lake Ontario	12.2	1.44	6.4	42.9	26.7	27.1	115.0
Lake Tahoe	6.1	1.7	2.5	9.4	1.9	2.5	40.0
Lake Nicaragua	11.0	2.0	6.0	16.0	19.0	0	65.0
World average fresh water	8	3	5	30	8	18	105
Sea water	10810	390	1300	410	19440	2710	140

n.d. = no data

eventually solidify to form the deposit known as tufa. Aquatic plants absorb carbon dioxide from the water for photosynthesis, so they sometimes have a similar effect: calcium carbonate comes out of solution and precipitates on to the surface of leaves, or on the bottom of the lake, or even forms crystals in the open water.

All the ions listed in the table above, and many others, are vital to the health of the living organisms in lakes. For example, calcium is vitally important to the cell processes of all plants and animals, it is a structural component of vertebrate bones, and it forms the shells of many invertebrate animals, especially molluscs. Magnesium is important for photosynthesis and, although it is not needed in great quantity, green plants cannot live without it. Iron is not only a major constituent of the red blood pigment haemoglobin (found in many lake invertebrates as well as ourselves) but is also essential to certain basic cell processes. These substances are obtained from the water; some organisms cannot survive if a certain minimum concentration is not present. For example, many molluscs cannot live in water with a calcium concentration below 20 mg/l. Silica is vital to the growth of diatoms (see Chapter 3) because their cells are surrounded by a hard outer casing, or frustule, which is composed entirely of it. As their populations grow they absorb silica from the water and when the supply runs out, the population crashes. The dead diatoms sink to the bottom of the lake and, because silica is relatively insoluble in water, their frustules remain in the mud and can be used thousands of years later to study the diatom community of the lake at earlier periods. Moreover, because many species of diatom are sensitive to particular pH levels in the water, their remains can be used to monitor changes in acidity of the water with time. Silica enters the lake almost entirely from the erosion of rocks and soils, so the concentration in the lake depends very much on the nature of the catchment area; relatively little is recycled from the sediments.

Oxygen

Oxygen is essential for almost all living things. Aquatic creatures are able to extract dissolved oxygen from their watery medium. Their survival in very cold water is facilitated because at lower temperatures cold-blooded organisms use up less oxygen in their life processes. They are also helped by the fact that more oxygen is able to dissolve in cold water than in warm. At 4°C, one litre of water is able to hold (i.e. is 'saturated' by) 12.7 mg of oxygen; if, for some reason, it contains only 6.35 mg oxygen per litre it is said to be 50% saturated. At 20°C the water is saturated by only 8.84 mg O_2/l. Thus lake water contains more oxygen in winter than in summer and more in the temperate zones than the tropics. At 100°C, boiling point, all oxygen is driven off and the water itself turns into a vapour.

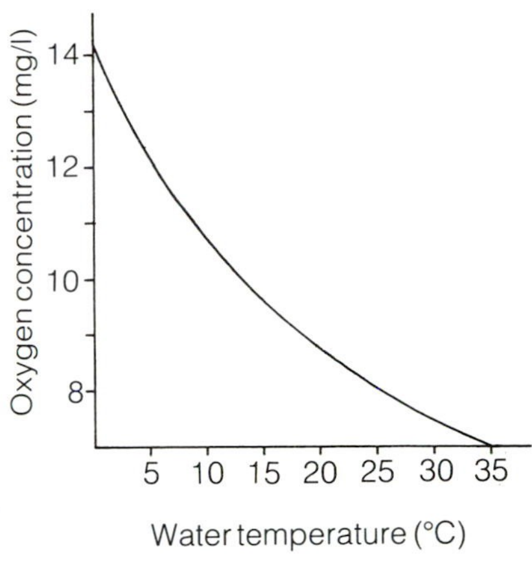

The amount of oxygen that pure water will hold at different temperatures when it is 100% saturated. Water at 5°C contains one third more oxygen than at 20°C.

Like all green plants, those in water give out oxygen as a waste product when they photosynthesise. This dissolves in the water around them, unless it is already saturated. Since plants need light for photosynthesis, oxygen is added to the lake water by this means only in the euphotic zone. Additional oxygen can come from the air, but only via the surface. All plants and animals take in oxygen for respiration, whether in the light or in the dark, so whereas oxygen is being added to the water as well as being taken out of it in the euphotic zone, in the deeper, darker water it is only being taken out. As a consequence the water at the surface usually contains more oxygen than that lower down, although this effect may be partially counterbalanced by the fact that warmer surface water cannot hold as much oxygen as the colder, deeper water. The distribution of oxygen in the waters of a lake is thus the product of these interacting factors.

Carbon

The carbon atom is like a piece from the centre of a jigsaw puzzle in that it can be linked to other pieces (molecules or atoms) on four sides. Thus carbon atoms can be linked in long chains to form complex **organic molecules**: carbon is the structural basis of all living material on earth. During the process of photosynthesis the energy supplied by sunlight is used to link carbon and carbon-based chemical units together to form sugars, starch and cellulose. Thus the plants take in carbon dioxide as a gas and 'fix' it into solid plant material. The amount of new material made by the plants in a lake is often measured as the amount of carbon fixed by the plants per volume of water. Alternatively, because the light essential for photosynthesis is measured per unit area of lake surface, it is also useful to express the amount of carbon fixed in the same way. The amount of carbon fixed *in a given time* is an important measure of the **productivity** of a lake (see below).

Plants obtain their supplies of carbon from the atmosphere in the form of carbon dioxide; animals obtain theirs from the carbon-rich bodies of plants and animals. Aquatic plants use carbon dioxide dissolved in the water but the amount available depends very much on the acidity of the water. Above pH 5 some of the carbon dioxide forms bicarbonate ions and by pH 9 it is all incorporated into bicarbonate plus some carbonate ions. Above pH 9.5, carbonate ions

Photosynthesis and respiration

All green plants make their own food by the process known as photosynthesis. They trap energy from sunlight with the aid of the pigment chlorophyll, which gives them their green colour, and use it to combine carbon dioxide and water to form sugars. They give off oxygen as a waste product of this process. The sugars may be stored and used as a source of chemical energy or they may be combined with other molecules to form proteins, fats, oils and carbohydrates.

The chemical energy stored as carbohydrates in the cells is released to power all the life processes through the mechanism of respiration. All plants and animals must respire in order to stay alive, so, unlike photosynthesis which can only occur in the light, respiration continues under all circumstances – if it stops, the organism is dead. The chemical process of respiration is the reverse of photosynthesis in that oxygen is taken in and combined with cell sugars to release energy; water and carbon dioxide are released as waste products.

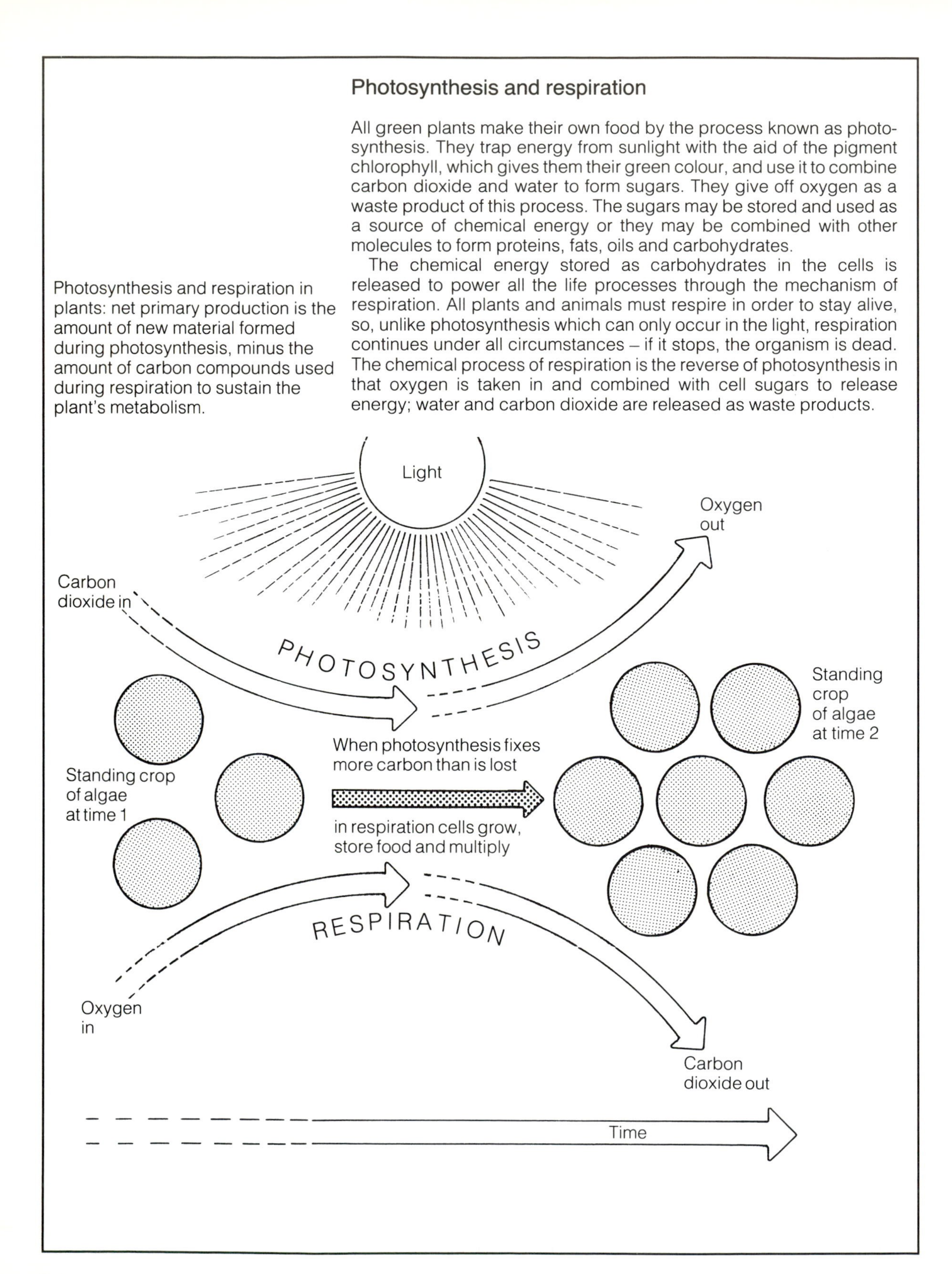

Photosynthesis and respiration in plants: net primary production is the amount of new material formed during photosynthesis, minus the amount of carbon compounds used during respiration to sustain the plant's metabolism.

assume increasing importance and no free carbon dioxide remains. As we have already seen, in calcium-rich waters crystalline calcium carbonate may start precipitating from the water and this, in effect, removes the carbon dioxide needed by the plants. Fortunately, in such alkaline water, when the plants take up any of the little carbon dioxide that is available, an equivalent amount of bicarbonate is converted to carbon dioxide and is replaced by the conversion of carbonate to bicarbonate. Thus, although the rate of photosynthesis may in some lakes be limited by the speed of these reactions needed to release carbon dioxide, there is unlikely to be a shortage of carbon for the formation of new living material. Not only is carbon dioxide available from the atmosphere but it is also released by the majority of living organisms as they respire and this includes many of the bacteria involved in decomposition. These bacteria thereby release carbon from the bodies of dead plants and animals back into the water for use by living plants.

Productivity

Lakes can be productive or unproductive in exactly the same way as farmland. Land which has poor soil, a cold climate, or not enough rain may be unproductive as a result of any, or a combination, of these factors. Some land can be made more productive by the addition of fertilizers or the provision of water by irrigation to make up for the natural deficit which limits productivity. Lakes seldom suffer from a shortage of water but they are often unproductive because they are cold and/or lack the nutrients necessary for rapid plant growth and reproduction. The addition of such nutrients (the same nitrates and phosphates that are in fertilizers on farmland) to lakes will usually increase their productivity: they will produce more plants and animals in the same area and period of time.

When the supply of nutrients is increased, whether through natural processes or human activity, the changes which take place in the lake are known as eutrophication. We use this word because lakes which have a naturally low supply of nutrients are called **oligotrophic** (oligo = few) while those that are more richly endowed are called **eutrophic** (eu = well); in fact there is only a clear distinction between oligotrophic and eutrophic lakes at the very extremes of the range. The use of these terms is always relative and frequently gives rise to confusion and misunderstanding. For example, we can say that Windermere in the English Lake District is eutrophic compared with Wastwater, but all of the Lake District lakes are oligotrophic compared with, for example, Loch Leven in Scotland, Lough Neagh in Northern Ireland or Lake Esrom in Denmark. In order to try and avoid some of the confusion associated with these terms we shall not use them but, where necessary, refer to lakes as productive or unproductive. Although these words can also be used relatively, they are preferable because they can be defined precisely, as the quantity of new organic material (i.e. plant and animal tissue) actually manufactured per unit area in a given length of time. This will be a consequence of the combined effects of availability of *all* nutrients,

Eutrophication

We have largely avoided the use of the terms **oligotrophic** (for unproductive lakes) and **eutrophic** (for productive lakes) because they are not clearly defined. However, the changes which occur in lakes due to the artificial addition of nutrients are so widely known as **eutrophication** that it would be unreasonable to abandon the word in this context. The additional nutrients (primarily nitrogen and phosphorus) most usually come from sewage, agricultural fertilizers or industrial effluents, and sometimes from all three.

Raw sewage is first decomposed by bacteria which use up oxygen from the water. This properly takes place in the treatment works so that the effluent does not deoxygenate the water, but treated sewage still contains large amounts of nitrates and phosphates. It has been calculated that an average human produces nutrients equivalent to 12 g nitrogen and 2.25 g phosphorus per day. Sewage always contains a much higher concentration of phosphorus in proportion to nitrogen than is found in natural waters, particularly since the use of phosphate-containing detergents has greatly increased over the last 20 years. In many natural waters the growth of algae is limited by a shortage of phosphorus rather than of nitrogen or carbon, so the addition of this essential nutrient greatly increases algal growth in the receiving water. While the algae produce oxygen during their daytime photosynthesis, at night they use it up; thus, dense algal blooms cause the suffocation of fish and other animals. Increased algal production initiates other changes in the aquatic ecosystem too and the whole process is known as eutrophication. Many lakes, particularly those situated among densely settled human populations, have suffered from these changes. They were first recognised, and steps taken to control them, in the Swiss lakes, whose importance for the tourist trade as well as for the local people made restorative action imperative.

whereas 'oligotrophic' needs to be defined in terms of the abundance (or lack of it) of principally two nutrients, nitrogen and phosphorus.

The productivity of a lake is manifest as the amount of plants growing in the water but this may be misleading. A miser hoarding a sack of money may be rich, but not actually earning very much each week. What there is in a lake at any moment is actually the **standing crop**; 'productivity' refers to the amount of new living matter **formed** in a period of time. Even if there are a lot of plants present, the lake may not actually be producing a great deal of new material each day. In contrast, an apparently thin population of plants may be vigorously photosynthesising and producing a lot of new material each day, and is therefore very productive. The apparent lack of plants could simply be due to rapid consumption by animals, which would themselves then grow and reproduce more quickly, contributing to the total organic production. The time factor is all-important in the concept of productivity.

Plants comprise the basic food material, directly or indirectly, for the whole lake system, so their abundance can be an important indicator of the lake's productivity. It is not very convenient to measure productivity in terms of weight of plants produced, however, and indeed it is more relevant to consider how much new carbon is collected from the air (as carbon dioxide) and 'fixed' in organic compounds by photosynthesis, so that it can be added to the lake's total food supply. Thus, productivity is often expressed as grammes of carbon fixed per cubic metre (g C/m^3) per unit of time, which is the intensity of production in a volume of water, or as g C/m^2 per unit of time, to allow comparisons to be made between lakes of different depths. Expressing production per square metre of lake surface allows comparison not only between lakes of different depths but also with the amount of solar energy arriving at the lake surface, which must always be expressed per unit of lake area.

The amount of small floating plants, or phytoplankton, present is an indication of the productivity; but again it is more convenient to express this not as a direct count of the number of algae in a litre of water but rather as the total amount of chlorophyll present. The chlorophyll is the green pigment which enables the plants to harvest light and to make the carbon compounds; it is also easier to measure (Chapter 4). Amounts of chlorophyll *a* per cubic metre (or per litre) are therefore used as a convenient index of phytoplankton abundance and *potential* productivity. As for carbon, chlorophyll concentration may also be expressed as the amount under one square metre of lake surface and, for a deep lake in which the water below the euphotic zone contains no algae, this figure will be very much smaller than the amount per cubic metre in the surface water.

Nutrients

An important group of compounds are those containing combinations of nitrogen and phosphorus. The most common are nitrates and phosphates, often referred to collectively as **nutrients**. They act as fertilizers of plant growth and their supply determines the quantity and quality of plants in a lake, just as in fields and gardens. Both nitrogen and phosphorus are essential to life, but while there are many sources of nitrogen, in general phosphorus is often in short supply and therefore limiting to plant growth. The availability of these two elements within a lake ecosystem depends not only on sources of supply from outside but also on physical and biological processes within the lake itself.

Nitrogen

Nitrogen is far more abundant in the atmosphere than is carbon dioxide or oxygen and it is also an essential component of the molecules of living material: for example, all proteins contain nitrogen. Some molecular nitrogen dissolves in water as a gas but most of the nitrogen in water is combined either with oxygen to form nitrate (NO_3^-) or nitrite (NO^{2-}) ions or with hydrogen to form ammonium ions (NH_4^+). There may also be numerous organic compounds of nitrogen present (such as proteins, amino acids and urea) which result from the decomposition of dead organic matter. These organic compounds must be further broken down to nitrates, nitrites or ammonia before the plants can use them, and this is done by bacteria.

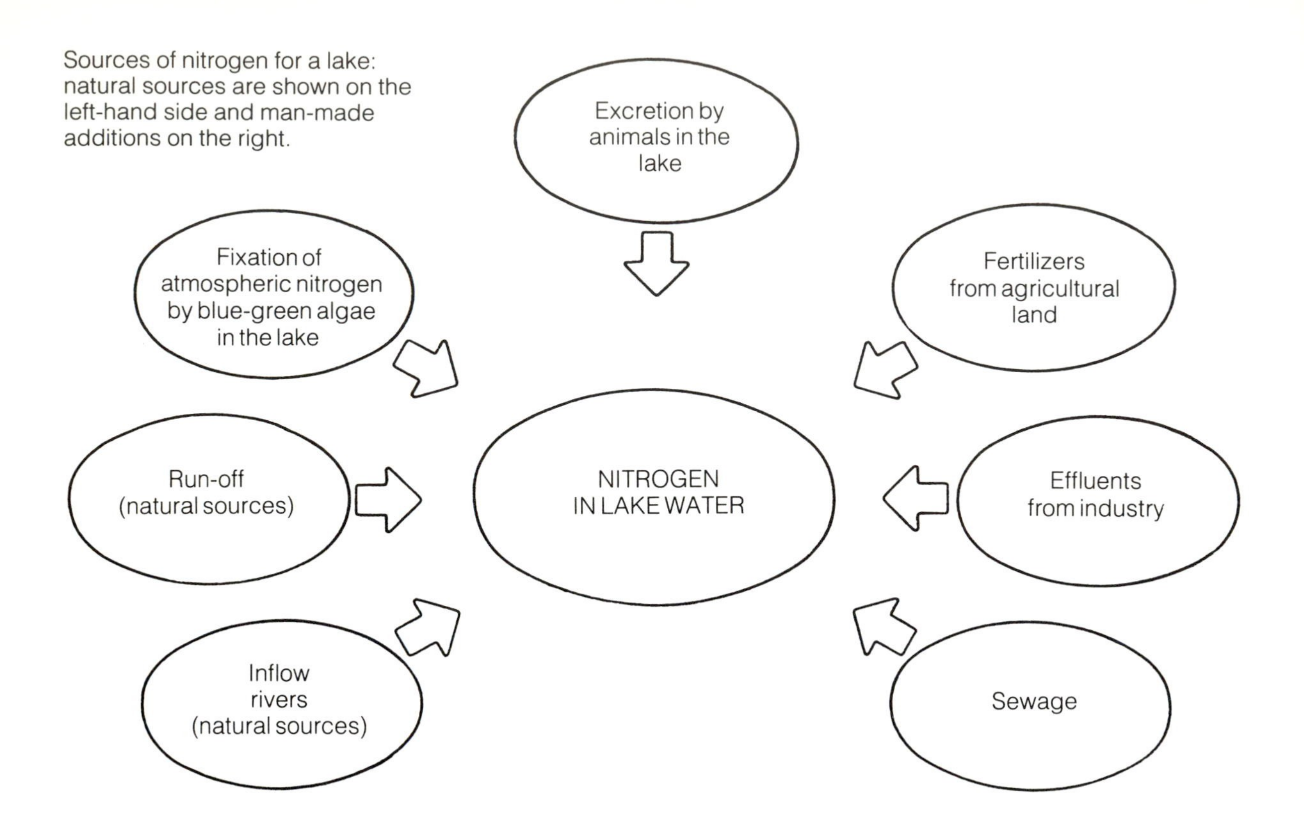

Sources of nitrogen for a lake: natural sources are shown on the left-hand side and man-made additions on the right.

Some plants and bacteria can capture gaseous nitrogen direct from the air and incorporate it into organic compounds; this process is known as **nitrogen fixation**. Terrestrial plants such as gorse and many other legumes rely on bacteria and fungi associated with their roots to fix atmospheric nitrogen. In lakes and ponds the blue-green algae are the most frequent nitrogen fixers; they have structures known as heterocysts in which nitrogen fixation occurs. The process is much encouraged if other forms of nitrogen are in short supply.

In many lakes the concentration of nitrogen compounds in the water is extremely low and any additional inorganic nitrogen is taken up immediately by the plants. This is why treated sewage effluent causes large growths of algae if put directly into a lake or river. Raw sewage contains complex organic compounds of nitrogen which must be decomposed by bacteria before they can be used by plants; the bacteria use up oxygen from the water and thus kill aquatic plants and animals. However, if this process of bacterial decomposition has been completed in the sewage works before the effluent is discharged, a 'clean' (from a health point of view) effluent is produced; but this effluent is still a very rich source of plant nutrients all ready to be used by the waiting algae. 'Clean' sewage is a major cause of eutrophication in lakes; the other is agricultural fertilizers washed off the surrounding land. These also contain soluble inorganic nitrates and have the same effect, vastly stimulating plant growth, but now in water instead of on farmland as intended.

Phosphorus

Phosphorus is one of several elements essential for all plant growth. Carbon and nitrogen are much more abundant than phosphorus in natural waters and are additionally available from gases in the atmosphere; so in the majority of lakes the availability of phosphorus is the limiting factor that controls the rate at which plants grow and, therefore, the productivity of the whole plant community. Phosphorus enters a lake via rainfall, from upstream lakes, from sewage, from industrial effluents and from the surrounding land. Phosphorus supplies to the plants are also enhanced by internal recycling within the lake itself, both from the sediments and through animal excretion. Phosphorus is much more readily lost from an ecosystem than nitrogen and carbon because it reacts with mud and chemicals in the water in ways that make it unavailable to plants.

There are many phosphorus compounds, both organic and inorganic, but plants can only absorb phosphorus as dissolved inorganic phosphate. In this form it is rapidly taken up by large plants (macrophytes) and algae as soon as it appears in the water. Thus concentrations of dissolved inorganic phosphate in unpolluted lake water are almost invariably low. Other phosphorus compounds are bound up in the tissues of living organisms or firmly attached to clay particles, either in the mud or suspended in the water.

In a shallow lake, phosphorus is frequently exchanged between the mud and water when wind stirs the water down to the bottom and disturbs the surface of the mud. The phosphorus released enables more algae to grow and when they die they use up more of the oxygen from the water just above the surface of the mud as they decompose. This creates conditions in which phosphorus is released by bacteria. The phosphorus is trapped for long periods in the still water of a deep lake but in a shallow lake the frequent reoxygenation of the bottom water and disturbance of the mud surface releases it more quickly back into the water and the cycle starts again. When more man-made nutrients (e.g. sewage and fertilizers) enter the lake, this process is accelerated. (After Biro, 1984.)

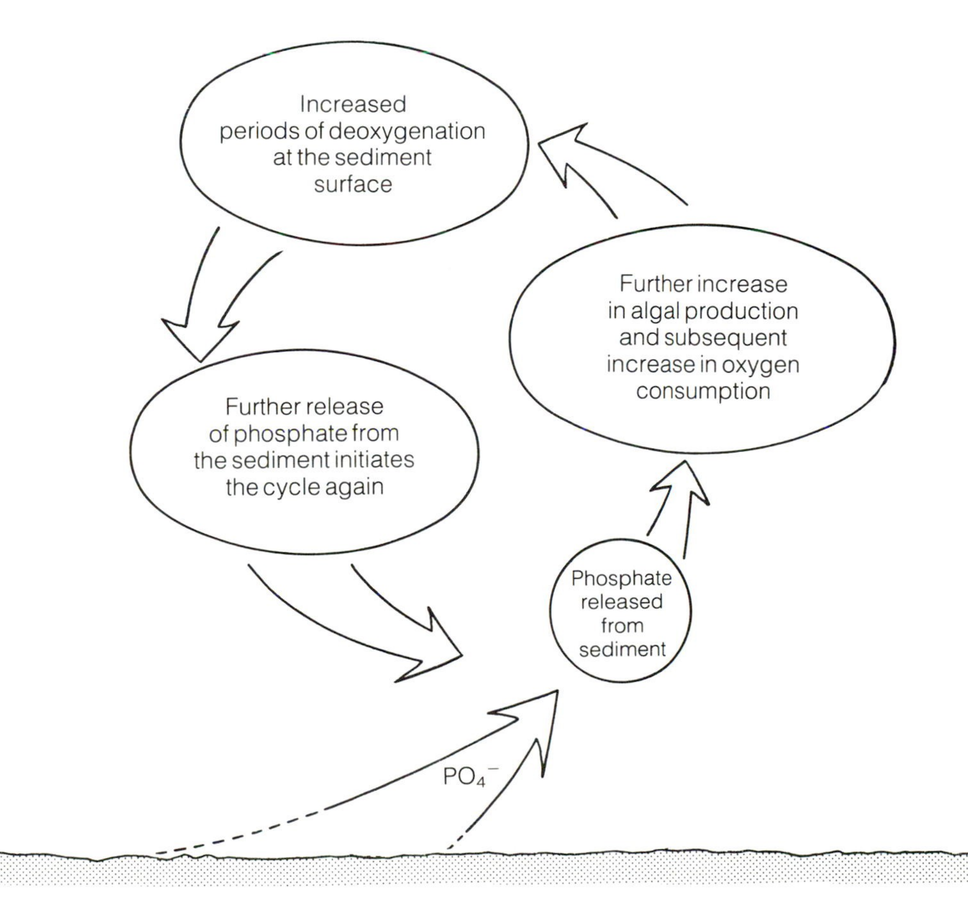

Phosphorus loading

The importance of phosphorus in determining the amount of plant growth in lakes has prompted a lot of research into the mechanisms by which it becomes available to the plants. More is known about lakes of the North Temperate Zone than those of other parts of the world and most of the work is based on lakes which have a Temperate Zone pattern of seasonal variation (see Chapter 4). It has been shown that for many such lakes the amount of algae may be predicted with reasonable accuracy if what is known as the **phosphorus loading** of the lake can be computed. The phosphorus loading is usually thought of as the total amount of phosphorus entering a lake each year per square metre of its area. There are two problems with this, apart from the difficulty of estimating the magnitude of all the sources of phosphorus, particularly those in run-off from the land. First, not all the phosphorus is in the right form to be available to the algae and, secondly, the amount available during a year depends on the depth and mixing regime of the lake. It has therefore been suggested that a better definition of phosphorus loading would be the yearly total of biologically active phosphorus per cubic metre of water in the August epilimnion of deep lakes (ie. the mixed layer), or of the whole depth in lakes shallow enough to mix to the bottom. Calculating phosphorus loading in this way allows very accurate prediction of the amount of algae which will be present in the summer and takes account of the differences due to lake depth. This is of interest to ecologists seeking to get to grips with the details of ecosystem function and is also potentially of considerable economic significance to reservoir managers concerned with controlling algae which clog up their filtration systems.

In lakes with a relatively low phosphorus loading, there are both large water weeds (macrophytes) and sparse populations of planktonic algae in the plant community. If the phosphorus loading is increased the macrophytes respond by growing more rapidly and prolifically, but if phosphorus loading continues to increase there comes a time when the water weeds decline and the algae take over as the dominant 'primary producers'. With very high phosphorus loadings, macrophytes may be completely eliminated; particularly those which have to grow from the bottom of the lake and are shaded out by dense populations of algae. Although this seems to be the mechanism causing water weeds to disappear, it has proved very difficult to establish the level of phosphorus loading at which the switch from macrophytes to algae will happen; some lakes with high loadings of phosphorus have algae dominant while others with equally high levels still have dense growths of water weeds. It is clear that other factors play their parts in determining the balance between these two sectors of the plant community. Studies on the Norfolk Broads in eastern England (described in Chapter 10) have begun to illuminate the complexity of the interrelationships between nutrient supplies, plants, animals and man. For one of the Broads (which are very shallow) it has proved possible to estimate past phosphorus loadings from measurements of phosphorus stored in the sediments, and to calculate the resultant phosphorus concentrations in the water. Comparison with concentrations and loadings measured recently show how greatly the phosphorus content of this lake has increased.

The history of phosphorus loading and concentration in Barton Broad calculated from analysis of the sediments at different levels, and recent measurements of phosphorus input and concentration (from Moss, 1980)

Date	Calculated loading of total phosphorus ($g\ P/m^2/year$)	Calculated concentration of total phosphorus ($\mu g\ P/l$)
1800	0.4	13.3
1900	1.55	52.0
1920	2.15	72.0
1940	3.55	119.00
1974–76 measured mean	10.83	329.0

Phosphorus is also present in the bodies of dead plants and animals, but it can only be of use for plant growth if the organic molecules which contain it are first broken down to separate the phosphate ions from the rest of their chemical associates. Thus, even if a lake contains a large quantity of total phosphorus, only a small proportion of it may be available for plant growth and then usually only at certain times of year.

Dead plants and animals are broken down by bacteria in the sediments at the bottom of the lake and phosphate is released into the water in the spaces between the sediment particles. This process is most rapid where the sediment is devoid of oxygen. In a lake with oxygenated water right down to the bottom, a thin surface layer of sediment (perhaps 5 mm) will also contain oxygen and this acts as a barrier, preventing the phosphate in the deoxygenated sediment below from diffusing upwards into the overlying water. If all the water becomes deoxygenated, as it does during summer stratification of productive lakes (see Chapter 4), then phosphate is gradually released into the water. However, in a stratified lake, this phosphate will not be mixed into the surface illuminated water (the euphotic zone) where the algae can use it, until the lake water is thoroughly mixed, right down to the bottom.

In a shallow lake which does not stratify for long, there is usually oxygen available at the sediment surface all the time and the phosphate might stay locked up in the deeper sediments for ever if it were not for the fact that shallow water allows wind mixing to extend to the bottom fairly frequently. This mechanical mixing breaks up the thin barrier of oxygenated sediment and releases phosphate from below into the water. During calm periods the barrier re-forms and breakdown starts again in the deoxygenated lower layers of the sediment. In the water the phosphate released is quickly absorbed by plants and their populations increase. This provides more phosphorus-containing material to die and decompose on the bottom, which in turn reduces the oxygen content at the surface of the sediment (because the bacteria use it up) and speeds up the release of more phosphate. It is the frequently repeated alternation of calm and mixing in shallow lakes that leads to the self-fuelling cycle of increased plant growth and helps explain why many shallow lakes are very productive.

The mud

Away from the shoreline, which often has a substrate of wave-washed stones, the bottom of a lake is covered by mud formed by the sedimentation of particles from the water. These particles comprise inorganic silt, derived directly from the surrounding land or via the inflows, plus the organic particles of dead plants and animals sinking down through the water column. The balance between these two types of particles determines the organic content of the mud and its value both as food for bottom-dwelling animals and as a source of chemicals for the water; we have already seen the importance of mud in the phosphorus cycle.

In many lakes the bottom mud is the principal site of decomposition (though this also occurs in the water to a certain extent, particularly in tropical lakes). Muds with a high organic content contain large numbers of bacteria which live on the organic molecules and

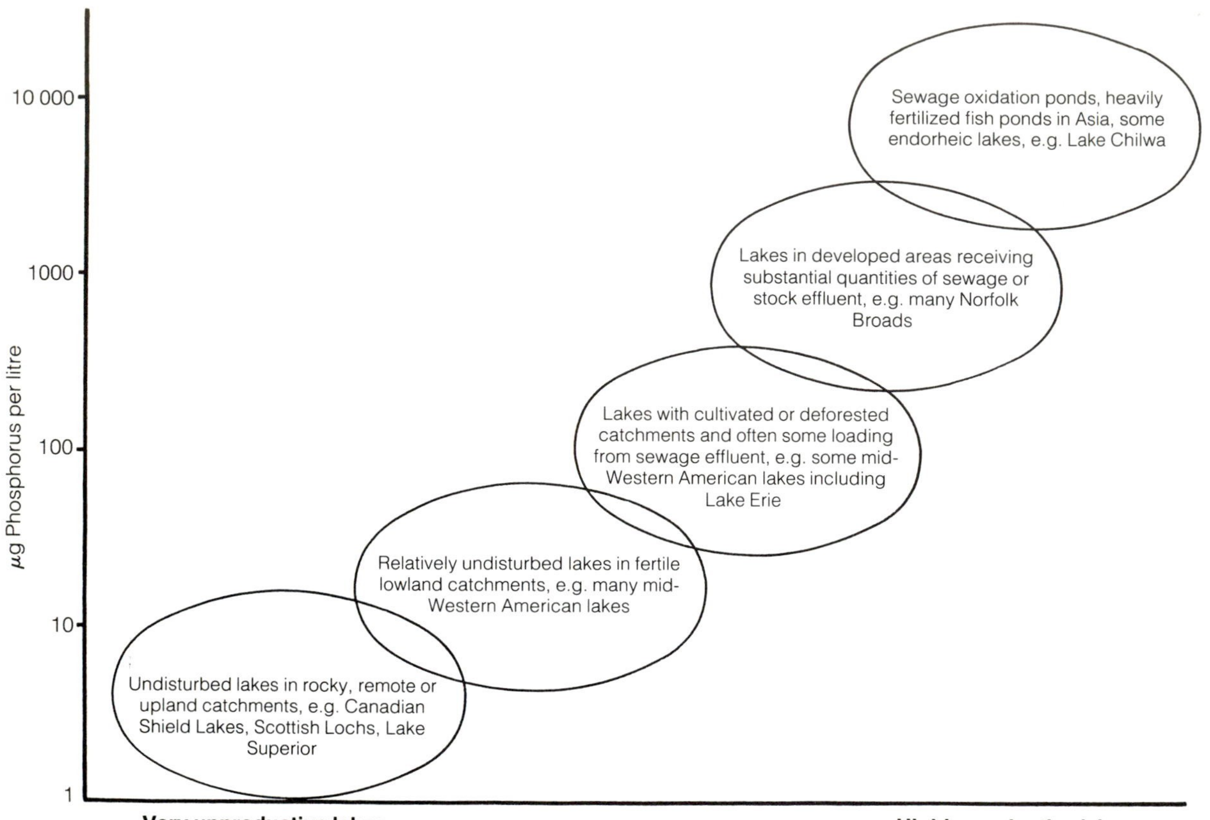

Various types of lakes whose water contains different concentrations of phosphorus. (After Moss, 1980.)

break them down into smaller, inorganic molecules. Many of the bacteria involved in these processes require oxygen and the more there are of them, the more oxygen they take from the mud until it may become totally devoid of oxygen and start to absorb it from the overlying water. If the lake is stratified there will be no mixing of the hypolimnion and thus no replenishment of the oxygen used from the lower water layer. This results in anoxic conditions and the suffocation of aquatic animals. In very productive lakes with highly organic mud, the hypolimnion may become completely deoxygenated before the stratification is broken down and the whole lake re-oxygenated by mixing. If the lake is small and shallow, animals may be killed during the oxygen famine and their populations take a long time to build up again. If the lake is larger and regularly has a deoxygenated hypolimnion, only those animals able to endure the period without oxygen will be found in the deeper parts.

While the mud is without oxygen some processes of decomposition cannot occur, but other chemical reactions, which only work in the absence of oxygen, take over instead. Anaerobic bacteria can operate without oxygen, and some of them break down large quantities of organic matter in the mud to produce methane gas, which has a simple molecule containing only carbon and hydrogen. Most of this methane dissolves in the water within and immediately above the mud and is normally used by other bacteria as a source of energy. These bacteria are aerobic and can only live in oxygenated conditions, so they are found in the hypolimnion just above the deoxygenated zone. Their activity uses up oxygen, so the more methane that is produced, the more they multiply and the more rapidly the hypolimnion becomes deoxygenated. If these methane-using bacteria are completely eliminated by deoxygenation of the water, but the production of methane continues in the mud, then methane will escape from the water as a gas, sometimes it ignites spontaneously to give 'will-o'-the-wisps' or 'jack o' lanterns' across the surface.

Other anaerobic bacteria can use sulphates in the organic matter of the mud and a byproduct of their activity is hydrogen sulphide. This gives the characteristic 'rotten eggs' smell when stagnant muddy water is stirred up. It is also toxic and is thus a further hazard to the life of lakes where conditions lead to its production.

The release of such noxious gases may occur naturally in small lakes and ponds which are overloaded with organic matter, but if they are evident in a lake of some size they are usually a sign that conditions are deteriorating. Many lakes have a layer of deoxygenated water at the bottom during the summer but maintain an overlying layer of oxygenated water. The balance between the volume of these two layers and between the intensity of aerobic and anaerobic activity is very important to the health of the lake ecosystem.

3

LAKE communities

Despite the varied characteristics of lakes and the variety of plants and animals inhabiting them, they have one feature in common: they are all basins filled with water. We can therefore treat them all in the same way by subdividing the whole watery habitat available to aquatic plants and animals into three major regions whose features are similar in most lakes. Where land meets water there is a **littoral zone** of gradually increasing water depth, sometimes colonised by large, emergent plants. Where the water is too deep for rooted plants to grow, both plants and animals must be adapted to surviving in the open water, or **pelagic zone**, free of both the bottom and the sides of the lake basin. The third component of the lake habitat is the **benthic zone** – the bottom of the lake. Usually this is covered by layers of fine mud, often very thick, in which the animals live. In the littoral zone there is enough light for plants to grow in the mud or on the stones and rocks, but beyond the littoral zone there are no rooted plants because the water is too deep for light to reach them. The deepest parts of the open water form the **profundal** or **abyssal zone** but this is only relevant in extremely deep lakes (perhaps more than 200 m deep).

There are also some rather specialised animals which live out in the air on the surface film of the water and this might be considered as another life zone of lakes. Each of these zones offers different opportunities to the species that live there, and many organisms have evolved appropriate structures or behaviour patterns which are characteristic of the particular life zone, no matter where the lake is sited. None of these zones, or their communities, is sharply defined; they just provide a useful way of subdividing and understanding what would otherwise be an even more complex situation.

In each community it is the green plants that make the food on which all the rest ultimately depend. That is why they are called **autotrophs** (= self-feeders) or **primary producers**. Animals which feed on green plants are known as **herbivores**; those which eat other animals are **carnivores. Omnivores** have a mixed diet and **detritivores** feed on dead plant and/or animal material. These different feeding habits link together into a **food web**, but distinctions between feeding types are often blurred and these terms are only a convenient short-hand. Although the food web will differ between regions of the lake, there will be a lot of overlap. For example, some fish which live in the open water of the pelagic zone come inshore to breed and their young feed in the littoral zone. Later these young fish move offshore to become part of the pelagic food web. The interlinking of food webs is important in parasite life cycles. For example, the tapeworm *Ligula* is often found filling the body cavity of sticklebacks (*Gasterosteus aculeatus*) and other small fish. It makes them bloated and unstable in the water – easy prey for fish-eating birds. The tapeworm matures in the bird's intestine and sheds eggs which are deposited back in the lake along with the bird's droppings. A parasitic larva hatches in a couple of days and invades a copepod. When this is eaten by a fish, the cycle is complete. Each stage of the parasite can only invade the correct host, there can be no short cuts; so the parasite is dependent upon one thing feeding on another in a predictable way.

The three main ecological regions of a lake for which plant and animal communities are described in this chapter.

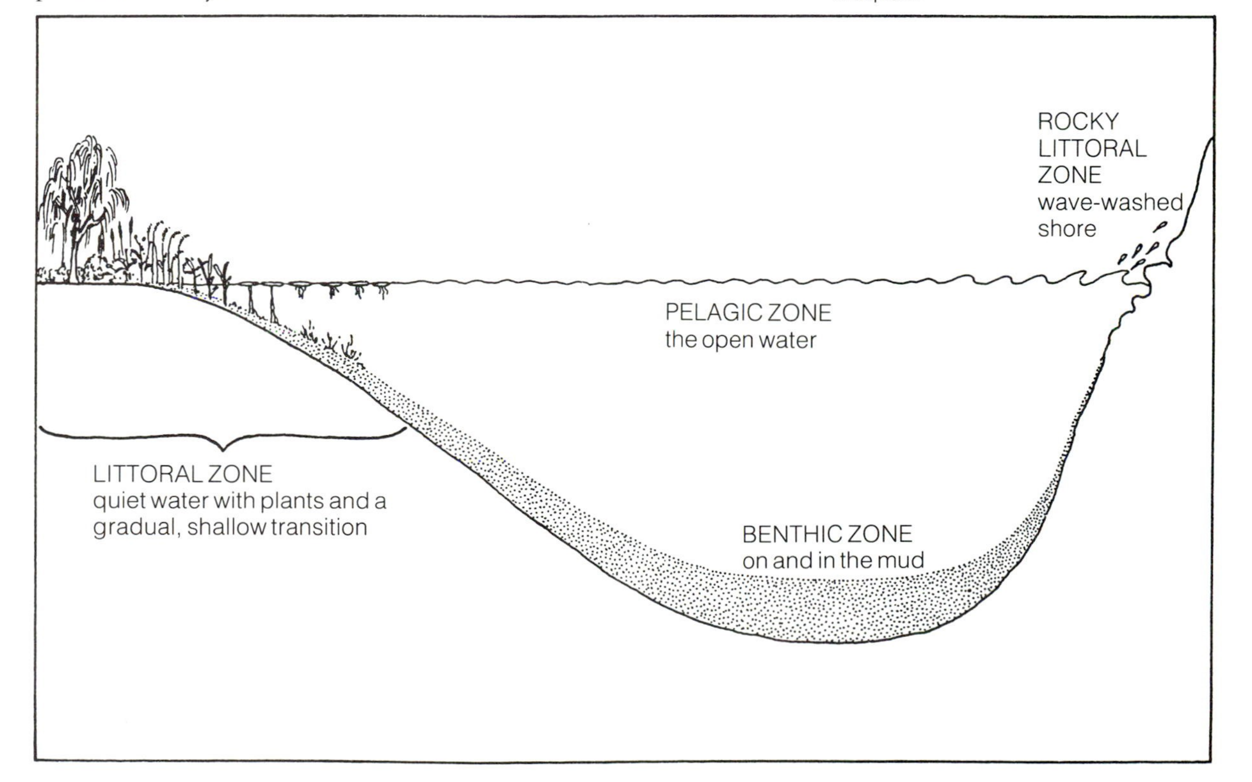

Table 3.1 *The number of freshwater fish species found in Europe, North America and Africa (from Maitland, 1977).*

Continent	Area (million km^2)	Species	Families
Europe	10	215	29
North America	24	687	34
Africa	30	1425	50

Fish are usually the largest animals in the lake community and between them can eat most kinds of invertebrates; some eat other fish, a few even survive by consuming their own young; very few are herbivores. Many have a mixed diet that changes in emphasis with the seasons or as the fish grows. There is no such thing as a typical fish and it should never be supposed that one fish is very like another; they all play different roles in the lake community. Because of their greater mobility it is more difficult to assign fish to particular life zones of the lake than it is for the smaller animals.

There are more than 20 000 living species of fish (more than all the reptiles, birds and mammals put together). Of these, about 7000 live in fresh water and the majority of these are found in tropical lakes and rivers, with fewer species at higher latitudes. The great variety of species found in fresh water is probably because river systems and their lakes are usually isolated from other such systems, allowing separate lines of evolution to take place in each. Some of the biggest lakes contain sufficiently varied life zones to offer many opportunities for evolution. They are also frequently old enough to have allowed plenty of time for new species to develop, some of which will occur in that lake and nowhere else (**endemic** species). Although some fish only live in lakes, others are found in rivers as well, migrating between the two. Many inhabit both, because the conditions in a large river are often not very different from those in a shallow lake.

Fish feed on plankton, plants and invertebrates and are in turn eaten by top carnivores such as birds, otters and man. At each step in the chain, food is eaten to gain energy and nutrients. Most of that energy is used to support daily activity and is dissipated into the environment as heat (even in 'cold-blooded' animals). Only the surplus is available for growth and reproduction, so the energy available to the next link in the food chain is always less than that which entered the present one. For this reason there can never be as many carnivores as there are herbivores and there will be even fewer top carnivores. For this reason too it is very inefficient for humans to use carnivores as food; more people can be fed if they use food lower down the food chain such as herbivores or plants. Unfortunately most algae are unsuitable as human food and must be converted into something more useful by being fed to animals first. Few large fish eat algae directly but in the tropics there are tilapia, which feed on blue-green algae (Chapter 4) and are themselves delicious. They are an important food for people who are then at the end of a relatively short food chain (algae → fish → man) and reap a higher proportion

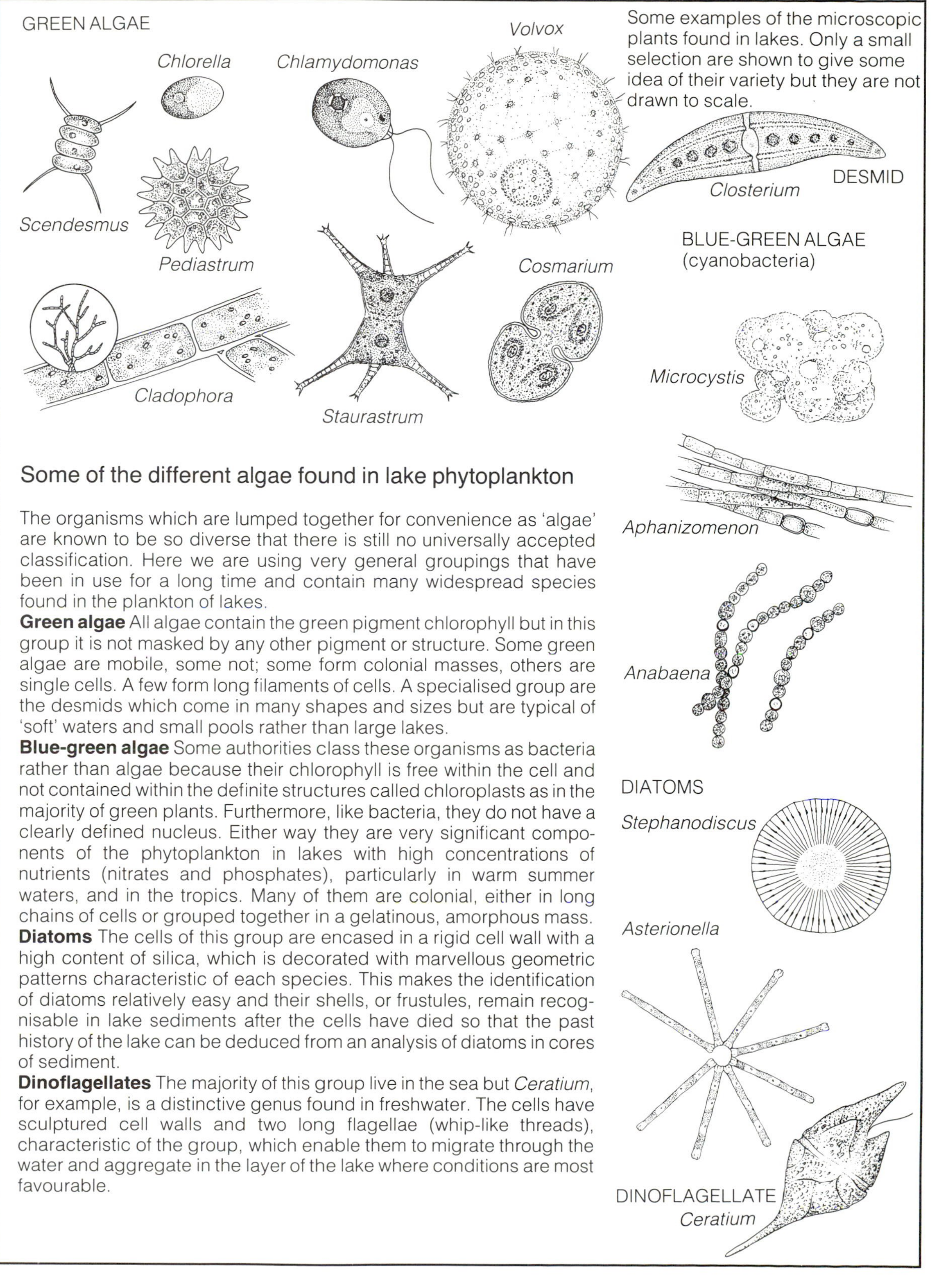

Some of the different algae found in lake phytoplankton

The organisms which are lumped together for convenience as 'algae' are known to be so diverse that there is still no universally accepted classification. Here we are using very general groupings that have been in use for a long time and contain many widespread species found in the plankton of lakes.

Green algae All algae contain the green pigment chlorophyll but in this group it is not masked by any other pigment or structure. Some green algae are mobile, some not; some form colonial masses, others are single cells. A few form long filaments of cells. A specialised group are the desmids which come in many shapes and sizes but are typical of 'soft' waters and small pools rather than large lakes.

Blue-green algae Some authorities class these organisms as bacteria rather than algae because their chlorophyll is free within the cell and not contained within the definite structures called chloroplasts as in the majority of green plants. Furthermore, like bacteria, they do not have a clearly defined nucleus. Either way they are very significant components of the phytoplankton in lakes with high concentrations of nutrients (nitrates and phosphates), particularly in warm summer waters, and in the tropics. Many of them are colonial, either in long chains of cells or grouped together in a gelatinous, amorphous mass.

Diatoms The cells of this group are encased in a rigid cell wall with a high content of silica, which is decorated with marvellous geometric patterns characteristic of each species. This makes the identification of diatoms relatively easy and their shells, or frustules, remain recognisable in lake sediments after the cells have died so that the past history of the lake can be deduced from an analysis of diatoms in cores of sediment.

Dinoflagellates The majority of this group live in the sea but *Ceratium*, for example, is a distinctive genus found in freshwater. The cells have sculptured cell walls and two long flagellae (whip-like threads), characteristic of the group, which enable them to migrate through the water and aggregate in the layer of the lake where conditions are most favourable.

of the energy originally trapped by the algae than is possible when carnivorous fish such as trout and salmon are eaten.

The open water or pelagic zone community

All living organisms, even the smallest, tend to sink in freshwater. To maintain a position in the open pelagic zone of a lake, plants and animals must therefore swim, have a flotation device, or be held up by water turbulence. Fish are powerful swimmers and can move regardless of water currents but many also have a swim bladder to keep them buoyant and reduce the energy needed to counter the

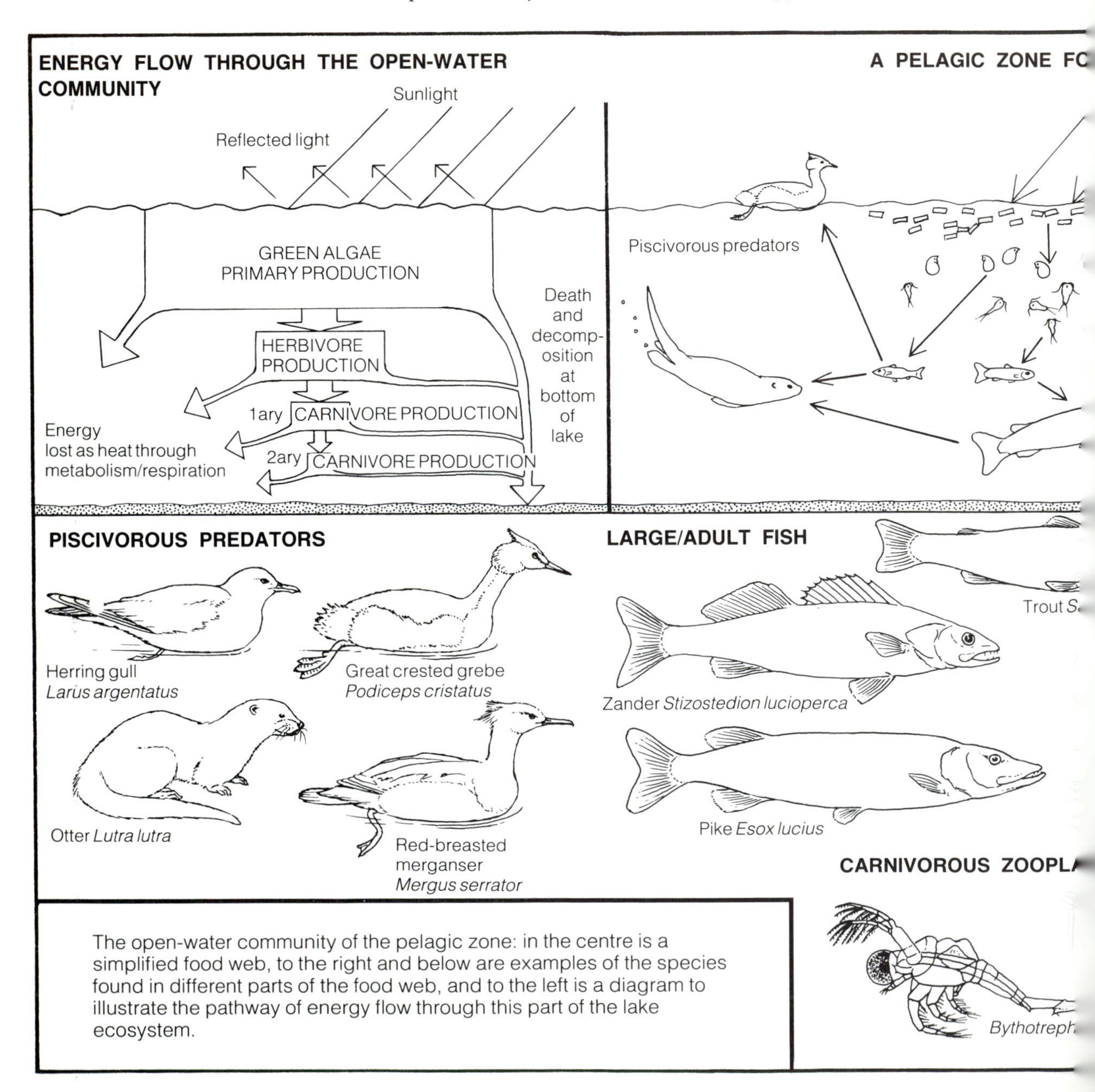

The open-water community of the pelagic zone: in the centre is a simplified food web, to the right and below are examples of the species found in different parts of the food web, and to the left is a diagram to illustrate the pathway of energy flow through this part of the lake ecosystem.

effects of gravity. The small planktonic animals (**zooplankton**) can also swim but not with sufficient strength to counteract water currents. This is also true of the microscopic algae (**phytoplankton**), although some of them can move independently and others have devices to keep them afloat. Consequently both plant and animal plankton populations tend to be swept away by running water in rivers and streams, but in the relatively still waters of lakes they can build up large numbers and achieve their full diversity. This is one of the most important biological distinctions between a lake and a river.

The pelagic zone is the home of the plankton – tiny organisms that swim, drift and twitch through their transparent, three-dimensional

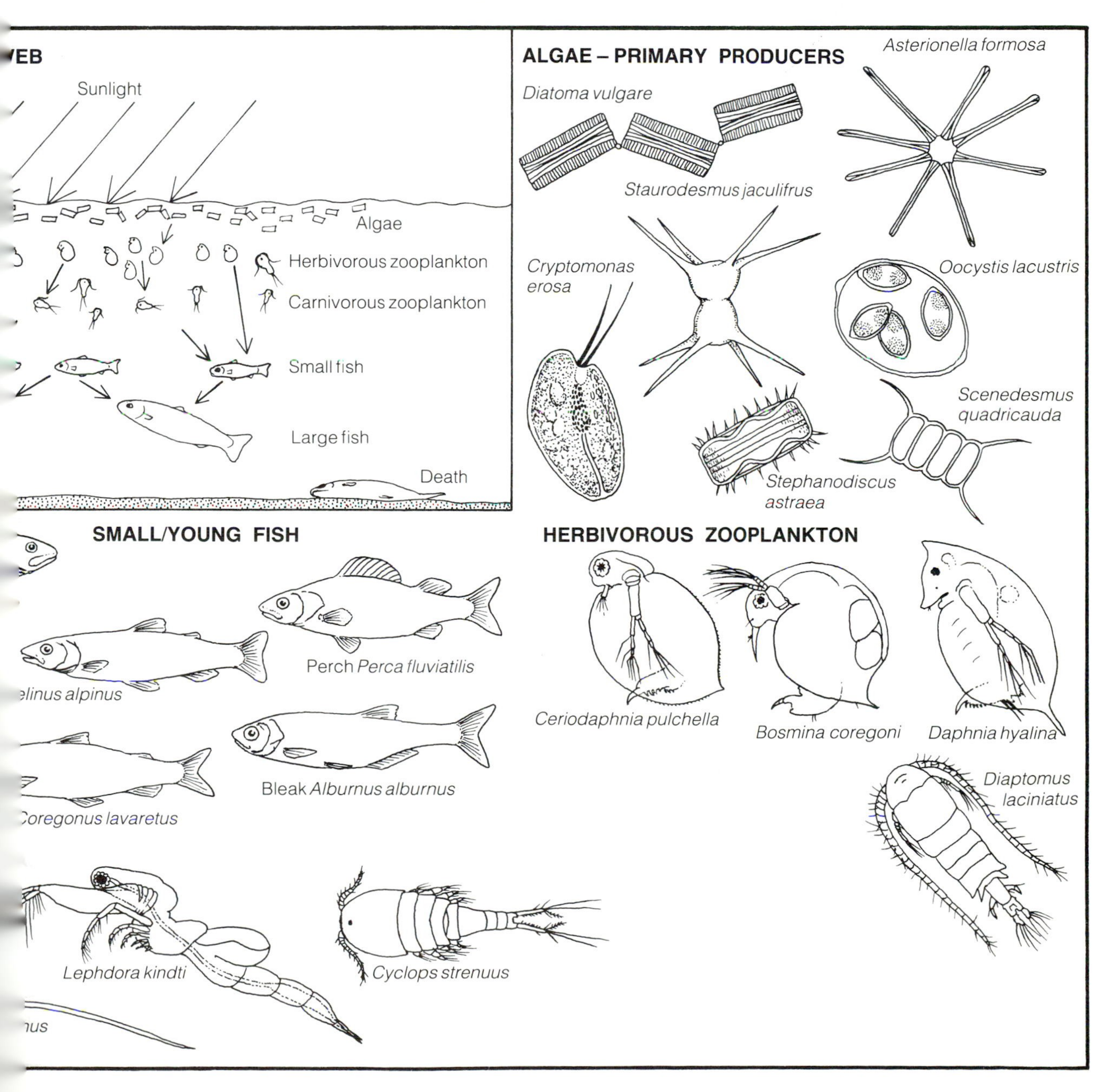

habitat. The phytoplankton consist of algae (singular = alga), ranging in size from single cells less than 10 μm in diameter (1 μm = 0.001 mm), such as *Chlorella*, to clusters of cells, like *Volvox*, which are large enough to be seen with the naked eye. The smaller algae impart a greenish tinge to the lake water but only become obvious when they proliferate to form 'algal blooms' at the surface. Some algae (e.g. *Spirogyra* and *Cladophora*) form thin filamentous strings of cells, up to 1 m in length. These are among the largest algae found in lakes. *Cladophora* is familiar to pond owners as 'blanket weed', which forms large clogging masses.

Several major groups of algae occur in the phytoplankton of lakes the world over, although the species vary. The sheer number of algal species, even in a single lake, is an apparent paradox: in a seemingly featureless environment, how can so many different species live together? In fact, although members of the pelagic community are all adapted to living in open water, there are innumerable differences between them: they live at different depths, reproduce at different times of year, need different amounts and types of nutrients, and grow at different rates.

The majority of planktonic animals are herbivores ('secondary producers') feeding on algae but there are also some carnivores. The most important species are crustaceans of two major groups: the Cladocera, of which the water-flea (*Daphnia*) is a common example, and the Copepoda. Comparatively few crustaceans are truly planktonic but the relevant ones are often widespread throughout the world. They range in size from 0.2 mm to about 4 mm long, with exceptional species reaching 1–2 cm. These small animals are a very important link between the algae and much larger creatures such as birds and fish, many of which are of great economic or aesthetic interest. A few rotifers ('wheel animalcules') are also planktonic inhabitants of the pelagic zone, but most creep and swim among the vegetation near the shore. The largest can just be seen with the naked eye.

Rotifers and cladocerans are parthenogenetic, i.e. the females can produce young without males being available. Cladoceran eggs are laid into a brood pouch under the back of the carapace and develop into replicas of the adult before being released into the water when the female moults. The speed with which young are produced and develop increases as the water gets warmer and when food is plentiful these creatures can increase their populations very rapidly. Under less favourable circumstances, some eggs develop into males and sexual reproduction occurs. This produces resting eggs which are enclosed in a horny pouch which resembles a saddle and is called an 'ephippium' from the Greek *ephippios* = for putting on a horse. When the female moults or dies, the ephippium sinks to the lake bottom and may remain there for many months before hatching. Ephippia can survive over winter or if the water dries up, and then hatch out when conditions improve.

In copepods, sexual reproduction is obligatory and the eggs hatch into small planktonic larvae called nauplii. Their life history is complicated and involves eleven moults before the animal is mature and able to reproduce. This contrasts with the four or five moults which precede maturation in cladocerans. It takes only about 10 days (at 15° C) for a cladoceran egg to develop into a mature female carrying more eggs; in the common copepod *Cyclops*, it may take as long as 3 weeks. Copepods are less able to respond to changing circumstances than are cladocerans and rarely manage to reach such dense populations.

In open water there is nowhere to hide, for either predators or their prey. Prey species have several possible strategies for survival, often used in combination. Some are transparent and almost invisible,

Animals found in the zooplankton of lakes

Rotifers are tiny but beautiful animals that move through the water by means of cilia which form a crown around and in front of the mouth. The beating of these cilia move the animal along and also create a current which draws particles towards the mouth at the centre. Many eat tiny bacteria and the smallest algae, rejecting particles that are too large by a brief reversal of the current. Others are able to eat much larger algae and some, such as *Asplanchna*, even eat other, smaller, rotifers. When food is plentiful rotifers reproduce very rapidly and their populations can increase markedly in a short time. Under favourable conditions numbers of some species may go from tens to hundreds per litre within a week.

The best known and most studied of the planktonic herbivores are the tiny crustaceans called **cladocerans**, such as the water-flea *Daphnia*. This is the animal that aquarists buy at the pet shop to feed their tropical fish. There are many species of *Daphnia*, and also a number of similar groups of cladocerans. All filter algae from the water by means of fine fringing hairs ('setae') on their legs. As they move jerkily through the water the legs beat rhythmically within the hollow carapace that covers them, and create a water current from which they extract food particles that come with it and also dissolved oxygen. The algae are gathered in a food groove at the base of the legs and carried forward to the mouth. As they pass down the gut and are gradually digested they can be seen, with a lens or microscope, through the transparent carapace.

There are two groups of **copepods**, the **calanoids** (including *Diaptomus*) which are usually herbivorous, and the **cyclopoids** (e.g. *Cyclops*), many of which are carnivorous. The former feed by sweeping algae into their mouth using their brush-like antennules. The latter use sturdier mouthparts to grab food and then chew it, so are often able to eat things larger than themselves. Some of this group have developed a parasitic way of life which may have been derived from holding on to animals such as small fish in order to feed off the surface of their body.

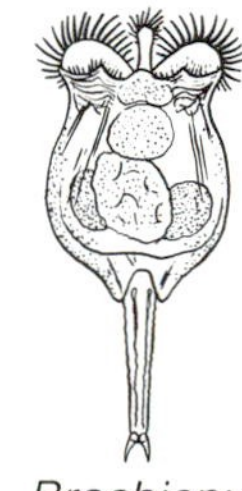

Brachionus

Asplanchna

Two rotifers found in the open water of lakes: *Brachionus*, a herbivore, and *Asplanchna*, a carnivore. The latter is about 1 mm long, and is just visible to the naked eye when it occurs in large numbers.

although their eyes need to be dark (to trap light) and their gut contents show up (very obvious in a well-fed *Daphnia* under the microscope). Potential prey can also be either too small or too big for particular predators to catch, but then they may fall foul of another predatory species; in any case most animals cannot stay the same size indefinitely. Prey might avoid predators by occurring at a different time of year, but there are limitations here too. In strongly seasonal lakes winter stops growth, and in aseasonal lakes there is no escape because predators can also live all year round. The best bet seems to be a nearly transparent body, coupled with the habit of swimming down to hide in the dark depths during the day. However, since phytoplankton do not live in the dark, it is then necessary to migrate back to the euphotic zone each night in order to feed. In view of the difficulties posed by life in clear open water, one might expect few species to manage it; the surprise is that the pelagic zone harbours such a diversity of organisms.

Fish are the most obvious predators upon the zooplankton. Most young fish start by feeding on rotifers and then move on to cladocerans. Some species such as the whitefish (*Coregonus*) of the European lakes and the alewife (*Alosa pseudoharengus*) of North America eat zooplankton all their life; others, like pike (*Esox lucius*) and perch (*Perca*), graduate to larger prey as they get older. Pelagic fish, particularly small ones, face the same problem in clear water as the zooplankton: there is nowhere to hide. This may be why many of them remain in the littoral zone, concealed among the macrophytes. However, pelagic fish can also hide effectively among themselves. So long as each individual lives along with many of its fellows, the chances of it being singled out and eaten are reduced, since its neighbours offer an equally tempting meal. Moreover, a shoal of individuals moving in different directions under attack offers a confusing target to a predator, which may end up not catching any of them.

Typical pelagic fish include the clupeids, a primarily marine family, which live in enormous shoals. A few species live in lakes where they play a role similar to that of their marine relatives (herring, sardine and anchovy) feeding on plankton, and are themselves eaten by many larger fish. They are important links in the food chains of the open water zone and are sometimes a significant source of human food.

The alewife *Alosa pseudoharengus* is an interesting freshwater clupeid. It occurs in the sea off the eastern seaboard of North America and migrates up rivers to spawn. This leads some to enter lakes, and certain populations, such as those in the Finger Lakes of New York, have become landlocked and isolated. The alewife hunts by sight and the most visible prey species and sizes are eaten first, so in lakes containing such predators the smaller zooplankton species come to predominate. This means that the community structure (the relative numbers of different sizes and species present) is partly determined by selective predation. This also applies to the phytoplankton: their population is at least partly governed by what the herbivores have left uneaten. The populations of food species in turn affect those who eat them, by influencing their rates of growth and reproduction. The European equivalent is the Twaite shad (*Alosa fallax*) which is also primarily a marine fish but enters rivers to spawn. There are landlocked, completely freshwater populations in Lake Killarney in Ireland and in Lakes Como, Lugano and Maggiore of the Italian Alps. Several African lakes contain clupeids and in Lake Tanganyika there are two endemic species known as 'Tanganyika sardines' (see Chapter 6) which have recently been introduced to man-made lakes, including Lake Kariba.

The Salmonidae, salmon, trout and their relatives, perhaps the best-known family of freshwater fish, are also pelagic. Unlike the clupeids, none is wholly marine and all spawn in fresh water, though some are anadromous, migrating to the sea to feed and returning to fresh water to breed. Salmonids are native to Europe, North America, North-West Africa and northern Asia but they have been introduced to suitable habitats all over the world, including India, Africa, Australia and New Zealand. They can only survive in cool, well-oxygenated water and in the tropics do not thrive except in highland regions. Spawning takes place at the coldest part of the year and eggs are laid in gravel beds, known as redds, and covered with gravel by the female after the male has fertilized them. The young feed on insects and other small invertebrates until big enough to tackle larger prey. Adults are also carnivores and in many places where they have been introduced are thought to have had a detrimental effect on the local fish fauna, not least because some of the

The perch

The Eurasian perch (*Perca fluviatilis*) is almost exactly the same as the yellow perch (*P. flavescens*) of North America and some authorities consider them to be the same species. They are usually about 20–35 cm long, but can grow to more than 50 cm and to exceed 4 kg in weight. In very dense populations they become stunted and even the oldest fish never reach maximum dimensions. They are generally found in lowland lakes and rivers, even in slightly salty conditions, but they cannot survive in water with an oxygen content less than 3 mg/l, so are confined to cooler places. In large lakes they are normally found in the top 50 m, out in the open water, but come inshore to breed. The females each lay 4000–300 000 eggs, winding them round vegetation or stones; the males follow and fertilise them. The eggs are 2–2.5 mm in diameter and coated by mucus which swells up and protects them from small predators until they hatch. Shoals of young swim in shallow water until they are old enough to move out into the pelagic zone. Survival of the young is very variable from year to year. This is evident when the whole population is sampled and unexpectedely large numbers of particular size groups are found. These are known as strong year classes and it has been shown for the perch in Windermere that they correlate with the number of degree-days (number of days × number of degrees) above 14°C in the year in which they were hatched. Warm years imply good crops of plankton on which the young perch feed and faster growth rates. In species like this, near the northern edge of their range, growth and survival rates are likely to be very responsive to even quite small climatic variations.

salmonids can grow to more than 20 kg. Today trout are delivered by tanker or aeroplane to remote lakes in Canada and the United States, often without hope of their breeding (for climatic and other reasons) but just to fatten up during the summer and provides sport for fishermen.

Apart from a few landlocked populations of the Atlantic salmon (*Salmo salar*) in northern Canada, all the salmon are anadromous and primarily found in rivers. The salmonids include trout and char, which are found in many lakes, but there is considerable confusion between the names used in Europe and North America. The brown trout (*Salmo trutta*) lives in lakes and rivers all over Europe. The sea trout is its migratory form which occurs in more northerly waters and grows much larger; it is silvery with only scattered spots compared to the more colourful, densely spotted brown trout.

The American population of brown trout came from Germany in 1883; more followed from Loch Leven (Scotland) in 1884, leading to a long-running controversy as to whether or not they were the same species.

The rainbow trout (*Salmo gairdneri*) is a native of North-West America but is now familiar in many other parts of the world through introductions and because it is cultivated on fish farms; the steelhead is its migratory form. The cut-throat trout (*Salmo clarkii*) of the Pacific coast also has migratory and non-migratory forms.

The char (*Salvelinus alpinus*) (see Chapter 5) is found in cool clear lakes of the Northern Hemisphere and is migratory around the Arctic seas. Its lake populations are probably relicts of Ice Age migrants. They mostly spawn in running water but some, such as those in Windermere, do so at the bottom of deep lakes.

The brook char or brook trout (*Salvelinus fontinalis*) is migratory and native to western North America but has been widely introduced in Europe. *Salvelinus namaycush* is known as lake trout in its native North America and as American lake trout where it has been introduced to Europe. In Europe it has been established in a few of the large deep lakes of Sweden and Switzerland.

The Coregonidae (including whitefish and cisco) is a family of fish closely related to the salmonids and similarly found in cool clear waters of the Northern Hemisphere. A few are anadromous. Some species are benthic but many are pelagic, moving through the open water in large shoals and feeding on the zooplankton, eating the largest first. In lakes with a large population of these fish the smaller zooplankton tend to be more abundant as a result. Coregonids are themselves eaten by many larger fish and are sometimes the basis of important commercial fisheries. The different *Coregonus* species are very difficult to distinguish, partly due to the fact that a single species may have slightly different forms depending on the environment in which it occurs.

Most birds feeding in the pelagic zone are fish-eaters and either dive for their prey (grebes, divers, sawbill ducks) or catch it close to the surface (osprey, pelicans, terns). There are also specialist insect feeders which take emerging midges and mayflies from the surface or snatch them from the air above. Birds are also attracted to the

The freshwater fauna of Australian and New Zealand lakes

Australia and New Zealand have been islands, separate from the other continents, for such a long time that they have a very distinctive fauna. This is very obvious when one considers their mammals: they have almost none of the familiar mammals of the Old World or the New World, except where these have been introduced, and instead the marsupials (of which very few are found elsewhere) have adpated to all the different habitats and foods available. Similarly, many of the animals found in their lakes are quite different from those found elsewhere. All the major groups are represented but many of the species or genera which are widespread in other parts of the world are missing and replaced by species found only in Australia and New Zealand. There is a good example among the calanoid copepods: the common genus *Diaptomus* does not occur and its place is taken by species of *Boeckella* which are found in many Australian lakes and also those of southern South America, including Lake Titicaca (Chapter 5).

The native fish fauna of Australia has very few species compared with other parts of the world. About 180 species of freshwater fish have so far been distinguished compared with, for example, about 2000 found in Africa. However, many of the species found in Australia are endemic and a lot of them are closely related to marine groups. Members of the family Galaxiidae (southern trout) are not found in the Northern Hemisphere and only a few species are known from Australia, New Zealand, the southern tips of South Africa and South America, plus a few of the Pacific islands. This is one of the most diverse families: a few species are migratory but most are confined to freshwater.

At least 25 species of fish have been introduced to Australian waters and many have successfully reproduced and established populations. They include many of the common European coarse and salmonid fish such as trout, roach and perch, as well as warm-water species such as the mosquito fish (*Gambusia*) and cichlids such as *Tilapia mariae*.

open pelagic zone of lakes as a refuge from terrestrial predators; huge flocks of gulls roost in safety on lakes during winter nights. Some species of geese and ducks seek the protection of open water while they are temporarily incapable of flight during their annual moult.

The benthic zone community

The term benthic applies to anything living on the bottom, and also to those species that live by burrowing in it. Algae attached to the surface of the substrate are also called benthic to distinguish them from the phytoplankton which live floating freely in the water. In some shallow lakes benthic algae may be an important source of food but most benthic animals beyond the littoral zone derive their nourishment from detritus raining down from the open water above.

Most of the bottom beyond the littoral zone will be covered in mud whose particle size and organic content depends on conditions specific to each lake. The larger inhabitants of this zone are mostly worms, the larvae of chironomid flies (midges) and, in some lakes, molluscs. In addition there are numerous smaller animals such as nematode worms and ostracods.

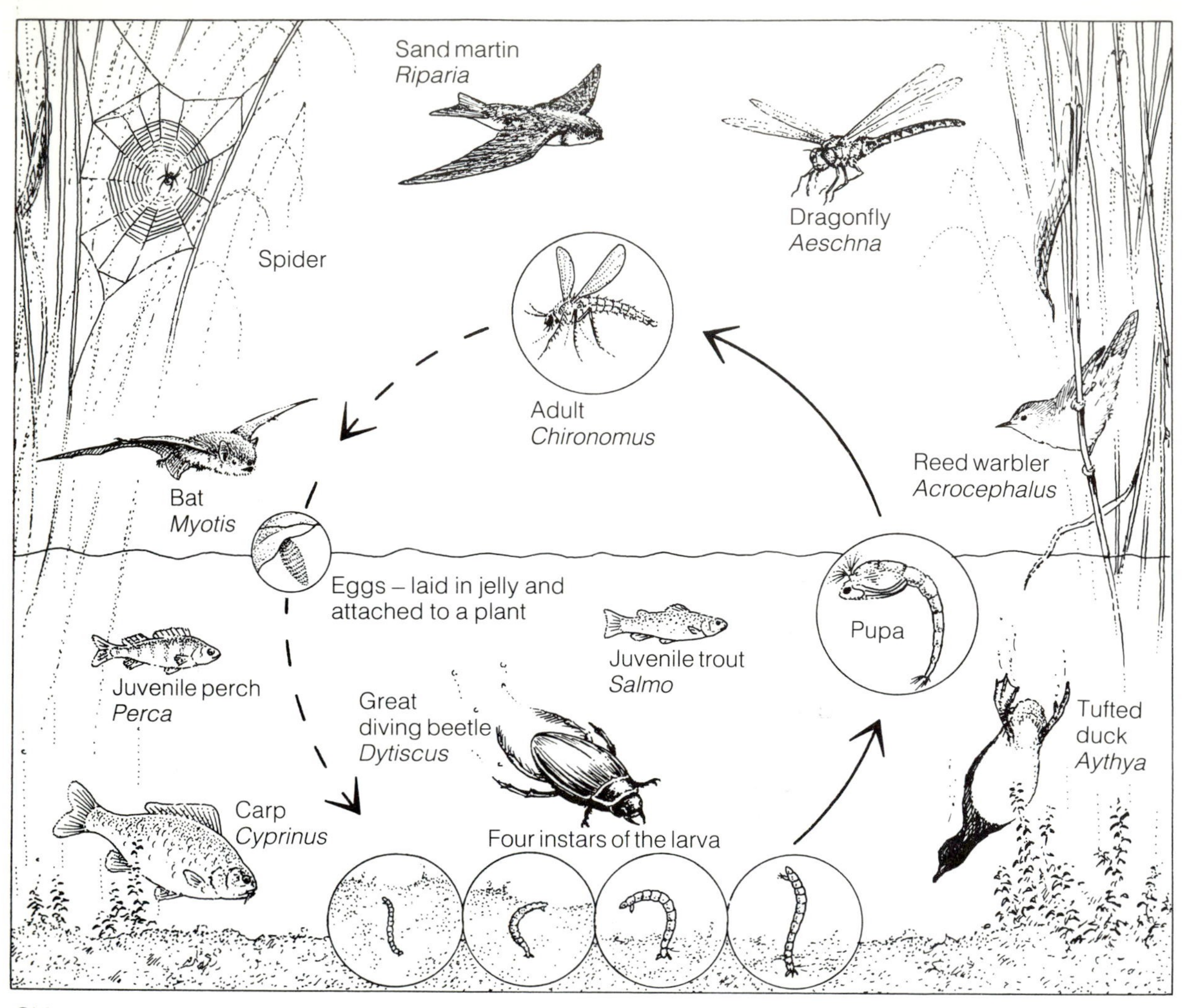

Chironomids are so prolific that they provide abundant food for other animals at every stage in their life history and are a vital link in the complex and varied food webs of almost every lake ecosystem. Chironomid (midge) larvae are usually among the most abundant animals of the benthic community and frequently many species will be found in one lake. The larvae live in the mud where they feed on detritus or benthic algae or are carnivorous, depending on the species. They moult four times before they form a pupa which does not feed but in which the grub-like larva is transformed into the adult. When this process is complete the pupa rises to the surface of the lake, moults its skin and the frail and delicate midge emerges from the water to fly away.

The chironomids are very characteristic lakeside inhabitants, familiar to visitors and fishermen. They are also typical of many lake insects, with larvae living and growing in the water (perhaps for several years) then emerging as adults into the open air. The larvae are essential food for many fish and ducks. The emerging adults provide rich pickings for insectivorous birds during the day and for bats at night, thus hitching terrestrial species to the aquatic ecosystem. Adult chironomids do not feed, which means that they do not bite, and they have only a brief life. Mated females lay eggs on the water which then sink to the bottom where they hatch into thin, caterpillar-like larvae. These increase in size at each moult and finally form a pupa which, when metamorphosis is complete, leaves the mud, rises to the surface and emerges as a free-flying adult. As the adults emerge they form swarms, resembling smoke rising from the lake. Blown inshore, they cluster around bushes or lights and die in their millions. So great are their multitudes that in some places the local people, such as the Sese Islanders of Lake Victoria, make protein-rich cakes from the drifts of tiny insects. At Lake Mývatn (see Chapter 7) in Iceland, swarms can be sufficiently dense to hamper drivers on the road around the lake.

The phantom midges, *Chaoborus* spp., also have aquatic larvae whose adults emerge from the water. They are particularly interesting because the larvae provide a link between the benthos and the pelagic zone. During the day, they rest in the substrate as members of the benthos and at night they rise through the water column, becoming pelagic to feed. They are voracious carnivores whose predation can have a marked effect on the other zooplankton. The thin body is about 15 mm long, and almost transparent except for two dark air chambers which are probably buoyancy regulators. The larvae hang motionless in the water and, against the night sky, must be as invisible from below as from above. Long grasping mouthparts seize any suitable prey that comes within range. They are themselves eaten by fish and their retreat to the bottom during the day may have evolved as a defence against this threat.

Many of the worms found in fresh waters (like the *Tubifex* used by aquarists to feed their fish) have red haemoglobin in their blood which helps trap oxygen, an advantage in the benthos where oxygen is often scarce. They are soft bodied and fragment easily when being sieved out of the mud, so they are more difficult to study than chironomid larvae (which are difficult enough) and less is known about them. They may be important components of fish diets but are readily digested in their entirety and therefore difficult to recognise in studies of fish stomachs unless their minute bristles (chaetae) can be found.

Molluscs are often important components of the benthic fauna. Some will be snails living on the substrate surface, though these are mostly found in the littoral zone. The molluscs particularly adapted to a benthic life are the burrowing bivalves such as the freshwater mussel (*Anodonta*). Smaller bivalves such as *Sphaerium* and *Pisidium* also occur, sometimes in great numbers. A mussel lies half buried in the mud, with the open edge of the shell uppermost. Water is drawn in through a siphon tube, passes over the gills and leaves via another part of the siphon. The gills are enormous and fill most of the cavity within the shell; their beating cilia maintain the water flow from which food and oxygen are extracted. The cilia also waft strings of mucus over the gills and it is this which traps the food particles. Only small phytoplankton are held in the mucus long enough to be transported to the mouth. Mussels are important filter feeders and are usually the largest members of the lake community to feed directly on the smallest algae.

The composition of the benthic community varies with the type of substrate both within and between lakes. In Loch Leven, for example, sand covers 42% of the lake bottom and is dominant inshore. It contains a very low proportion of organic material as measured by the quantity of carbon present (less than 1% by weight). Fine-grained muds cover the deeper, offshore areas (57% of the lake bottom) and their organic carbon content is up to ten times greater. The same chironomids are found in both substrates but the dominant genera differ. In the sandy parts *Glyptotendipes* and

The swan mussel (*Anodonta*) is a typical inhabitant of mud in the littoral and benthic zones of temperate lakes. Concentric rings on the shell indicates age in years and the shells may grow up to 15 cm long.

Stictochironomus predominate, while in the mud *Chironomus* is most abundant. Among the nine species of segmented worms, tubificids occur commonly in both substrates but the sand contains more enchytraeid (and nematode) worms than the mud. Almost three times as many animals are found per square metre in the sand as in the mud, though their total dry weights are similar (12.7 g/m^2 in the sand and 10.0 g/m^2 in the mud). This suggests that although the sand contains more individuals, they are smaller animals.

Lake Esrom in Denmark has been studied for over a hundred years and the animals living on the bottom of it are probably better known than any other benthic community. The lake area is 17.3 km^2, 60% of which has water more than 10 m deep. Its maximum depth is 22 m and the bottom is covered with a thick layer of highly organic mud. The benthic community is very diverse but the number of species varies with the type of substrate and depth of water. At the edge of the lake there are about 150 bottom-dwelling species, increasing to 300 about 100 m from the shore in the zone of submerged macrophytes. Further out, under about 7–8 m of water, there are only fifty species, reducing to twenty in the deepest parts of the lake. Herbivorous animals are only represented in the shallower water; thereafter detritivores dominate the community. Some, such as the midge larvae, feed from the surface of the mud, but others, including the tubificid worms, live on bacteria within it.

The animals which live in the deepest parts of the lake are able to extract oxygen even from water with very little in it. At about 20–25% saturation their respiration rates decrease, but they can survive periods of very low oxygen content, although feeding and growth are slower. This often means that they take longer to complete their life cycles than in more favourable circumstances. They frequently have to tolerate long periods in summer when not only the mud but also the overlying water contains no oxygen at all.

Benthic animals, particularly chironomid larvae, are important food for many bottom-dwelling fish. Unlike pelagic fish near the surface, these are feeding in the gloomy depths, so they cannot find their prey by sight and many feel and taste for it with barbels ('whiskers') around their mouth. Carp are typical benthic fish. They have large, protrusible mouths which they use for sucking up surface mud from the bottom; they retain any food material and spit out the rest. They eat whatever is available – molluscs, crustaceans, insects and plants – consuming more at higher temperatures, but becoming very sluggish and hardly feeding at all below 8°C.

Carp are easy to keep and have been taken all over the world to be reared as a popular food fish. They mature at 3–4 years old but in the wild rarely breed successfully in northern Europe because the water does not reach sufficiently high temperatures (they need 17–20°C). They grow and breed best in warm, productive waters with a lot of vegetation but they are extremely tolerant of adverse conditions, especially low oxygen concentrations (down to 0.5 mg/l) and will survive where many other species die. They are also very long lived (perhaps 20 years in the wild; 40–50 years in captivity) and in Britain the largest specimens are highly prized by anglers.

Carp are heavy-bodied fish belonging to several genera: the Crucian carp and the goldfish are in the genus *Carassius*; the grass carp is *Ctenopharyngodon idellus*. In the natural state carp are fully scaled; selective breeding is responsible for the varieties known as mirror carp and leather carp – these are forms of the common carp *Cyprinus carpio*. This species comes originally from the borders of Europe and Asia but has been widely introduced. The population in the United States, for example, is derived from a mere 345 specimens landed at Boston in 1877. These formed the basis of a burgeoning introduction programme, where seemingly everyone wanted stock. From them a quarter of a million were distributed only 5 years later. Today the American carp are accused of eating so much aquatic food that they leave too little for ducks like the canvasback. They also stir up the mud, decreasing light penetration and reducing productivity.

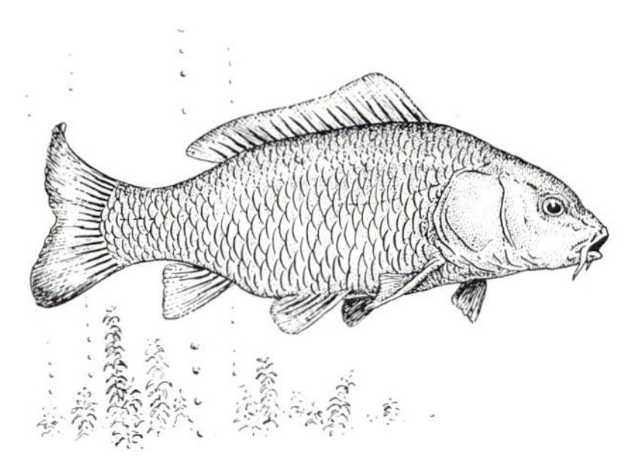

A typical cyprinid fish, the common carp *Cyprinus carpio*

Another abundant and typical group of bottom-dwelling fish is the catfish: there are fourteen families confined to South America, eight found only in Asia and three only in Africa. Most catfish live in rivers but there are some which inhabit lakes. Although they vary in size from the fragile glass catfish of Africa (*Physailia pellucida*), about 10 cm long, to huge species thirty times bigger, they all have the characteristically flattened head and long barbels which have inspired their popular name. Many catfish are nocturnal and can also survive in very poor conditions, even in waters that periodically dry up. Some, such as *Clarias*, are able to breathe air (see Chapter 7).

The European catfish or wels (*Silurus glanis*) is a native of eastern Europe and the southern Soviet Union. It lives in lakes and slow-flowing rivers, where it can reach a weight of 200 kg. Fish of this size will eat large prey, including water voles and ducklings, but the wels mainly feeds on eels, tench, bream and roach. It rarely spawns in water below 20°C so it is hardly surprising that the population introduced to southern England shows little sign of spreading.

The littoral zone and its inhabitants

The main groups of benthic invertebrates are also found in the littoral zone but are often represented by different species. The inshore area is a much more complex habitat, due mainly to the presence of macrophytes, and consequently supports many more animal species.

Stony shores have a more varied fauna in sheltered places than where the wind blows onshore unimpeded. Waves splashing over the rocks stop mud accumulating and prevent macrophytes becoming established. The only substrate for animals is then the rocks themselves and the only foods are the algae in the water or attached to the rocks, plus what little detritus becomes lodged among the stones. There are similarities between this type of habitat and that of a rocky stream. Both have the advantage that oxygen is never likely to be limited, because the water is constantly moving and dissolving more from the air. But animals living in these situations must hold fast to the rocks or hide underneath them to avoid being swept away or battered about. The disadvantages of living on exposed shorelines are sufficient that relatively few animals do so. Those with a means of attachment are at an advantage. Robust snails with a heavy shell and strongly muscular foot are common, particularly if the calcium content of the water is high. Sponges encrust the rocks and there are leeches which have suckers at both ends of the body. Among the organisms which escape the waves by hiding are worms and the freshwater shrimps (*Gammarus* spp.) The larvae of caddisflies and chironomids either attach themselves under stones or burrow among the smaller pebbles at their base. An animal that specialises in living on wave-washed stones of the lakeshore (and in rivers) is the freshwater limpet *Ancyclus*. Limpets are well known on rocky seashores but the freshwater ones are much smaller and are easily overlooked.

In sheltered places, where sediments can accumulate, plants can grow and they make a tremendous difference to the numbers and types of animals found in this marginal area. The large plants along the shore are not just living in a habitat, they are creating it. Reed beds, for example, deflect the wind and play a vital role in damping down waves to provide a still-water habitat in which fine mud settles and delicate organisms can live. The plants are not only interesting in their own right, but also a structural component of the littoral zone, providing a three-dimensional habitat for animals both above and below the water.

The large plants in a lake are referred to as macrophytes, as distinct from the other major plants, the algae, most of which are microscopic. Most macrophytes need to be rooted in the bottom silt and

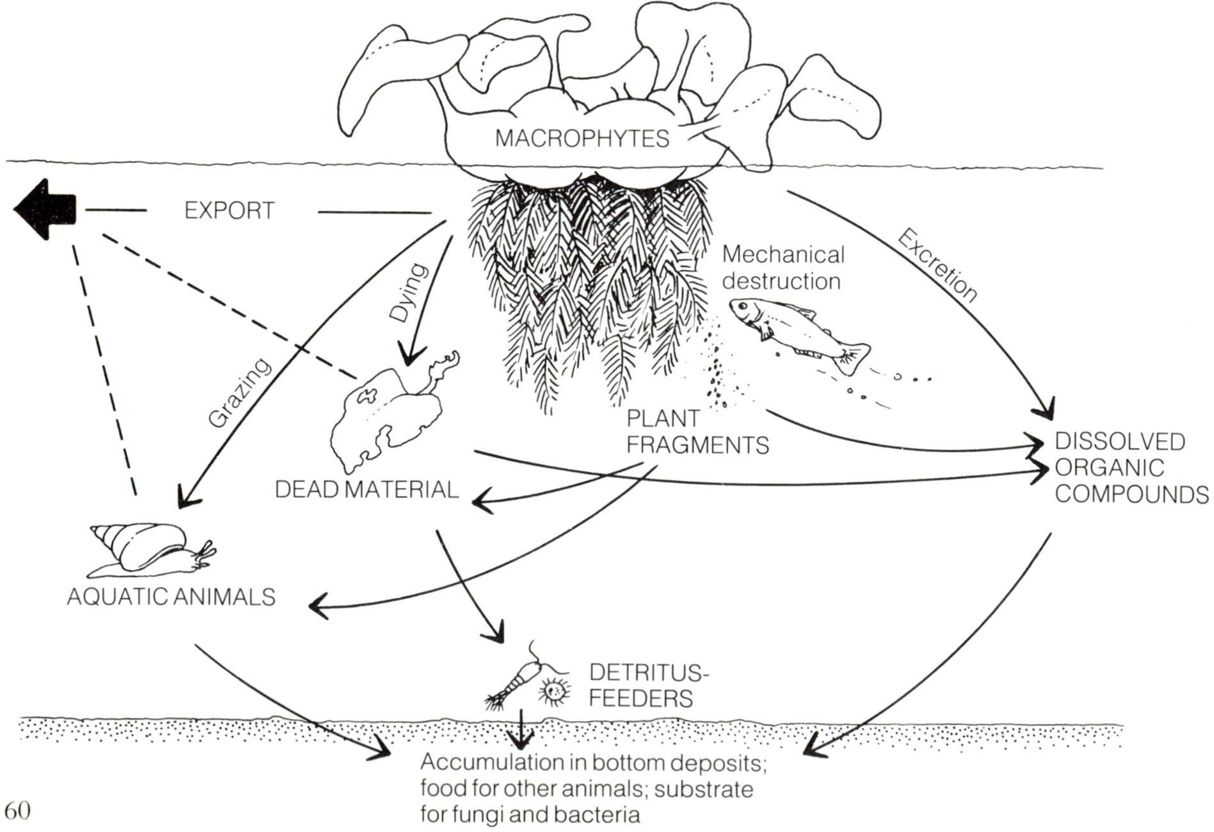

The role of floating plants in the lake ecosystem (based on Pieczynska, 1976).

therefore grow in relatively shallow water. Hence they are characteristic of the littoral zone. Those at the lake edge normally grow tall and stick out of the water. They are called **emergent macrophytes** to distinguish them from floating and submerged species more characteristic of deeper water. Most macrophytes are flowering plants (angiosperms) but a few large algae (such as the *Chara* species) and mosses are included among them.

On gently shelving shores the most obvious macrophytes are tall aquatic relatives of grasses, notably reeds (*Phragmites*) and reedmace (*Typha* – known as cat-tail in North America). On the landward side the macrophytes are often rooted in apparently dry ground but on the lakeside they spread into the water. In front of them are the flat floating leaves of plants such as water lilies (*Nymphaea*) and pond weeds (*Potamogeton*), which are still rooted on the bottom but have long stems. Sometimes there are also underwater leaves and only the flowers protrude above the water. Further out into the lake are wholly submerged plants, some of which are rooted on the bottom, but many, such as water milfoil (*Myriophyllum*) and hornwort (*Ceratophyllum*), float freely in the water. In sheltered places are plants which float on the surface with roots trailing down into the water. The smallest are the duck weeds (*Lemna*); the largest include the water hyacinth (*Eichhornia*) and floating fern (*Salvinia*) (see Chapter 9).

Canadian pondweed, *Elodea canadensis* which was introduced to British waters in the mid-nineteenth century. It has since spread very widely in rivers and lakes.

This gradation of plants from land into water represents a transition from one environment to another and is sometimes known as the *hydrosere*. It is also a zone of succession. As the emergent macrophytes die back each year their remains accumulate. They also trap silt and debris and the level of the substrate is raised until the plants are growing too high above the water to be comfortable. By then trees, such as alder (*Alnus*) and willow (*Salix*), which like to grow out of water but in soggy places, have invaded their domain. Eventually terrestrial species dominate and what was the lake edge has become land supporting a typical terrestrial community. A similar process is happening among the reeds and floating plants: as the water depth decreases the reeds are able to colonise further out into the lake. The lilies also creep forward as the years go by. The open-water part of the lake is thus gradually strangled by the tightening noose of macrophytes around its edge. Eventually the whole water body becomes invaded by macrophytes and ultimately the lake disappears, leaving a characteristically flat area to mark its former site. In some cases, particularly with very shallow lakes, this succession runs its full course, but in others an equilibrium develops between the further colonisation of open water and factors which inhibit the process. Then the shoreline community remains stable for long periods.

Emergent macrophytes are similar to terrestrial plants except that they are nearly always rooted in deoxygenated mud rather than well-aerated soil. Like other parts of a plant the root cells need oxygen, and if this cannot be obtained from their immediate surroundings it must be transferred down the stem from those parts of the plant which are out in the air. For this purpose the roots, leaves and stems of macrophytes contain large air-filled spaces along which oxygen

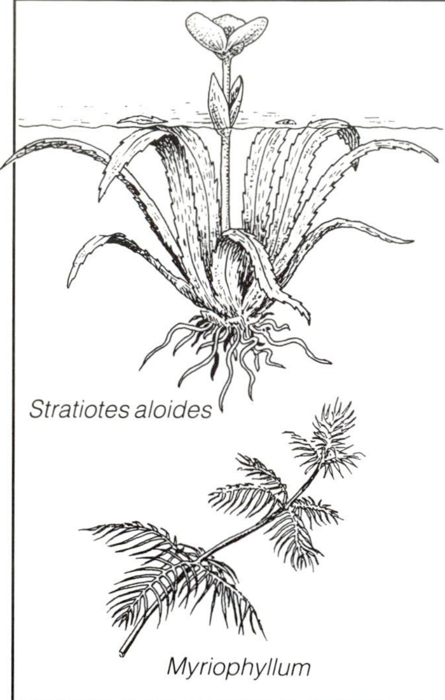

Water flowers

Increased pressure under water seems to inhibit development of flowers in many species so that stems must reach the surface before flowers will form. Many species are wind-pollinated but the hornwort (*Ceratophyllum demersum*) has flowers underwater and gets them pollinated without reaching the air. When they open, the anthers detach, float to the surface and burst open to release their pollen. Being heavier than water, the pollen sinks back to pollinate the flowers. The water soldier (*Stratiotes*) has its flowers pollinated by flies which are attracted by the nasty smell of the flowers when they appear briefly above the water. This plant, like many others, does not rely only on seeds for the production of new individuals, but produces stolons from which small new water soldiers develop. In winter they accumulate chalk in their tissues and sink to the bottom of the lake, safe from ice and storms. In spring this is released and they float back to the surface. Almost all the rooted macrophytes can reproduce vegetatively by means of rhizomes or stolons. These structures also enable them to overwinter after the green parts have died back. Plants which lack them (such as *Myriophyllum* and *Utricularia*) often overwinter by means of turions, which are bundles of partially developed leaves that grow very close together among the lowest leaves on the stems grown the previous year. Turions separate from the parent plant and sink to the bottom of the water to await improved conditions.

can travel to relieve the shortage in the roots. These spaces make macrophytes feel typically spongy when squeezed and also dictate the maximum depth of water in which they can grow. At about 10 m down the water pressure is increased equivalent to one extra atmosphere and the air channels collapse. Macro-algae can grow at greater depths (if there is enough light) because their thin-walled cells do not contain air spaces; they obtain sufficient oxygen directly from the water.

Wholly submerged plants cannot obtain oxygen from the air and must extract all they need from the water. They have very thin leaves, and have lost all the strengthening that is found in terrestrial plants since they are supported by water. Much of the oxygen released during photosynthesis goes to alleviate the shortage in their roots, so although bubbles of oxygen sometimes rise from the plants on sunny days their quantity cannot be used to measure the rate of photosynthesis because they do not represent the whole output of the plant. Submerged plants also need other adaptations. Photosynthesis must be possible in poorly lit water, floating leaves need to be very flexible to avoid damage by water movements, and the stems must be long enough to allow for changes in water level.

Littoral zone animals

Animals in the littoral zone mostly live on the surface of leaves and stems or burrow among the plant roots. There are also tiny swimming animals living in the still water provided by the protective tangle of stems and leaves. Although aquatic plants may be a nuisance to fishermen and to boating, without them the lake would have a much less varied fauna and there would be less variety of

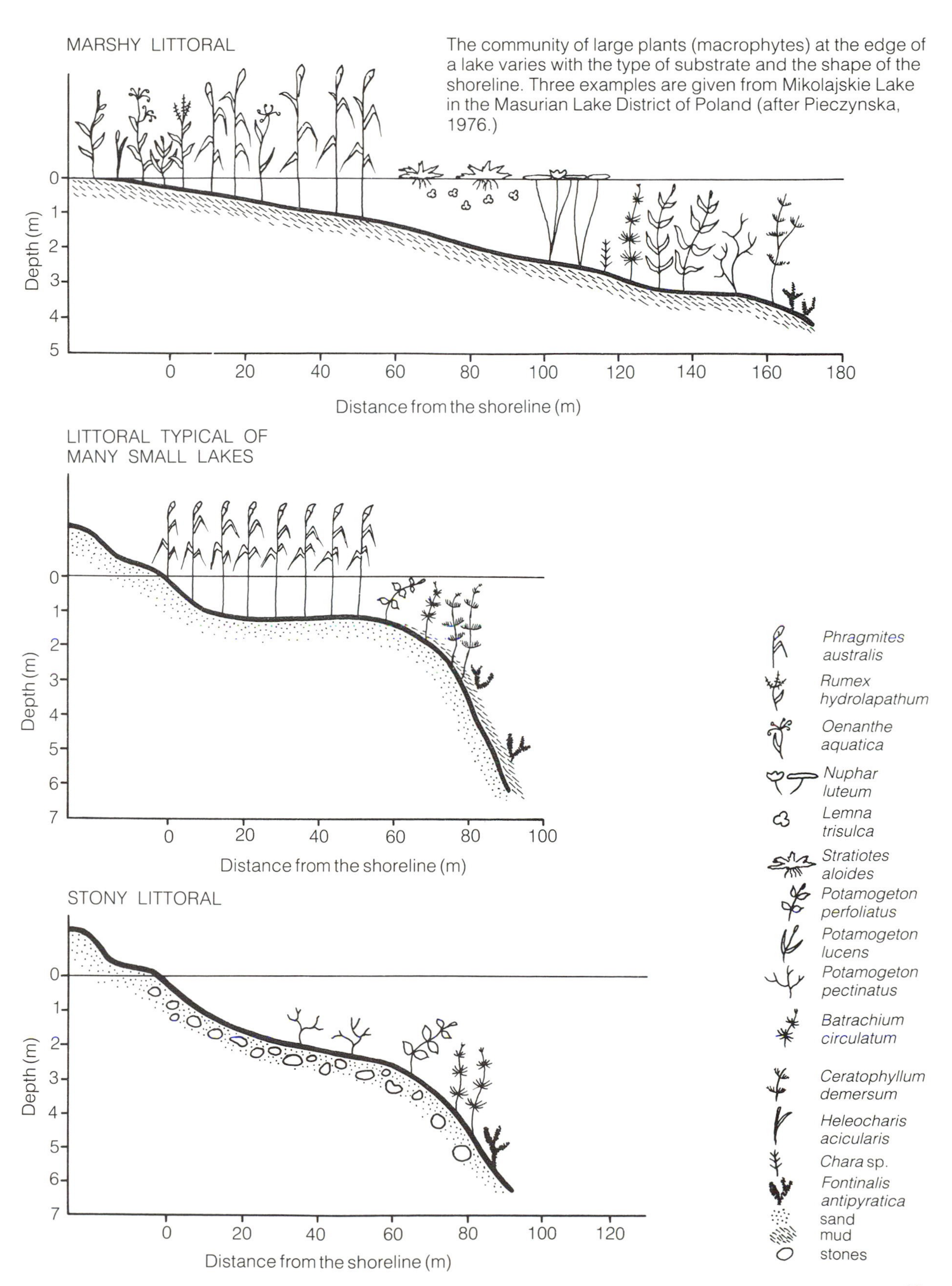
MARSHY LITTORAL
The community of large plants (macrophytes) at the edge of a lake varies with the type of substrate and the shape of the shoreline. Three examples are given from Mikolajskie Lake in the Masurian Lake District of Poland (after Pieczynska, 1976.)
Depth (m)
0
1
2
3
4
5
0
20
40
60
80
100
120
140
160
180
Distance from the shoreline (m)
LITTORAL TYPICAL OF MANY SMALL LAKES
Depth (m)
0
1
2
3
4
5
6
7
0
20
40
60
80
100
Distance from the shoreline (m)
STONY LITTORAL
Depth (m)
0
1
2
3
4
5
6
7
0
20
40
60
80
100
120
Distance from the shoreline (m)
Phragmites australis
Rumex hydrolapathum
Oenanthe aquatica
Nuphar luteum
Lemna trisulca
Stratiotes aloides
Potamogeton perfoliatus
Potamogeton lucens
Potamogeton pectinatus
Batrachium circulatum
Ceratophyllum demersum
Heleocharis acicularis
Chara sp.
Fontinalis antipyratica
sand
mud
stones

'Aufwuchs': the community of algae, bacteria and small animals which form a slimy layer around the underwater stems of aquatic plants (after Pieczynska, 1976). Within the circle is an enlargement of the details which would be seen if the stem were scraped clean and then examined under a microscope a few days after the plants and animals had started to recolonise.

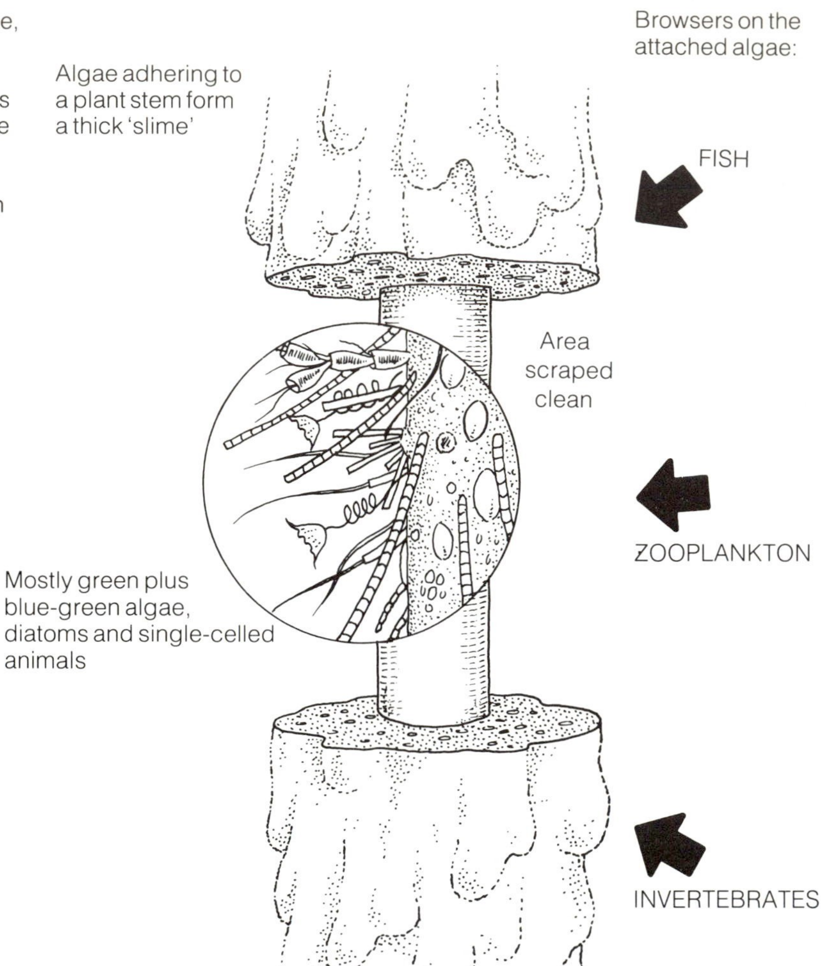

places for animals to live. Complexity of habitat structure enables a much more diverse community to live in harmony than would be possible in a lake with only open water and bare mud.

Surprisingly, perhaps, few animals feed directly on the living macrophytes. Those that do make very little impression and this leaves a lot of plant material to die and provide food for the detritivores. Many animals feeding on and among the macrophytes are actually exploiting the layer of living material which builds up on the stems and leaves. This is a slimy mixture of plants, animals and micro-organisms, best described by the splendid German word *Aufwuchs*. Diatoms and blue-green algae are common components and the latter secrete a gelatinous covering which attracts and holds other tiny organisms. The aufwuchs needs surfaces on which to grow, so there will be more food for animals in places offering greater surface area. The submerged surfaces of plants available for colonisation by animals may be equivalent to forty times the area of the lake bottom.

Pond snails are a conspicuous group of littoral zone residents which scrape aufwuchs off plants using a tongue (radula) which bears rows of tiny teeth, arranged in patterns characteristic of the species, which rasp off particles, including bits of plants, algae, bacteria or dead animal material. The snail shells are composed of calcium carbonate and, consequently, snails are more abundant and varied in 'hard' waters; few snails live in very 'soft' water. Some pond snails breathe by means of gills but others have a simple lung chamber which they can fill with air. If the water freezes they can obtain enough oxygen from the water, especially as they need much less at low temperatures.

Many species of Crustacea, relatives of shrimps and crabs, live among the plants of the littoral zone. The largest are the water woodlouse *Asellus*, various freshwater shrimps and the crayfish. In some parts of the world there are also freshwater crabs. The small Crustacea are mostly copepods and cladocerans like those living in the pelagic zone, but some are particularly adapted for life in the weed beds and do not survive elsewhere. *Sida crystallina* is a large cladoceran (4–5 mm long) whose movements are slow and which is very vulnerable to predators such as dragonfly nymphs and small fish. It is made more visible by a large black eye. Its survival depends upon being able to hide among the macrophytes; it can also attach to the underside of floating leaves by means of a sticky secretion produced by a gland on the back of its head.

Insects of the littoral zone

Many of the littoral zone animals are young stages of insects including dragonflies (Odonata), mayflies (Ephemeroptera), stoneflies (Plecoptera) and caddisflies (Trichoptera), as well as the midges (Diptera). The first three of these groups have young called nymphs which resemble the adults fairly closely except that they have no wings and obtain their oxygen from the water. The nymphs are also dull in colour and inconspicuous. As in all insects, the body grows by moulting its hard cuticle to allow expansion to a larger size. Each stage between moults is known as an instar and the later instars have wing buds on the thorax which increase in size with each moult. The last instar crawls up the stem of an emergent plant and out into the air. When the skin splits for the last time the adult insect emerges to fly away as soon as its wings unfold and harden. Adults do not survive the winter; it is their nymphs which provide the continuity from year to year.

Dragonfly and damselfly nymphs are among the larger insects in the littoral zone. Damselfly nymphs are much longer in relation to their girth than the short, stout dragonfly nymphs and their rear ends bear three long, flattened 'tails' which help extract oxygen from the water. Dragonfly nymphs do not have these 'tails' but instead breathe by pumping water over gills which line the rectum. Dragonflies are carnivores, both as nymphs and as adults. They have prominent compound eyes and catch their prey by shooting out a 'mask' which has grasping claws on the end to seize their prey. Normally this structure is folded up under the head. The larger nymphs sometimes tackle tadpoles and even small fish.

Mayfly nymphs have three long hair-like 'tails' and an elongate

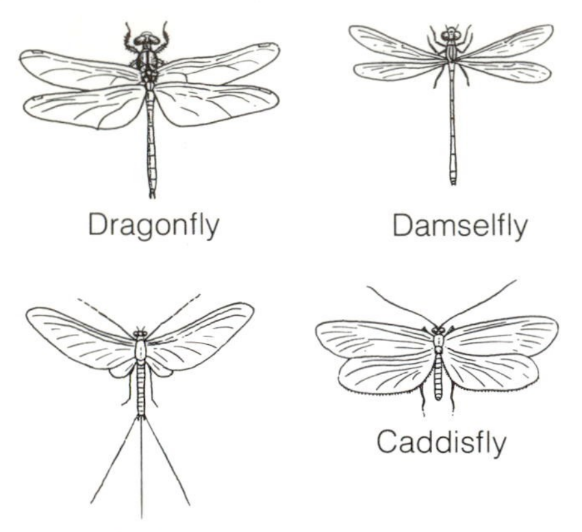

Some of the adult insects which lay their eggs in water and whose juvenile stages feed and grow in the littoral zones of lakes. (Not drawn to scale.)

body. Some, like the species of *Baetis* and *Chloeon*, breathe with seven pairs of plate-like gills arranged along the abdomen and eat mainly algae and plant detritus. They have up to twenty-seven moults during a life cycle, which takes from a few months to a couple of years depending on the species and the temperature. Stonefly nymphs are similar in shape to those of mayflies but have only two 'tails' and lack the plate-like gills. They are confined to cold, unpolluted, well-oxygenated waters and in lakes are only found on stony shores.

Caddisfly young are called larvae because, unlike the nymphs of dragonflies and mayflies, they are quite different from the adults and look more like caterpillars. They must undergo a complete transformation (metamorphosis) before they become adult. The head and thorax of the larva are covered in a hard brown cuticle, but not the abdomen. There is no sign of wing buds but the thorax bears the usual three pairs of legs with which the larva walks about and clings to vegetation. Many caddis larvae protect themselves by building a 'case' of sticks or stones, glueing together the components with a sticky silk thread secreted by a gland near the mouth. Different species use different building materials and patterns of construction, so some can be identified on the basis of their case structure alone. They walk about carrying their shelter with them. Species found on stony shores differ from those found in more sheltered situations. On stony shores the cases are made of sand grains or of small stones and are heavy enough to avoid being constantly rolled about by the water. In some species the case is cemented to a rock. In sheltered places, cases can be made of lighter material such as wood and vegetable fragments. The tubes are open at both ends and, typically, water is drawn through by undulations of the body, so that oxygen can be obtained by feathery gills. The larva moults several times and then forms a cased pupa which metamorphoses into the adult.

Other insects found in the littoral zone spend their whole life in the water, even though many of the adults need to breathe air and have to store it when they dive. The bugs (Hemiptera) show a variety of ways in which this can be achieved. They are primarily a terrestrial group but a few live in water. These include the backswimmers (*Corixa*), waterboatman (*Notonecta*), water scorpion (*Nepa*) and water stick insect (*Ranatra*). They are all lighter than water so bob up if they stop swimming, unless they cling to plants. At the surface the waterboatman puts the tip of its abdomen through the surface film and the row of hairs with which it is fringed channels air to the openings (spiracles) of the internal system of tubes (the tracheae) which carry air to all parts of the body. A bubble of air also collects among the water-repellent hairs on the underside of the body. This bubble is carried down when the waterboatman leaves the surface and can be seen as a glistening layer covering the body. The bubble provides buoyancy, an embarrassment to a diving insect, and it has to work harder to keep itself submerged. The almost frantic paddling action of these insects underwater is quite distinctive. The water scorpion and water stick insect have, instead of fringing hairs, a long tube at the rear of the abdomen which is thrust through the surface film and

This ornamental lake was created in the eighteenth century to enhance the landscaped park of Blenheim Palace in Oxfordshire.

Above *A new lake forming behind the terminal moraine of a glacier in Iceland.*

Right *Buttermere in the English Lake District lies in a typically round-bottomed valley gouged by a glacier during an Ice Age.*

Facing Page Top *A typical example of a crater lake in an extinct volcano in Iceland.*

Facing Page Bottom *A lagoon formed behind coastal sand dunes in Australia. A black swan sits on its nest among the reeds in the centre of the picture.*

Above *Australian black ducks (*Anas superciliosus*) in the thick algal soup of a highly productive lake whose water is enriched by fertilizer run-off from a pineapple plantation.*

Left *Birds like these pintail (*Anas acuta*) face a difficult time when lakes freeze in winter.*

Facing Page *The clear, unproductive waters of a high-altitude lake in the Sierra Nevada, California.*

Above Left *The water flea (*Daphnia*) with eggs visible in its brood pouch and digested algae in its gut.*

Above *A male frog* Hyperoleus viridiflavus *calling loudly at night among the water lilies fringing Lake Awasa in Ethiopia.*

Below *The discus (*Symphysodon*) is one of many fish from the Amazon basin often kept in ornamental aquaria.*

The sacred lotus Nelumbo nucifera *has floating leaves but carries its magnificent flowers high above the water.*

Right *An adult midge (*Chironomus*), showing its typical feathery antennae. Midges do not bite, which is just as well since swarms of them emerge from lakes during the summer.*

Below *In winter the hollow dead stems of the common reed (*Phragmites australis*) shelter stem-boring insects and the seed heads provide food for many small birds.*

Facing Page Top *The moose* Alces alces *often wades into shallow water to eat the aquatic vegetation.*

Facing Page Bottom *The clear waters of Lake Tahoe are over 500 m deep and span the border between California and Nevada.*

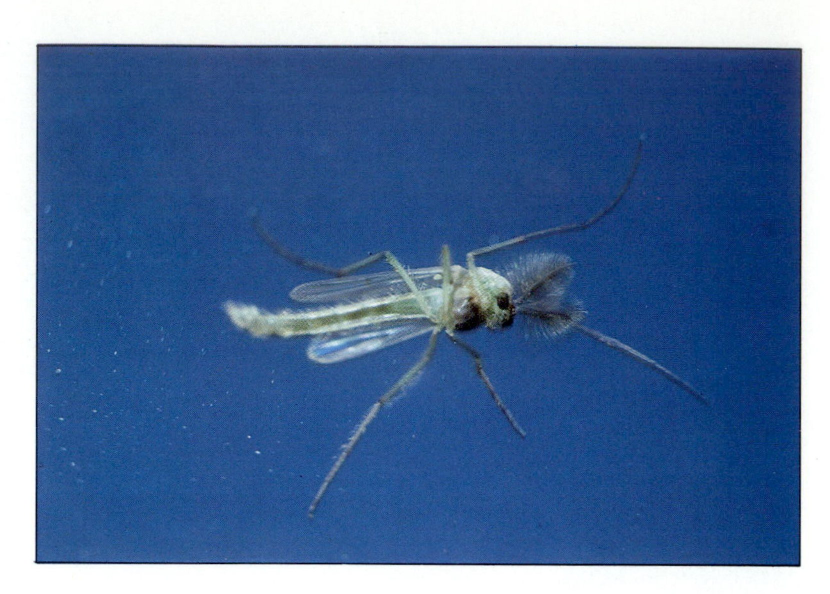

KASSA
BOOTE

Some examples of Lake Tanganyika cichlids (Lake Malawi has a quite different set of species).

Above *Lamprologus brichardi*

Left Top *Cyphotilapia frontosa*

Left Centre *Haplochromis burtoni*

Left Bottom *Limnochromis auritus*

Facing Page Top *Hallstatt in Austria, a typical picturesque alpine lake resort.*

Facing Page Bottom *Fishing boats on Lake Tanganyika; behind them rise the precipitous slopes of the Western Rift Valley escarpment.*

Above *An American alligator (*Alligator mississippiensis*) basking at the edge of a shallow freshwater lake in the Florida Everglades.*

Facing Page Top *Massed flamingos at Lake Nakuru, Kenya, with the Rift Valley wall in the background.*

Facing Page Bottom *Clumps of papyrus* (Cyperus papyrus) *break free from the swamp fringe and float across the surface of Lake Naivasha in the Eastern Rift Valley.*

Green-backed Heron (*Butorides striatus*) An opportunist feeder that may even pursue prey into the water, diving from a perch or the air

Goliath Heron (*Ardea goliath*) A typical 'wait and see' hunter, standing motionless until it strikes at passing prey

Indian Pond Heron (*Ardeola grayii*) Wading slowly through shallow water, this small heron darts forward to catch its prey

Reddish Egret (*Egretta rufescens*) With seeming abandon this species dashes around in shallow water, wings flapping, presumably to disturb fish from cover only for them to be snapped up before they can hide again

Grey Heron (*Ardea cinerea*) Long legs allow this bird to wade into deeper water than most other species. It will wait motionless or stalk slowly and purposefully

Little Egret (*Egretta garzetta*) One method used by this species is to agitate or rake the lake bottom with a foot and snap up any prey that is disturbed. Fish are possibly attracted to its yellow feet

Black Heron (*Egretta ardesiaca*) Unique in forming an 'umbrella' with its wings which attracts small fish to the shade provided. The bird can probably see better too

Herons are important predators in the littoral zone, various species having different feeding strategies.

Top *A small shallow lake drying up in the fierce sun of the Hattah lakes region in New South Wales.*

Bottom *Saplings of the Australian Red River Gum growing where floating seeds were stranded at the edge of a temporary lake as it dried up.*

Above *Inspecting the salt crust, coloured pink by salt-tolerant microorganisms, at the edge of a lake in Nevada.*

Right *The water in what remains of this Nevada lake is a saturated salt solution; as evaporation continues a raft of salt crystals forms at the surface.*

allows air to diffuse into the tracheal system.

There are more species of beetle in the world than of any other group of animals, although only a small proportion of them are aquatic. Many species live in the littoral zone of lakes. Some have larvae quite different from the adults – pale, soft and sausage-shaped 'grubs' with only a hard head capsule. Other beetle larvae have legs and hardened cuticles but most larval and adult beetles have biting mouthparts, enabling them to be plant or animal eaters. The great diving beetle (*Dytiscus*) is a fierce carnivore and will attack small fish and frogs as well as other insects and worms. The larva has long pointed jaws projecting from the head which enable it to pierce its prey and suck it dry. There are innumerable smaller water beetles, none more obvious than the whirligigs (*Gyrinus*) which swim round and round in dizzy circles at the surface, catching small insects. They are shiny, black and very smooth. Their eyes are divided into two sections: one for seeing above the water surface, and one for seeing while submerged.

Larger animals in the littoral zone

Some fish spend their whole lives in the littoral zone and many favour the weedy areas: the roach (*Rutilus rutilus*) is a good example. The feeding activities of fish contribute to the breakdown of plants, helping to form detritus which forms food for many invertebrate animals, particularly those living on the bottom. Sticklebacks (*Gasterosteus aculeatus*) eat small crustaceans swimming among the plants, tench (*Tinca tinca*) feed on the bottom, and pike (*Esox lucius*) prey on other fish. All are members of the littoral zone community over much of lowland Europe and have their equivalents elsewhere.

The pike are a small family of predators which lurk among the littoral vegetation of lakes in Europe and North America. They ambush prey rather than chase it, and for this their coloration provides effective camouflage among the underwater macrophytes. Their diet changes as they grow: from Cladocera and insect larvae in the first year, through young fish and tadpoles, to large fish including salmon and trout. Occasionally they take water voles and aquatic birds. Large pike can have devastating effects in a lake populated by smaller fish and they are sometimes introduced to lakes where there are too many small fish. Predation by the pike leaves more resources for the survivors to grow larger and become more attractive to anglers. Care must be taken to use pike of appropriate size and to remove them before they eat too many of the other species. Pike are themselves prized as sport fish and also relished for eating in mainland Europe. Full-grown specimens can exceed 20 kg and records over 30 kg exist. Females are larger than males and spawn in spring among the littoral vegetation. The growth of the young depends on the water temperature, but they can be 20 cm long at one year old.

Many pelagic fish come inshore to breed in the littoral zone. Perch, which are so familiar to fishermen in both North America (*Perca flavescens*), Europe and Asia (*P. fluviatilis*), feed in open water but lay their eggs in ribbons, about a metre long, among the littoral vegetation. When the young hatch they feed first on copepods and cladocerans among the plants but soon graduate to insects and, when they are big enough, eat other fish. By this time they have moved offshore into the pelagic zone. In the London reservoirs (see Chapter 9) the littoral zone is reduced to a steeply shelving concrete slope, with only a sparse covering of attached algae and lacking macrophytes. Perch populations survive, but only just. There are no weeds to provide habitat for the food on which the young depend, nor places for them to hide from their voracious predatory parents.

In tropical waters cichlids make nests in the littoral substrate as part of their courtship ritual. In some species the eggs are picked up by the female and incubated in her mouth; in others the parents guard the nest until the eggs have hatched. This sort of breeding behaviour needs shallow water and suitable substrates, so even if the fish feed in deeper water for most of the year they come to the littoral zone during the breeding season. Even in herbivorous species, such as *Oreochromis niloticus*, the young feed on small crustaceans inshore until they are old enough to digest the algae of their adult diet and exploit the whole lake area. For these fish, and many others, the littoral zone is as vital as the open water and both must be suitable if their populations are to prosper.

4

LAKES and the seasons

Lakes are very much affected by seasonal changes and in most regions of the world the weather varies to a greater or lesser extent from one part of the year to another. On the Equator, seasonal changes in sunshine and temperature are less pronounced than they are at higher latitudes but wind and rainfall are often markedly seasonal. These profoundly influence lake functioning; seasonal cycles of physical events in lakes, whether dictated by sun, wind or rain, lead to seasonal changes in the chemistry of the water and in the plants and animals in the lake. The activities of the living organisms may also determine changes in the water chemistry and chemical changes may affect the biological elements of the community.

In the temperate zone the annual cycle of solar radiation dominates all other features of the climate: the short, cold days of winter lengthen in the summer, the temperature of air and water rises, and more light energy is available for photosynthesis. The response of the terrestrial vegetation is apparent to everyone and is parallelled by a similar response in the lake's green plants, both large and small.

Stratification

In the spring, water at the surface of the lake heats up more rapidly than that lower down and may eventually become so much warmer and lighter than the colder, denser water beneath, that the lake is divided into two layers which cannot mix because the light summer winds are not strong enough to overcome the difference in density. The lake is then said to be **stratified**. When a thermometer is lowered into the water, a narrow layer can be distinguished between the warm **epilimnion** at the top and the colder **hypolimnion** below. This zone is called the **thermocline**, because there is a sharp decrease in temperature between the bottom of the epilimnion and the top of the hypolimnion. (See diagram p. 28.)

The water of the epilimnion is illuminated by the sun, mixed by light winds, and oxygenated both from the surface and by the algae which photosynthesise actively throughout its depth. In the cooler water of the hypolimnion there is usually no light, and the decomposition of dead plants and animals raining down from above gradually uses up oxygen, which cannot be replaced either from the surface or by photosynthesis. How much of the total oxygen trapped in the hypolimnion by the thermocline is used up depends on the volume of the hypolimnion and the intensity of decomposition processes. If the lake is productive and a lot of organic material enters the hypolimnion from above, to add to already highly organic mud at the bottom, then there will be a lot of very active bacteria present, needing oxygen for their respiration. The deepest water will become deoxygenated first and as the season progresses the boundary between oxygenated and deoxygenated water will move upwards. In relatively shallow, productive lakes the hypolimnion may be devoid of oxygen by the end of the summer. In very cold, unproductive lakes the whole of the hypolimnion may remain well oxygenated throughout the period of stratification. Every lake is different in the details of its seasonal pattern.

As the air temperature declines during autumn so does that of the surface water, until there is much less difference in density between the water of the epilimnion and that of the hypolimnion. This enables the first strong wind to mix the layers of water and the lake is said to have overturned. The temperature is now more or less the same from top to bottom and all the water contains oxygen once again, but by now the temperature is cooler and the amount of light energy has decreased, so there is much less photosynthesis. The lake starts the winter season and its inhabitants merely 'tick over' until production starts again in the spring.

When a lake freezes there is a period of calm in the water below the ice. The cold water can hold a lot of oxygen and the plants and animals use very little because their metabolism is greatly slowed by the low temperature. If ice cover is prolonged there may be an inverse stratification – the water immediately under the ice is colder and less dense than that lower down. Snow on top of the ice acts as insulation and also prevents light reaching the water, but clear ice is transparent; in cold but sunny climates the water may be warmed by incoming radiation and species of algae which are adapted to photosynthesise at low light intensities may start to grow even while the lake is still frozen. With the coming of the spring thaw the surface of the lake is again exposed to the wind and the water is mixed until its temperature is the same from top to bottom. As the air temperature increases the yearly cycle starts again. The cycle mixes the water and takes oxygen down from the surface, and also brings up essential nutrients from the bottom, to start the system working again.

When the lake is frozen in winter, fish-eating birds such as herons and kingfishers are unable to feed. Others, such as ducks and coots, become much more vulnerable to predators, e.g. foxes, than when able to swim freely on the water.

It is generally true that there are more lakes in the temperate zones of the world than at lower latitudes, and the majority of them probably experience some form of the basic annual cycle just described: they stratify in summer, freeze at the surface during winter, and are completely mixed *twice* in the year. In natural systems there are always nearly as many exceptions as there are typical examples, and this certainly applies to lakes. In mid-continental regions of North America, Europe and Asia, this pattern of mixing twice each year will be found in most lakes except the very shallow and some of the very deep. Nearer the maritime margins of the continents, where the winters are milder, lakes often do not freeze and winter winds continue to mix the cold water, which remains at the same temperature at all depths until the spring. Such lakes therefore have only *one* period of mixing per year instead of two.

Some deep lakes with steep sides have a rather small surface area in relation to their volume. In these cases, particularly if they are sheltered by high mountains, it may be almost impossible for the wind to mix the lake water right down to the bottom. Such lakes can remain stratified for more than a year; then the epilimnion may itself become stratified due to differences in water density, thus adding secondary stratification to the water above the original thermocline. Wind mixing readily disrupts the secondary stratification but leaves the main thermocline intact. Meanwhile the water at the bottom seldom contains any oxygen. The depth of the boundary between the oxygenated and the deoxygenated zone varies from year to year, depending on the weather. In a year of cool temperatures and strong winds the overlying water may be stirred more deeply than usual,

Depth–time diagrams

One of the most useful ways of describing lake conditions which change with both depth and time is to plot them on what is known as a depth–time diagram. The depth of the lake is shown along the vertical axis and the time of year (or day) along the horizontal axis. Each measurement of, say, temperature is then marked in the appropriate place for depth and time and labelled with its value. When all the measurements taken have been plotted, it is possible to join up all the points of equal value to make something resembling a contour map. In the same way as contour lines close together on a topographical map indicate steep slopes and lines far apart indicate flat land, so depth–time diagrams show periods of time and layers of the lake in which conditions (such as temperature) are changing most rapidly and periods where conditions are the same throughout the water column and for lengths of time. The thermocline is shown by lines lying horizontally and close together as a period of time during which the temperature changes abruptly across a narrow layer of water. Where the lines are vertical and far apart the water is the same temperature throughout the depth of the lake for some time. Similar diagrams can be used to show the seasonal pattern of oxygen concentration or pH, or any other characteristic of the water, including the concentrations of plants and animals.

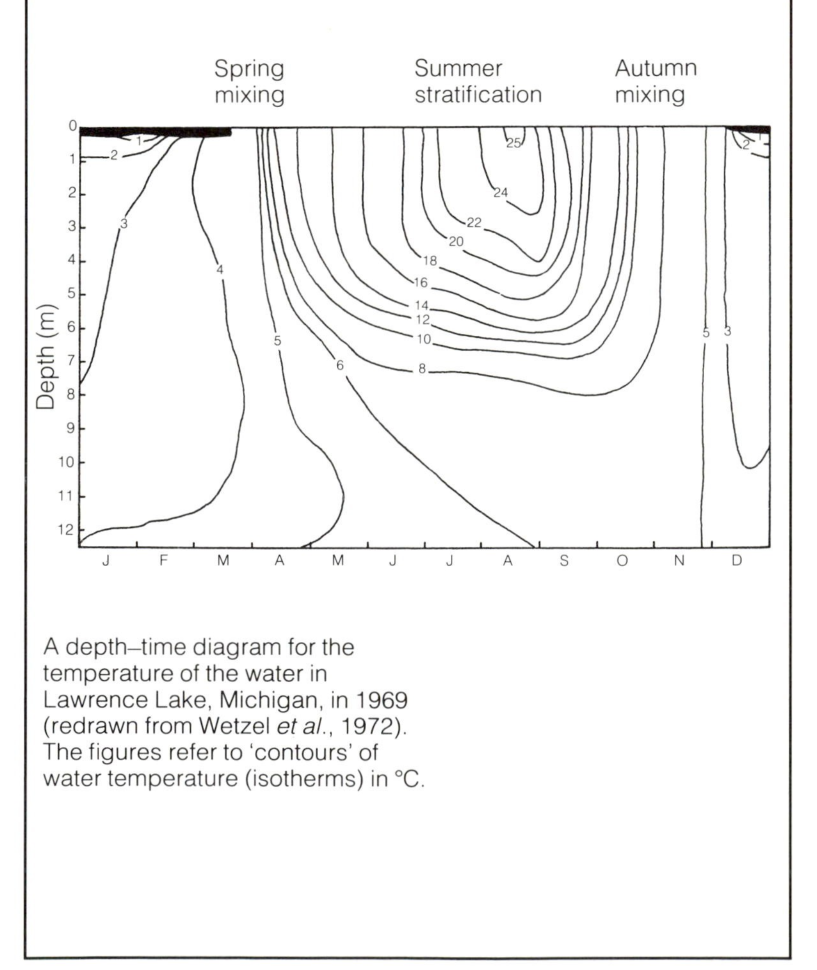

A depth–time diagram for the temperature of the water in Lawrence Lake, Michigan, in 1969 (redrawn from Wetzel *et al.*, 1972). The figures refer to 'contours' of water temperature (isotherms) in °C.

mixing oxygenated water into the top of the deoxygenated zone and pushing the boundary downwards. In a warm summer the difference in density between top and bottom water will be greater. Production in the epilimnion will be higher and more dead material will sink to the bottom, to use up oxygen in its decomposition, raising the level of the deoxygenated zone. Such lakes will only mix to their full depth once in 5 to 10 years, if at all. Such *infrequently mixed* lakes occur throughout the world but are more common in the tropics than at higher latitudes. Apart from the very deepest lakes (Chapter 6), the only ones that are *never* mixed by the wind are those with a permanent cover of ice found only in Greenland and parts of the Antarctic.

In a very shallow lake it is possible for even light summer winds to stir the lake water right down to the bottom, so it may never become stratified for more than a few days at a time. Such lakes contain relatively warm water throughout their depth and the mixing ensures that all the algae spend at least some of the day in the light. It also means that nutrients released by decomposer organisms from dying algae and from the mud are mixed into the water and quickly made available for re-use. These circumstances combine to ensure that many shallow, unstratified lakes are highly productive compared with deeper lakes. If the weather is calm for a few days the water column of even the shallowest lake may become stratified. In these circumstances the bottom water rapidly loses its oxygen, but usually the lake is mixed again before the whole depth of water becomes deoxygenated. If this disaster should occur however, the plants and animals would die; this sometimes happens in warm-water fish ponds, with great loss of stock, it may also happen if a very shallow lake is frozen for a long time and the water below the ice becomes deoxygenated. Shallow lakes which mix *many* times in a year are also found throughout the world and factors other than mixing, such as temperature and nutrient supply, are more important in determining the patterns of events in these lakes (see Chapter 7).

Temperate and tropical cycles

In the temperate zones the annual cycle of stratification and mixing that occurs in most lakes, except those less than about 10–15 m deep, is primarily driven by the sun and its effect on lake temperature, though wind strength also plays a significant role. Nearer the Equator the number of sunny days in the year increases but the day-length becomes more uniform throughout the year. The relative lack of sunshine further north or south is partially compensated for by the long summer days. Even so, it comes as a surprise to learn that the amount of solar energy available per year on the Equator is only twice as much as it is in Scotland: but even this is far more than the plants can use. A more significant contrast between tropical and temperate climates is that in the tropics more than enough radiation is available to plants every day of the year and the temperature never drops so low that plant metabolism is too slow to use the radiation energy. The combination of 'too much' sun and very high tempera-

Chlorophyll *a* and seasonal variation in quantity of phytoplankton

All green plants contain the green pigment called chlorophyll *a*, which enables them to trap the energy in sunlight and use it to combine carbon dioxide with water to form sugars during the process of photosynthesis. This is the only chlorophyll in blue-greens but most other plants contain at least one other similar pigment, called chlorophyll *b*. This too traps light energy but has to pass it to chlorophyll *a* before it can be used in photosynthesis. In general, the bigger or more numerous the plants are, the more chlorophyll *a* they contain, so it is a convenient indicator of the total quantity of plant material present in water.

In order to obtain an indication of the total quantity of phytoplankton present in a lake, the algologist takes a measured volume of water, passes it through a very fine filter so that the algae are left on the filter paper, and then grinds up the algae and the paper in either acetone or methanol. The grinding breaks up the algal cells and the chlorophyll is released from inside them to dissolve in the acetone or methanol. The fragments of paper are then compacted into the bottom of the tube by spinning the sample in a centrifuge and the green liquid can be poured off. The intensity of the green colour is then measured in an instrument known as a spectrophotometer and a simple calculation converts this to the quantity of chlorophyll *a* per litre of the sample. Since the volume of the sample taken from the lake is known, the concentration of chlorophyll *a* in the lake water can be found.

This is a much quicker way of estimating the total quantity of phytoplankton in the lake water than counting all the cells in the sample under a microscope, but it does not give information about the species of algae that are present. Nevertheless it is a very useful technique and is widely used. In very unproductive lakes, such as Millstättersee in Austria, the quantity of chlorophyll gradually increases during the spring from a very low winter level to a maximum in summer and then declines again. The level of the summer maximum is primarily determined by the quantity of nutrients available and is never very high. Rather more productive lakes in the temperate zones frequently show two peaks in the concentration of chlorophyll. The first, as in Lake Erken (Sweden), comes in spring just after the ice has melted and the second in the autumn. The dip in the middle, during summer, may be due to the fact that the spring population of algae has used up all the nutrients, leaving insufficient to sustain such a high population until the autumn mixing temporarily releases more nutrients from the hypolimnion. Some of the reduction in the summer population of phytoplankton may also be due to grazing by the zooplankton, whose populations built up more slowly during spring than the early algal species.

In more productive, shallow lakes, such as Crose Mere (England), the two peaks of spring and autumn are still evident and a good deal higher because of the larger quantity of nutrients available. The summer trough is occupied by a number of lower peaks which are due to different species of algae becoming dominant at different times. These species are adapted to use whatever nutrients are available at that time and, perhaps because the general level of nutrient supply is relatively high, have been able to replace those species which were dominant in the spring. In a very productive, tropical lake such as Lake George in Uganda, where both physical and nutrient conditions seem to be almost constant throughout the year, the concentration of chlorophyll *a* remains high and almost constant.

tures can actually inhibit production, but year-round warmth and sunshine means that plant and animal populations never drop back to very low levels from which they must rebuild their numbers each year. This enables many of them to achieve higher levels of annual production than in the higher, colder latitudes where there are seasonal cycles in production. Levels of production *per day* in temperate zone lakes during the summer may be equal to, or even higher than, those in tropical lakes, but such levels cannot be sustained for more than short periods before temperatures drop.

Conditions are not constant throughout the year in tropical lakes – far from it. In Lake Victoria there is a distinct period of stratification each year which breaks down when the surface water is cooled by the advent of the South-East Trade Winds from May to July. Although there is no appreciable change in daylength or solar radiation during the year, the wind is distinctly seasonal and the annual cycle in the lake is linked to this.

In many parts of the world the main seasonal change is between the wet and dry seasons. Rainfall, either on the lake or within the catchment area, increases the flow-through of the lake water, mixing a water body which was static and stratified throughout the dry season. Other lakes are flushed out when melting snow in the catchment area increases an inflow that has been frozen to a standstill during the winter. In a small lake the water may be completely changed within a matter of days and any organisms not attached to the bottom are swept out of the basin by the flow.

The seasonal cycles, whether driven by sun, wind or rain, result in changes in the chemical composition of the water. These changes, plus those in light and temperature, influence the growth of phytoplankton. The algae use up some chemicals as they grow and release some as they die: these processes also change the chemical composition of the water. The algae are eaten by small animals and some are more suitable as food than others: this influences the composition of the algal populations, which modifies the chemical composition of the water, in turn affecting algal growth. All the interactions work in both directions. The small animals are eaten by larger animals and thus further complexities are added to the interactions which take place throughout the year among organisms living in the lake. Unravelling these complexities in order to discover what are the most critically influential factors is the work of many research scientists, but even casual observers can see some of the changes which are the end results.

Seasonal changes in the lake community

Algae

Algal density in a lake is often measured in terms of the concentration of chlorophyll, and if this is done regularly throughout the year the seasonal variation in the total quantity of algae can be seen. What is not shown by this method is the change in the *species* of algae whose chlorophyll has been extracted. Individual algal cells are very short lived compared to larger plants found around the lakeshore.

The annual cycle of total phytoplankton biomass (as indicated by the concentration of chlorophyll) in the waters of different lakes: (a) Crose Mere in Shropshire, England, in 1973, which had two periods with very dense phytoplankton in spring and autumn; (b) Lake Erken, Sweden, in 1957, which also had two peaks (that in the spring later than in Crose Mere) but less dense populations; (c) Millstättersee, Austria, in 1935; this unproductive mountain lake had very small populations of phytoplankton, compared with the other two examples, and only one small period of increase in the summer. (Redrawn from Reynolds, 1984.)

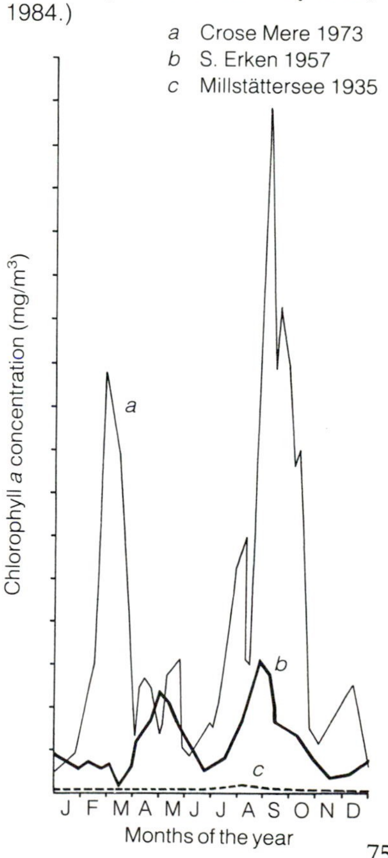

Depending on the species, algae divide to form new individuals two to ten times per month; one individual of the most rapidly dividing species may give rise to more than one million million individuals during one growing season. This means that, in a lake, we are observing the build up and decline of whole populations, instead of the growth, flowering and reproduction of individual plants as seen on land where populations build up over years rather than days. Consequently it is possible for populations of algae in lakes to exhibit a succession of species during the course of one year which, in some ways, is parallel to the sort of succession that takes place over a number of years on land, for example when a patch of bare earth is colonised first by grasses and herbs, then by woody shrubs and finally by trees.

In lakes which stratify during summer the succession of species starts again each spring and is interrupted each autumn when the lake mixes. During the period of stratification the replacement of one dominant species by another may be determined by chemical factors or by the grazing effects of animals. In lakes where secondary stratification occurs within the epilimnion but breaks down intermittently before complete mixing occurs, the succession may be repeatedly interrupted and forced to start again. How far it proceeds each time depends on the duration of secondary stratification between the mixing episodes.

Only detailed identification of algal species and cell counts will reveal details of the succession. Some species are present more or less all the time, but they fluctuate in numbers; many other species show a clear seasonality and disappear from the plankton for the remainder of the year. These appear regularly at more or less the same time each year, but the size of the population varies from one year to another. The time of appearance each year is characteristic for the different species. In general there is a tendency for diatoms to dominate in the spring, followed by green algae and then blue-greens later in the year. This pattern of succession is seen in lakes all over the world but the details are highly variable and each lake is unique in this respect, as in so many others.

Insects

The lake animals also respond to seasonal cycles. Many of these changes are not readily seen by the non-specialist, but clues to what is going on can be gained from some of the more obvious seasonal events. Clouds of insects dancing over the vegetation at the lake's edge on a warm summer evening in the temperate zones mark the end and beginning of the insect's life cycle. Adult midges, mayflies and stoneflies emerge from the water, where they have spent their juvenile life. They mate over the water and lay their eggs in it, starting the cycle again. If the population of one species is sampled from the lake at intervals throughout the year, first one size will be most numerous and then another. Adults of different species emerge at different times of year so, like the algae in the phytoplankton, there is a seasonal succession of species. As the insects grow larger, they provide food for fish; the fact that each prey species reaches maximum size at a different time means that there is always food available for predators that switch progressively from one species to another.

Zooplankton

Zooplankton populations also exhibit seasonal changes in their dominant species. Rotifers can increase their populations rapidly and in strongly seasonal lakes are often the first to respond to the increasing supply of algal food in the spring. They are followed by the cladocerans or the calanoid copepods. At the present time there is continued controversy over whether or not the populations of the larger species in the zooplankton are controlled by food supply or by predation. Where seasonal cycles are very marked, as at high latitudes, zooplankton with complex life cycles, whose development involves a number of instars, are frequently represented at any one time by only one instar. Their development is highly synchronised and as each instar follows the one before, it is the stage of development rather than the species composition of the population which changes.

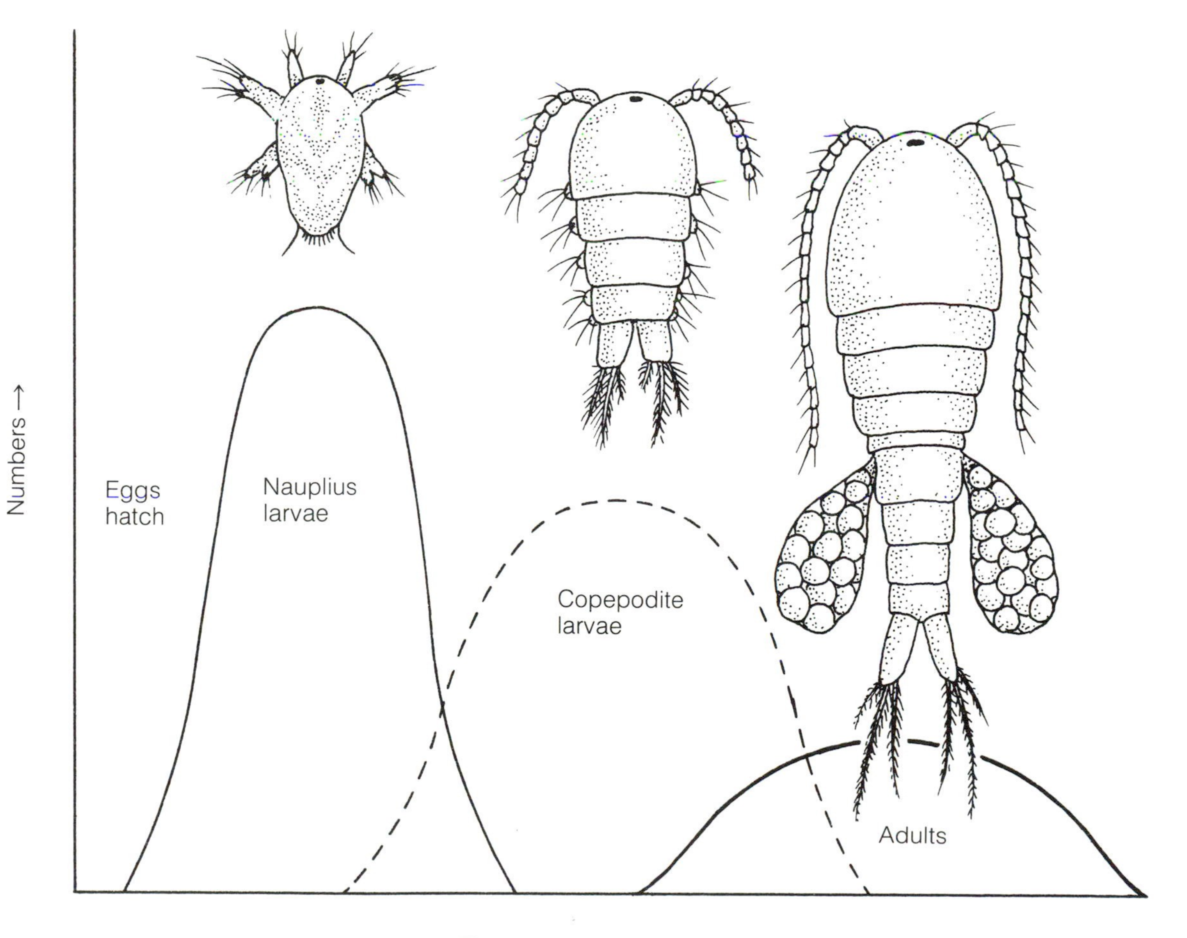

Seasonal succession in the stages of a copepod life cycle. In all copepods the eggs hatch into nauplius larvae; in a cyclopoid copepod (which is the type shown here) there are six nauplius stages followed by five, increasingly segmented, copepodite stages, and finally the adult. A female is shown here in dorsal view. A succession of stages such as this is most clearly seen in the coldest lakes; in the warmest lakes several generations may overlap and all stages of development may be present in the lake at any one time.

A species of *Barbus* There are many species in this genus of tropical cyprinids. They are primarily river fish but some are found in lakes although they return to the inflowing rivers to breed and usually do so in the flood season.

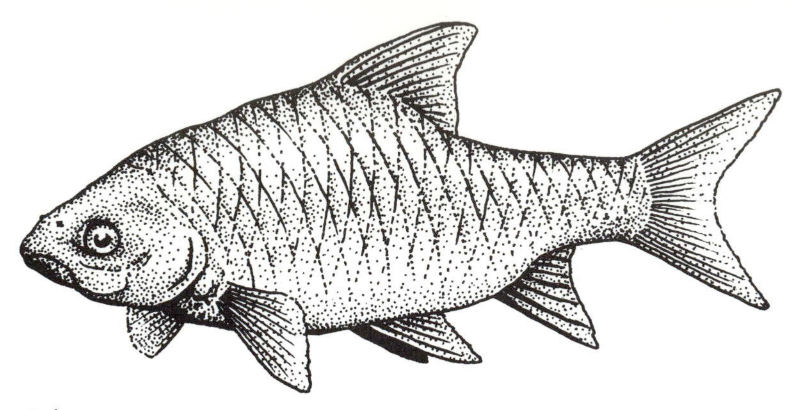

Fish

Fish mostly have a clearly defined breeding season and there is one period of the year when larval fish are abundant in the lake. Again, in most lakes where there is more than one species, their breeding may be separated in time so that the larvae of one species do not compete with those of another for the food available. In temperate zones many species of fish breed in the warmest seasons of the year, but others depend on the flow of river water into the lake so that they can swim upstream to breed in running water. In tropical lakes some fish breed all year round while others confine breeding to periods of flood, when they can breed either in shallow inshore areas or up rivers which flow into the lake. The return of young fish from inflowing rivers, or from floodplain areas, to feed in the open lake probably has as sudden and marked effects on the lake communities as physical events such as mixing.

Øvre Heimdalsvatn

Seasonal changes in solar radiation and temperature are most marked at higher latitudes. Lakes of the true Arctic and Antarctic contain animals and plants which have to tolerate particularly difficult conditions of cold and prolonged darkness, so it is hardly surprising that they are relatively few in number. The behaviour of ice and snow are the principal determinants of the physical conditions in such lakes and the development of their living communities. These will be discussed in the next chapter but here we can look at a small lake which illustrates the sharp seasonal changes associated with long winters and short summers found at higher latitudes.

Øvre Heimdalsvatn lies on the eastern slopes of the Jotunheimen Mountains in Norway at a latitude of 61° 25′ North and 1090 m above sea level, which is just below the upper limit at which trees will grow (1200 m). The lake is only 0.75 km^2 in area but its catchment area is 26.3 km^2 and rises to an altitude of 1843 m. The maximum depth of the lake is only 13 m, so during the ice-free period, from early June to mid-October, the water is well mixed with only occasional brief stratification. In winter the air temperature falls as low as −30°C but the lake is covered in a layer of ice (70–90 cm thick) plus a blanket of snow up to 1 m deep. The water level under the ice is maintained by groundwater supplies and its renewal time varies from 100 to 400 days. In May and June, when the inflow streams are in spate, renewal time may be as short as 2 days and during this period supplies of organic matter from outside the lake reach a peak. They are also high after autumn rainstorms.

During the ice-free period, wind mixing keeps the water temperature uniform throughout its depth. The water temperature rises to a maximum of about 15°C during July. The water contains very low levels of chemicals (conductivity 10–30 μS/cm) and, although there is a relatively diverse phytoplankton, the algae are not very productive. They start to increase in quantity, from very low winter levels, before the lake is free of ice, and between the end of April and the end of June the biomass may increase to four to eight times the winter level. This seems to be triggered by increasing spring sunshine which melts the overlying snow and penetrates the ice.

Both the planktonic and benthic animals have very clearly defined life cycles and show marked seasonal changes. The copepod *Heterocope saliens* hatches from resting eggs and has one generation per year, during which the numbers of each instar have peaks of abundance in rapid succession with seldom more than three stages present in the plankton at one time. Such synchronised development is fairly typical of animals in environments with such sharp differences between periods when food is available and when it is not. The cladocerans outnumber the copepods and because they have a less complex life cycle (see Chapter 3) have several generations and increase their numbers more rapidly in the same time period.

The benthic fauna is dominated by insects (74%) which is typical of unproductive lakes. Mayflies predominate, followed by caddisflies, midges and stoneflies. Each group is represented by a number of species; the majority are detritus feeders, reflecting the importance of dead organic matter brought in from outside the lake basin compared to the sparse growth of benthic and epiphytic (growing on other plants) algae actually in the lake. All these insects have highly synchronised life cycles which culminate in mass emergence of the adults from the water. The records of emergence dates illustrate the seasonal succession of some of the dominant species.

The four most abundant species of mayfly have one generation each year. They include *Leptophlebia marginata*, which manages to grow while the lake is ice-covered, and *Baetis macani*, whose growth is restricted to the ice-free period. *L. vespertina* and *Siphlonurus lacustris* are intermediate in this respect. Where there are two species of the same genus, they always emerge at different times: *L. marginata* emerges at the end of June and beginning of July, while *L. vespertina* appears from the end of July through the beginning of August.

The ten stonefly species are dominated by *Diura bicaudata*, which is found among the stones along the shoreline, and *Nemoura avicularis*, which lives in deeper water where it is warmer in winter and detritus is always available. *D. bicaudata* takes at least 2 years to complete its life cycle in this lake. The eggs are laid in July after emergence and diapause (remain dormant) during the first winter. They hatch the following spring and the larvae grow throughout that summer and the following winter, to emerge as adults the summer after.

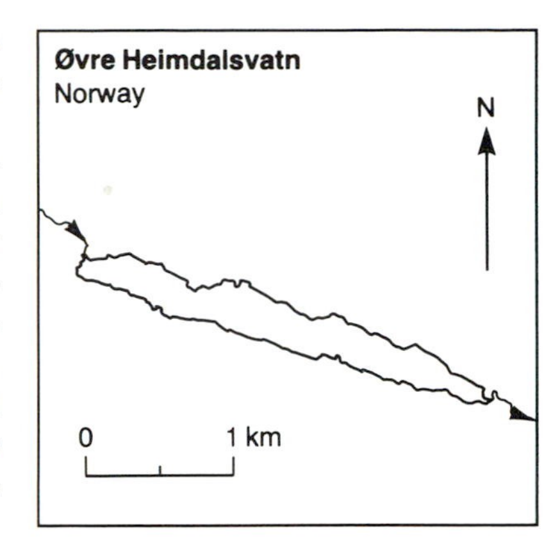

Location: 61°N; 9°E
Altitude: 1090 m
Catchment area: 26.3 km^2
Principal inflow: Brurskardbekken
Lake area: 0.75 km^2
Lake volume: 3.6 million m^3
Maximum depth: 13 m
Mean depth: 4.7 m
Outflow: River Hinøgla

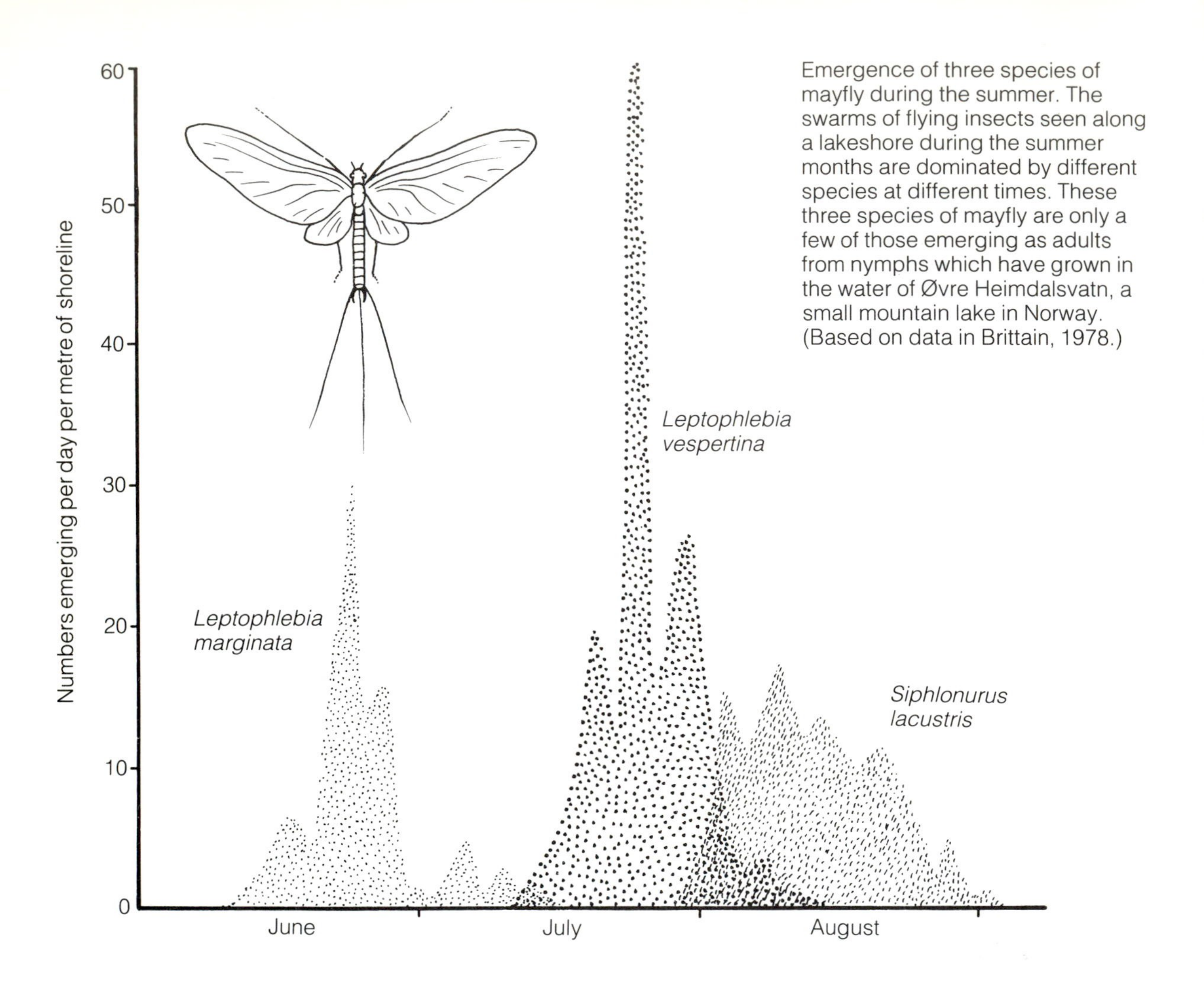

Emergence of three species of mayfly during the summer. The swarms of flying insects seen along a lakeshore during the summer months are dominated by different species at different times. These three species of mayfly are only a few of those emerging as adults from nymphs which have grown in the water of Øvre Heimdalsvatn, a small mountain lake in Norway. (Based on data in Brittain, 1978.)

The chironomids are the most numerous and diverse group of insects in the lake, with sixty-seven species recorded in emergence traps. As individuals they are much smaller than the mayflies and stoneflies. The fact that their emergence periods are spread out through the ice-free period means that there is always some food available for the fish. The growth of fish depends on the insects and thus, indirectly, on the detritus washed into the lake by its inflow streams.

In lakes like Øvre Heimdalsvatn the volume of water in the lake basin is so small in comparison to the volume of inflow during a spate that it can be totally and rapidly flushed out. So the seasonal variation in solar radiation affects the seasonal cycle in the lake, not only through the availability of energy for primary production, but also via its effect on run-off as a result of melting the snow lying on the catchment area. In deeper lakes the effects of seasonal floods and meltwater are usually lessened by the larger volume of water in the lake basin.

Windermere

Further south seasonality becomes increasingly muted, although mid-continental climates are more extreme than those of the continental margins. Windermere is the largest lake in the English Lake District and is subjected to a North Temperate maritime climate. This means that the lake only freezes over in exceptional years and normally has only one period of complete mixing and one of stratification per year. The dimensions of the lake basin and its catchment area disguise the fact that the long narrow lake actually lies in two basins which are separated by a stretch of relatively shallow water (<5 m) and several islands where the lake is at its narrowest, about half-way along its length.

The temperature of the lake was among the first things investigated when research work started there. In late 1931 the water temperature, in the middle of the North Basin, was 7°C from the surface down to 60 m. By early April 1932 it had fallen to 5.5°C but after that the temperature at the surface began to rise quite rapidly until, by the beginning of July, it had reached 18°C. The water at the bottom of the lake was still only at 7°C. Stratification lasted from the end of June until the end of September; the surface reached a maximum of 19°C and at the bottom the water temperature crept up to 8°C. The thermocline was established between 10 and 20 m and in that distance the temperature fell from 15°C to 9°C. Mixing occurred in October. Between 1947 and 1964 there were only five years in which the surface temperature reached 20°C and only three where this temperature was reached at a depth of 5 m.

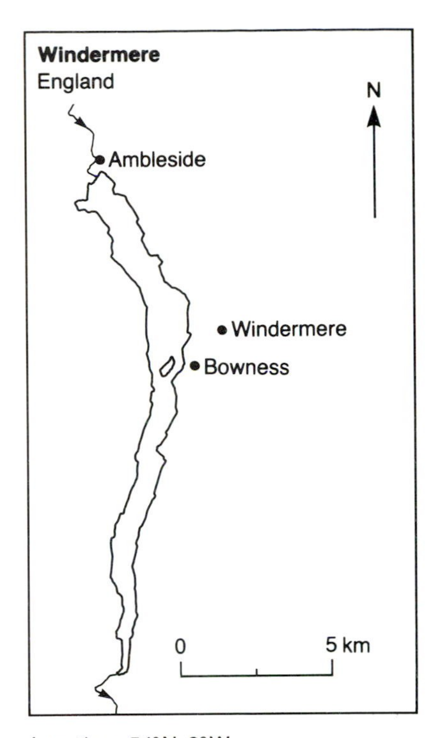

Location: 54° N; 3° W
Altitude: 39 m
Catchment area: 230.5 km^2
Principal inflow: River Brathay
Lake area: 14.8 km^2
Lake volume: 314.5 million m^3
Maximum depth: 64 m
Mean depth: 21.3 m
Outflow: River Leven

Windermere is among the more productive of the Lake District lakes. The concentrations of the ions listed in Table 2.1 show little variation through the year but this is not true of the nutrients, nitrate and phosphate, nor of silica, all of which follow a regular seasonal pattern. The natural concentrations of nitrate and phosphate in Windermere are low compared with those of many other lakes elsewhere (although higher than those in the other lakes of this district), despite the fact that they are augmented by the effluent of a sewage works which drains into the north basin. Nutrients are always present in sufficient quantity to support the growth of *Asterionella formosa*, the alga which dominates the phytoplankton at the beginning of the year. By June, however, *Asterionella* is rapidly being replaced as the dominant species by others; what has caused its population to decline?

Asterionella is one of the algae known as diatoms, which are characterised by their rigid cell walls (frustules) composed of silica. Cells of this species, and of other diatoms, are always present in the lake water, even during winter, but they only begin to grow and divide rapidly in the spring. As the population expands, the cells absorb more and more silica from the water until the supply is exhausted and population growth ceases. When the diatoms die their heavy frustules fall to the bottom of the lake. Some of the silica may be released and returned to the upper layers of the water at the autumn overturn but most of it is unavailable until the next spring.

Diatom numbers depend on silica concentration in the water. In Windermere the population of the diatom *Asterionella formosa* increases greatly in spring but only until the supply of silica, which it needs for building new cell walls, is almost used up. The concentration of nitrogen in the water also falls but it is the low level of silica that limits the growth of *Asterionella* once they can no longer extract it from the water. No more new cells can be made, the old ones die and the population crashes, to be replaced by other species of algae which take advantage of the remaining nitrogen as the year goes on. (Redrawn from Macan, 1970.)

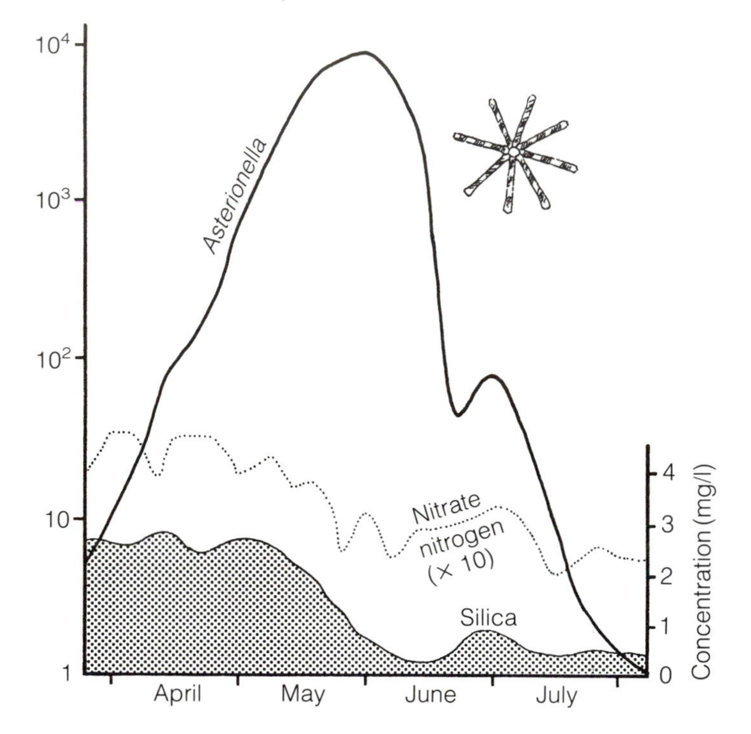

During winter, fresh supplies of silica are brought in by the inflows from erosion of the catchment area and decomposition of some diatoms on the lake bottom releases more silica to the water, ready to support the spring bloom of *Asterionella* once again.

Melosira italica is another diatom which forms significant populations in Windermere during the spring, but it sinks in the water much faster than *Asterionella* and can only remain in the upper layers during periods of strong winds which stir the water vigorously. It does have the advantage of being able to survive long periods lying on the bottom mud and usually has its spring population peak before that of *Asterionella*.

As the year proceeds, the diatoms are replaced as the dominant phytoplankton, first by green algae and later by blue-green species. The latter reach their peak towards the end of summer. If by then the other algae have used up all the nitrates in the water, some blue-greens have the advantage of being able to use nitrogen directly from the atmosphere. Most blue-green algae grow best at warm temperatures, so as the water cools in autumn they are replaced by other species; frequently there is a second brief outburst of diatoms. None of the other species of algae reaches the population numbers achieved by the diatoms in spring.

This seasonal succession of algal species is common to all lakes which undergo a marked seasonal change in physical conditions. The dominant species vary in different lakes and, in some cases, from year to year, emphasising the individual character and sensitivity of a lake system.

The phytoplankton succession is paralleled by seasonal changes in the dominant planktonic animals. One reason why algal populations are so much larger early in the year is because the herbivorous animals cannot increase their populations until there is enough food available. Once the algae start to grow, the grazing animals soon follow: first the rapidly reproducing rotifers, then the cladocerans, and later the copepods.

The zooplankton populations are influenced by their food supplies and also by the seasonal cycles of their predators, including fish. Typically in Windermere, fish eggs hatch towards the end of May and the young fish, such as perch, begin feeding a few days later. They start by eating rotifers and, as they grow, turn to larger prey: cladocerans, then insects, and finally other fish. Depending on the numbers and growth rate of the fish larvae (which partly depends on the water temperature), different sizes and species of planktonic animals are abundant or scarce in the water as a result of predation. The populations of herbivores in turn affect the abundance of different sizes and species of algae. This whole intricate series of interrelationships is closely bound up with the equally complex physical and chemical events which follow the seasonal cycle in the lake.

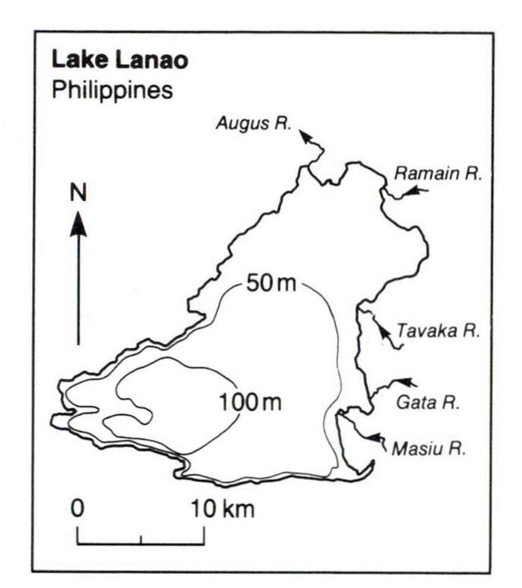

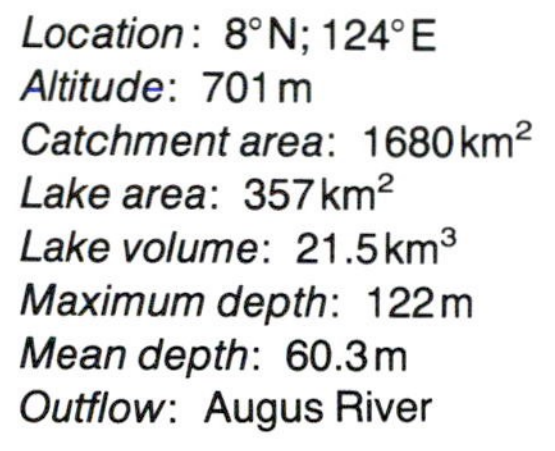
Location: 8°N; 124°E
Altitude: 701 m
Catchment area: 1680 km^2
Lake area: 357 km^2
Lake volume: 21.5 km^3
Maximum depth: 122 m
Mean depth: 60.3 m
Outflow: Augus River

Lake Lanao

Many tropical lakes of moderate depth (except those at very high altitude which may freeze) have one period of mixing per year. Lake Lanao in the Philippines is a good example which has been well studied. It is one of the few natural lakes in South-East Asia and was largely unaffected by human activities until the early 1970s. It lies in a sparsely populated area, 702 m above sea level, and has an area of 357 km^2 with steeply sloping shores. The maximum depth is 122 m and the mean depth 60.3 m. For most of the year it is stratified with a very thick epilimnion of 40–60 m depth, but in January and February the sunshine is curtailed by thick cloud cover and this reduces the surface temperature sufficiently for the stronger winds of that season to mix the whole lake. Even then the water temperature is 24°C. When stratified, the temperature of the epilimnion varies around 25–27°C and the surface water temperature may rise as high as 29°C.

Although the whole depth of the lake is only mixed once a year, secondary stratification occurs within the epilimnion. This forms, breaks down during windy weather, and re-forms repeatedly at irregular intervals without disturbing the main thermocline. The overall seasonal variation in temperature is so small that these non-seasonal events assume greater importance in the functioning of the lake than they would in more seasonal climates. Within the epilimnion the alternation of shallow and deeper mixing facilitates the return of nutrients to the euphotic zone and this increases primary production. Similar enhancement follows the complete breakdown of stratification and the mixing of the whole lake.

Each of these abrupt increases in nutrients within the euphotic zone initiates a new succession sequence in the phytoplankton. It starts with the diatoms and cryptomonads, which are the first to respond as the turbulence dies down and they are able to stay in the light long enough to photosynthesise. As they use up the nutrients and conditions get calmer green algae take over and these are followed by the blue-greens. This sequence is similar to that in temperate lakes but how far it proceeds in Lake Lanao depends on the time between mixing events. It is interrupted when stratification breaks down in January.

The total quantity of both phytoplankton and zooplankton in the lake declines during the seasonal mixing, which extends to such depths that much of the phytoplankton community is deprived of light and cannot grow. Nevertheless the amount of phytoplankton is never so low as in a temperate lake during winter mixing. The zooplankton consists of seven rotifer species, four cladocerans and two copepods, all of which are herbivores and mostly present throughout the year. There are always sufficient algae to support some of the rotifer and crustacean population, and food is not scarce for sufficiently long that their numbers are drastically reduced. This emphasises the contrast between a constantly warm tropical lake and the wide seasonal oscillations in temperate lakes, which include a prolonged period when growth and development of most species is severely curtailed.

Lake George

Lake George is a shallow, freshwater lake spanning the Equator in Uganda, where daylength never varies and there is no seasonal variation in temperature of either air or water. This means that the amount of energy entering the ecosystem and available to the algae is almost constant throughout the year. Indeed, there is so little *seasonal* variation that it is the *daily* pattern of events which becomes the dominant regulator of the plant and animal communities.

There are two rainy seasons and two dry seasons each year, so, although evaporation rates are also very high all through the year, the income from rain is more evenly spread than in regions of the world that have only one wet season. The location of Lake George, at the foot of the Ruwenzori Mountains, also ensures that it receives plenty of river inflow all year round. The Ruwenzoris rise to over 5000 m above sea level and 4000 m above the floor of the Western Rift Valley in which Lake George lies. Water from the eastern side of these mountains pours down their steep slopes in innumerable fast-flowing streams that flow only a very short distance over level ground to enter the extensive papyrus swamps at the north end of the lake. Water from the eastern escarpment of the Rift also drains to the lake. Water leaves the lake via the broad shallow Kazinga Channel which is 35 km long and in that distance drops only about 1 m before entering the much larger and deeper Lake Edward. This drains north to Lake Mobutu Sese Seko (formerly Lake Albert) and the River Nile.

The Ruwenzori Massif is crowned by seven permanently snow-capped peaks so during the rains some of the precipitation is in effect stored as snow. When the dry season comes, cloud cover decreases and the snow melts to fill the rivers and maintain the inflow to the lake. Thus, despite its tropical location and a mean depth of only 2.4 m (maximum about 3 m), the lake level varies by only 20 cm during the average year. There is some year-to-year variation in the water balance but, in comparison with many other lakes, particularly those in the tropics, Lake George is remarkably stable in its physical conditions and could reasonably be considered to lack seasonal variation.

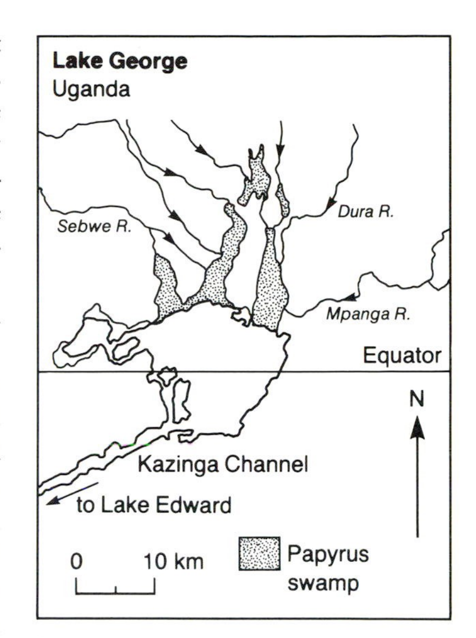

Location: 0°; 30° 01′ E
Altitude: 913 m
Catchment area: 9955 km^2
Principal inflow: Mpanga River
Lake area: 250 km^2
Lake volume: 600 million m^3
Maximum depth: about 3 m
Mean depth: 2.4 m
Outflow: Kazinga Channel

Another major seasonal factor in lakes is wind, but Lake George is partially shielded by the bulk of the mountains to the north-west and by the escarpment close to the lake. Prevailing winds are light and usually blow most strongly towards the end of the afternoon. Thus the lake starts the day with the same temperature from top to bottom and then the shallow column stratifies as the surface warms up until mid-afternoon, when the wind rises and is normally sufficient to disrupt the stratification. Sometimes the wind is also strong enough to stir the bottom mud into the water. Occasionally there are periods of calm when stratification may last for several days. Thus almost every day Lake George goes through a pattern of events similar to those experienced by a temperate lake in a whole year. Many shallow lakes have a similar pattern of daily stratification, in summer or in the less windy season, but not many experience it throughout the year.

These remarkably constant, warm conditions seem to allow the blue-green algae (particularly *Microcystis* spp.) not only to gain, but to maintain, dominance. Although more than 150 species of algae have been recognised in the Lake George phytoplankton, the blue-greens comprise more than 70% of the algal biomass all through the year. The total concentration of algae is very high (on average about 200 μg chlorophyll *a* per m^3 or 400 μg/m^2) which gives the water of Lake George the appearance of thick pea soup. When the water is calm, rafts of *Microcystis* can be seen forming at the surface. Consequently, light does not penetrate beyond about half a metre down into the water and many of the algae spend much of the day in darkness. The fact that they can form gas vacuoles which bring them back to the surface gives the blue-greens an advantage over other species. Species that have no flotation devices must wait until stirring by the wind brings them back into the lighted water where they can photosynthesise.

It was thought for many years that animals could not digest blue-green algae such as *Microcystis* and that even if fish swallowed them they passed through the gut unharmed and that nourishment was obtained from bacteria and other algae taken in at the same time. One of the most interesting aspects of the Lake George community is that it is dominated by herbivores that can and do utilise *Microcystis* as food. The small copepod *Thermocyclops hyalinus* is one, and it dominates the zooplankton. The others include two species of fish, the tilapia *Oreochromis niloticus*, which is harvested commercially, and the much smaller cichlid *Haplochromis nigripinnis*. These two herbivorous fish comprise 60% of the total fish biomass in the lake.

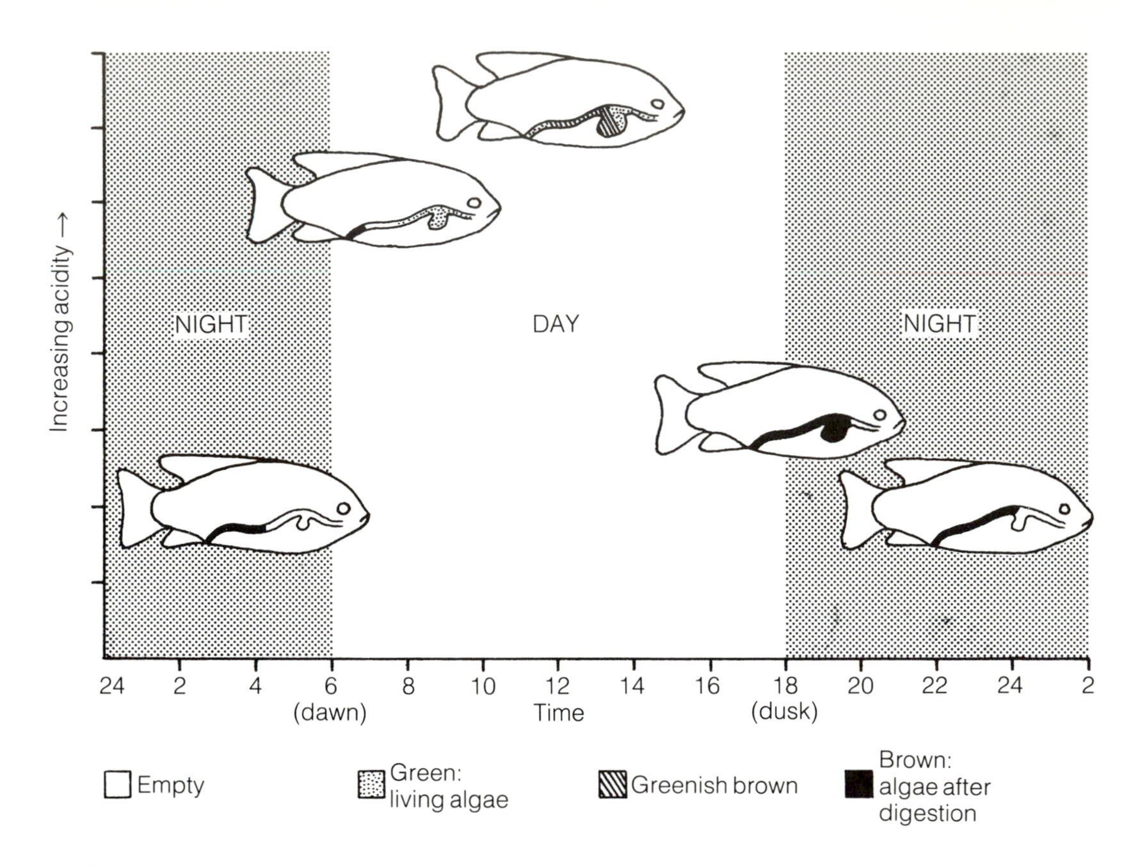

The herbivorous fish *Oreochromis niloticus* (formerly called *Tilapia nilotica*) is found in several African lakes. It feeds on blue-green algae and has a daily digestive rhythm of acid secretion in the stomach to cope with feeding during the day. It does not feed at night, despite being surrounded by a thick 'soup' of algae. By dawn the stomach is empty and there are only dead, brown algae in the intestine, (left-hand fish). It begins to feed and the passage of living algae into the stomach stimulates the secretion of strong acid that aids digestion of the algae which turn from green to brown as they pass into the intestine. After feeding ceases at dusk, digestion continues through the night until the stomach is empty (Based on information from Moriarty, 1973.)

They digest the *Microcystis* by secreting very strong acid in their stomachs. This breaks down the mucilage in which the cells are enveloped and releases the cells for digestion in the intestine. When feeding starts in the morning the first food to enter the stomach stimulates secretion of the acid so at first the algae do pass through to the intestine undigested and the faeces appear green. Very soon, however, the faeces are brown as enough acid accumulates in the stomach and digestion is able to proceed. When the tilapia are harvested there is thus a very short food chain from algae through fish to man, which must be a very efficient way of exploiting natural resources; eating the algae direct would be more efficient but less enjoyable.

Looking at the green water of Lake George from the shore, and knowing that large numbers of fish are caught each year, might lead one to assume that this is a very productive lake. However, although large algal populations in warm water are able to photosynthesise very rapidly and form new cells, they also respire very rapidly and thus use up oxygen very quickly. In addition they shade each other from the incoming light so that only a small proportion of the population can be photosynthesising at any one time. This means that the surplus of new living material formed *per day* (the net production) is in fact not as high as might be expected. It is often less than on a summer's day in Britain. Moreover, the days are short compared with those at higher latitudes in the summer, so the time available for photosynthesis is reduced. However, because conditions are favourable for photosynthesis throughout the year, the *annual* net production is greater than that in a similar lake at higher latitudes, although not enormously so.

High levels of primary production require large supplies of nitrogen and phosphorus. The waters entering Lake George from the swamps contain very low concentrations of these nutrients and so do the surface layers of the soft bottom mud. Presumably this mud is so frequently stirred that only small amounts of nutrients accumulate and these are used immediately they are released. The escape of nutrients from lower layers of the mud is irregular and less frequent. Overall, the mud seems to play rather a minor role in supplying the nutrients needed by the algae. A large proportion of what the algae need comes from the excretions of herbivores, particularly the tiny copepods of the zooplankton. Although the fish are much bigger animals, their total population excretes only a fraction of the total amount supplied by the millions of tiny copepods. It has been shown that if the copepods are put in water on their own, the concentrations of both nutrients steadily increase, but if *Microcystis* are kept in water with the copepods there is never any nitrogen or phosphorus free in the water because the algae absorb the nutrients as fast as they are released.

Lake George contained a large population of hippopotamus at the time of these studies (1966–72) and it was thought at first that they must be responsible for the fertility of the lake, but even 3000 hippos don't excrete enough nitrogen to supply a fraction of that required by the dense population of algae. It seems more likely that it is the constancy of the environment in Lake George which allows the dense algal population to persist. We have seen that in our other examples, from Øvre Heimdalsvatn in Norway to Lake Lanao in the Philippines, the plant and animal communities are dominated by different species at different times of year and that it is only when there is a change in conditions, brought about either by the species themselves or by the physical environment, that the normal sequence is interrupted. In Lake George the events which take place within 24 hours are within the lifetime of single algal cells (particularly the slow growing blue-greens), so the algal populations do not have time to respond to the changes. Provided the algae are well adapted to the conditions occurring during the daily cycle they can maintain their populations indefinitely. Longer-lived animals such as fish do not usually show seasonal succession of species and the lack of diversity in the fish fauna of Lake George is probably also due to the relatively uniform structure of the habitat. The lake is also relatively young (having been formed only about 3500 years ago); this too reduces scope for evolutionary diversification.

Hippopotamus

The common hippo (*Hippopotamus amphibius*) can grow up to 4 m long and weigh 4 tonnes. It is a truly amphibious species, dependent upon and linking together the aquatic and terrestrial habitats. Hippos spend their days in lakes and rivers and their nights on the savanna grasslands. During the 1960s they were particularly abundant in Uganda along the Semliki River and the shores of Lakes Edward and George, where they caused considerable stress to the local ecosystem.

The outer layer of skin is very thin and in dry air hippos lose a great deal of water through it: up to 12 mg/cm^2 of skin in 10 minutes, three to five times the equivalent rate in man. To reduce such loss and keep cool they must spend most of the day in water.

The hippo is adapted to aquatic life by having its eyes, ears and nose placed along the top of the head and for much of the day these may be all that is visible above the water. When completely submerged, the ears and nostrils can be closed. Hippos can stay under for at least 5 minutes, longer in an emergency. They can swim and are also able to walk on the bottom of shallow water while totally submerged but most of their time is spent quietly lying about in family groups, averaging ten to fifteen individuals, though sometimes comprising more than a hundred. Dominant bulls each hold sway over a length of shoreline 250–500 m long; other bulls that are submissive may join the group within that range but only the dominant bulls breed.

Every evening as the sun goes down and the air cools, these massive animals wade ashore and follow well-trodden traditional pathways through the bush and across the grass, grazing as they go. They have four stout toes on each foot which support their ponderous body and tread a path of beaten earth about half a metre wide. On either side of the route hippos leave dung, scattered by rapid beating of the flattened, bristle-fringed tail as it is voided. Perhaps it serves as a scented marker to help the hippos find their way in the dark. The average hippo needs to eat about 40 kg dry weight of grass each night – a smaller proportion of its body weight than the requirement of most other hoofed animals. The digestible parts of the grass are broken down in a three-chambered stomach; much is not in fact digested and is still clearly recognisable in the dung. The hippos can probably get by with this low intake and inefficient digestion because, apart from the five or six night hours spent feeding, their energy expenditure for the rest of the 24 hours must be relatively low, as they lie in the water doing very little.

Much of the daily output of dung and urine goes into the water and effectively transfers nitrates and phosphates, as well as organic matter, from the terrestrial to the aquatic ecosystem. It was assumed that the main reason why Lake George contained a 'pea soup' of algae and plenty of fish was the large population of hippos that enriched its inshore waters. At one time the population in the lake was estimated at about 3000, clustered in the shallow bays around the edge. Their dung provided food for detritus-feeding invertebrates in those places and its decomposition must have enhanced nutrients and algal populations; but careful calculations indicated that their contribution to the total nitrogen in the lake as a whole was very small.

Because of their size, weight and behaviour hippos affect their environment markedly. In the water their movements stir up sediments and enrich water; they keep open paths through the swamp vegetation allowing access for other, terrestrial animals; their grazing promotes areas of productive short grass which are good for other grazers too. However, hippos are limited to feeding within 3–4 km of the lakeshore, so if their numbers increase greatly they destructively overgraze this zone and their paths become eroded gullies. By 1960 their grazing density around the shores of Lakes Edward and George exceeded 30 animals per km^2, equivalent to 31 000 kg/km^2 of hippo flesh alone, without the elephant and buffalo which were also locally abundant. From 1962–66, 1000 hippos were shot each year to reduce their population and allow the vegetation to recover. It was the first time such a culling programme had been undertaken in a National Park. Where hippos were completely eliminated, the bare earth once more become covered by thick grass and the numbers of several species, particularly buffalo, increased. The hippos themselves responded to the reduction in their population by maturing younger, at 10 years instead of 12, and the proportion of calves in the population increased from 6% to 14%. When the control programme stopped they soon recolonised the areas from which they had been eliminated, but in more supportable numbers.

During the rainy season, of which there are two in the Ruwenzori National Park, the hippos increase their grazing range by not returning to the lake and instead wallow in temporary pools during the day. As these pools start to dry up the hippos are forced back to the permanent lake and river waters and their populations become concentrated again. It is during the dry season that they mate in the water. Most young are born during the rains when females leave the herds to give birth to single calves weighing about 42 kg. Mother and young return to the herd after about a week.

Although the adult hippo has no natural enemies the young are vulnerable to lions and to crocodiles. There are no crocodiles in either Lake Edward or Lake George, but in other places when crocodiles are about the calf may ride on its mother's back and she will fiercely defend it. Other animals take advantage of the hippo's bulk: one little cyprinid fish (*Labeo velifer*) grazes on algae growing on its skin; terrapins and young crocodiles use half-submerged hippos as basking platforms; and birds like egrets and the hammerkopf frequently perch on a hippo to peer into the murky water for prey. In many ways this strange mammal epitomises the interdependence of a lake and its surroundings.

5
Polar and mountain LAKES

As one climbs a mountain, the air, even on a bright sunny day, gets colder and colder, by −1°C for every 150 m extra height. Even in the tropics the effect is the same, and the transition from the lowlands to the summit of very high mountains is like passing from the Equator towards the North or South Pole. As a result, there are similarities between the conditions found in lakes at high altitudes and those at high latitudes. In both cases the water is very cold, and covered in ice and snow for long periods. The main difference lies in the pattern of sunshine: near the poles there is almost continuous light for several weeks in summer and continual darkness for a similar period in mid-winter. The pattern of sunlight falling on high mountain lakes depends, as for all other lakes, on their latitude. Lakes surrounded by steep, high mountain peaks may also be in shadow for longer than many lowland lakes and this reduces the total amount of light they receive.

Ice, snow and cold water are the principal features these lakes have in common; otherwise they are as varied as other lakes. They may be shallow or deep, fresh or saline, or even fed by hot springs. In this chapter we are only concerned with the majority of the lakes, which contain fresh water and show the most characteristic effects of high latitude and high altitude on lake ecosystems.

Arctic lakes

The red-necked phalarope *Phalaropus lobatus* is a typical summer visitor to lakes in the Arctic. The long summer days enable them to feed for 24 hours a day and the lakes provide an abundance of insect food during the short summer when the birds breed.

The truly Arctic areas of the Northern Hemisphere are those with a mean air temperature of the coldest month less than −3°C and of the warmest month between zero and 10°C. This includes the northern land masses of Alaska, Canada and the Soviet Union as well as Greenland, Spitzbergen and the northern fringes of Scandinavia. Here there are enormous areas of land where meltwater collects in every hollow on the bare ground and the landscape is dotted with thousands of lakes. Most of the lake basins were formed by glaciers and the lakes are relatively recent; some, on high ground, are still fed directly by the glaciers which scoured out their basin, but the majority are fed by the melting of snow which fell the previous winter.

Winter comes very early in the high Arctic. Shallow lakes are all frozen over by the end of September, while the greater amount of heat stored in deeper lakes keeps them open for a week or two longer. Inflow rivers cease to flow even earlier, before the lakes freeze, but outflow rivers may continue to drain water from beneath the ice for some time and thus lower the level of the lake water sufficiently for the ice to sag in the middle and be dragged down the sides of the lake basin during the course of the winter. Ice scraping the shore removes any vegetation growing there, and the shorelines of Arctic lakes are usually bare down to about 2 m or more.

Lake ice usually has a maximum depth of about 2.4 m but greater depths have been recorded. Anguissaq Lake, a deep (maximum 187 m) lake in North-West Greenland, is permanently frozen. The depth of ice reaches about 3.4 m in winter and about 1.4 m of that melts off the top each summer. This is exceptional; most Arctic lakes have an ice-free period for a few weeks in summer. Melting begins when the air temperature above the ice exceeds 0°C and the sun reaches the surface of the ice (after first melting any overlying snow). The ice melts only at the surface but it does so first at the boundaries between the crystals and this eventually makes the ice porous so that overlying water drains down through the thickness of the ice which then resembles a mass of candle-like columns. When its thickness is reduced to about 1 m and when all of it is 'candled', it becomes unsafe and may break up as soon as the ice columns become separated and tip over. It then melts very quickly.

During the Arctic winter, when the sun does not rise above the horizon, there is no energy for primary production, and living organisms have to subsist on what remains of the previous year's production or spend the winter in a dormant phase. Light is available for much longer than the lake is free from ice. So life can go on beneath the frozen surface, but the quantity of energy reaching the water depends very much on the clarity of the ice and the thickness of the overlying snow. Only 10–20 cm of snow will absorb or reflect up to 99% of the available light. Similarly, although most deep Arctic lakes have very clear ice, if the ice is bubbly this too will reduce light penetration. In very windy locations much of the snow cover may be blown away or drifted to one area of the lake so that sunlight can reach the ice and warm the water beneath it. Because water reaches its maximum density at 4°C, the initial warming above zero results in heavier water which sinks into the colder water below and causes some circulation, even under the ice. The sediment, too, may slowly transmit stored heat to the deep water during the winter, thus raising its temperature a degree or so above that of the overlying water.

When the ice is thick, the lakes are not free of it until after the annual peak of solar radiation has passed. Cloud cover is also more frequent in July and August than during the rest of the year, reducing light at what would otherwise be a favourable time. All these factors result in Arctic lakes receiving very low levels of solar radiation annually and their surface water temperatures seldom rise as high as 15°C and then only in very shallow lakes. Such low temperatures ensure that the water of Arctic lakes is usually rich in oxygen and may become supersaturated if photosynthesis starts under the ice. Only very shallow lakes become deoxygenated if there is no photosynthesis, though a pool of deoxygenated water may form in the very deepest water as the sediment uses up oxygen during the winter when the water is very still. During the ice-free period most Arctic lakes are well mixed by the wind and stratification is uncommon.

Char Lake gets its maximum exposure to sunshine in the early part of the brief Arctic summer. However, at this time the lake is still frozen and covered with snow, which restricts the amount of light reaching the phytoplankton. Thus the peak of algal productivity comes later, after the peak of available sunshine, when the ice and snow cover has melted away. (Based on data given by Kalff and Welch (1974) and Welch and Kalff (1974).)

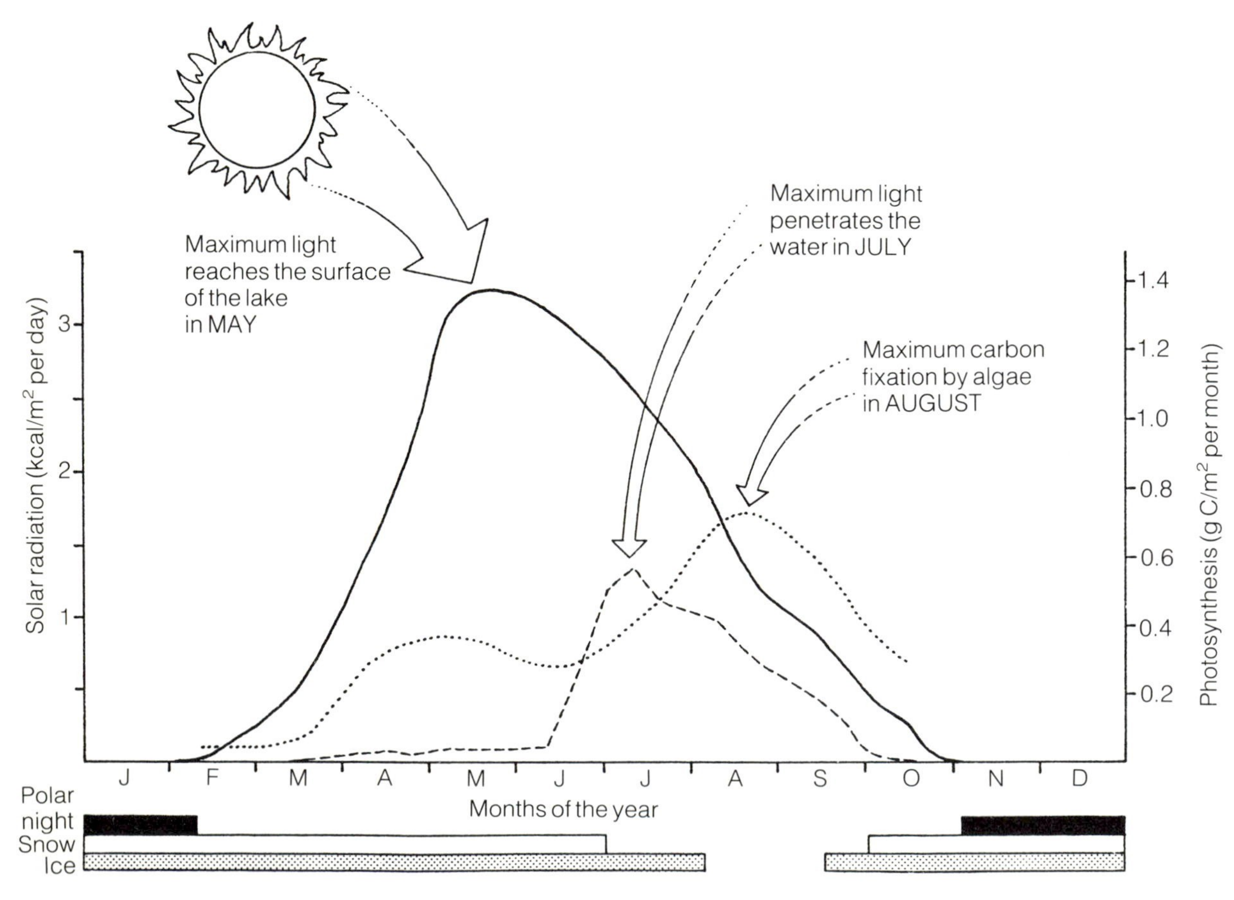

Arctic lakes are generally very unproductive and contain communities composed of a very limited number of species. Simple animals and plants seem to occur in Arctic lakes in similar numbers to those in temperate lakes, and are sometimes of the same species, but the diversity of higher plants and animals is more restricted. Only those which can function in continuous cold thrive and they have very low rates of metabolism and growth. The very low levels of production also limit the total number of animals able to exploit the small amount of food produced.

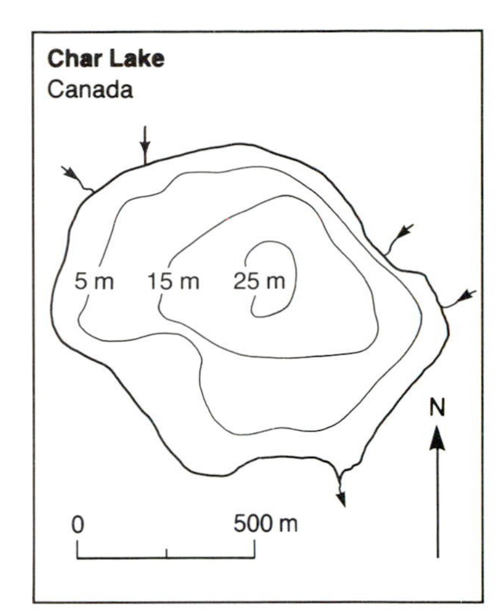

Location: 74° 42′ N; 94° 50′ W
Altitude: 34 km
Catchment area: 43.5 km^2
Inflows: 4 small streams
Lake area: 0.53 km^2
Lake volume: 5.4 million m^3
Maximum depth: 27.5 km
Mean depth: 10.2 m
Outflow: a small stream

Char Lake

Char Lake is a small, well-studied example of an Arctic lake. It is situated at 74°N on Cornwallis Island in the Canadian part of the Arctic Ocean. At the nearby village of Resolute the mean annual air temperature is −15°C and in the coldest month, February, the mean falls to −33°C. The highest monthly mean, in July is only +4.5°C.

The lake is almost circular with an area of only 52.6 ha and a maximum depth of 27.5 m. Although most of its catchment is low lying and undulating, all three of the main inlets flow from a plateau about 120–150 m above sea level. The water is hard and has a conductivity of about 195–240 μS/cm which is very little influenced by the proximity of the sea because the ocean around the island is frozen for most of the year. As in other Arctic lakes, the conductivity of the water increases when ice forms because ions are frozen out into the underlying water.

Eighty percent of the light reaching the water of Char Lake in one year is received during the 11 weeks from 15 June to 31 August. The lake is free of ice for about 5–6 weeks in almost every year but in some years, such as 1972, the ice never completely melts. This prevents the water mixing and cooling, so the lake starts the next winter rather warmer than usual; this does not prevent the ice from becoming thicker than usual, however, as the new ice is added to the previous year's residual layer. The ice is normally 2.2–2.3 m thick and occupies 18–20% of the lake volume. Over winter, the lake water steadily loses heat and the temperature beneath the ice sinks to 0.5°C, while that at the bottom remains at about 1.5°C: this is an example of a temperature inversion. Even in summer the temperature of the water never rises above 4–5°C but the water is always well mixed and rich in oxygen.

The phytoplankton in most Arctic lakes is dominated by very small cells whose high surface-to-volume ratio may be advantageous in nutrient-poor waters. Many are motile, allowing them to select the best among the rather poor light conditions available. In Char Lake the phytoplankton biomass begins to increase in February and continues to do so until May, but then declines until September. Even at the peak the concentration of chlorophyll *a* is only about 0.69 mg/m^3 and primary production is very low. The productivity of plants growing on the bottom of the lake (mostly mosses) is considerably higher, perhaps because they are able to obtain nutrients from the substrate.

The red-throated diver *Gavia stellata* is a typical fish-eating bird of small Arctic lakes. It often nests on lakes that are so unproductive that the birds have to fly elsewhere to feed. In winter the lakes are frozen so the birds migrate south to coastal waters and unfrozen lakes.

Apart from single-celled ciliates, the zooplankton contains only two species: the copepod *Limnocalanus macrurus* and the rotifer *Keratella cochlearis*. These are both present in the water throughout the year and have only one generation annually. The eggs of *Limnocalanus* hatch in January–February and the numbers in the population decline from then onwards as the animals grow, until only adults are present by September. Neither they nor the rotifers show any sign of being able to develop more rapidly than their temperate zone relatives would at these very low temperatures and the birth rate of *Keratella* in Char Lake is only about one tenth of that even in sub-Arctic lakes. Clearly the cold and lack of algal food slows everything down.

The benthos includes more animal species, 18 altogether: but this is still very restricted compared with that of lakes further south. Apart from four species of nematodes, seven species of chironomid larvae and a few oligochaete worms, there is only one representative of each of the other major groups such as ostracods and copepods.

There is only one species of fish in this very simplified community, the Arctic char (*Salvelinus alpinus*). It is the most widely distributed fish at these latitudes, occuring in all Arctic waters deeper than 3–4 m, and in the most unproductive waters it is the only fish present. In these cold waters growth is restricted to the short summer and fish may live for a very long time and reach large sizes. In Char Lake they continue growing until they are 16–18 years old and 30–50 cm long. By then they weigh up to 1 kg and may continue to live for many more years.

There are many lakes in the world that achieve as much primary production in a single day as Char Lake manages in a whole year. This very low level of production is not only due to cold and lack of light, but is also a consequence of the lowest inputs of nitrogen and phosphorus that have ever been recorded. It seems likely that the algae are not even able to use what light there is very efficiently because the supply of phosphorus is so low. Maretta Lake, another small lake close to Char Lake, receives the sewage from the nearby transport base and its reaction supports the idea that Char Lake is limited by lack of nutrients. The phytoplankton production in Maretta Lake is three times, and the concentration of chlorophyll *a* more than 20 times, that of Char Lake.

Antarctic lakes

The climate of Antarctica is even more severe than that of the Arctic at comparable latitudes; the winters are longer and the summers colder. The main plateau of the Antarctic Continent is at high altitude, in contrast to the Arctic regions, and also remote from other land masses. Surrounded by ice-covered sea, it is isolated from possible sources of colonisation for lakes uncovered by recently retreating ice and this results in a very impoverished fauna. There are relatively few lakes in Antarctica and the majority are found on the nearby islands, such as the South Orkneys, or on the northern end of the Antarctic Peninsula where there is a more maritime climate. In

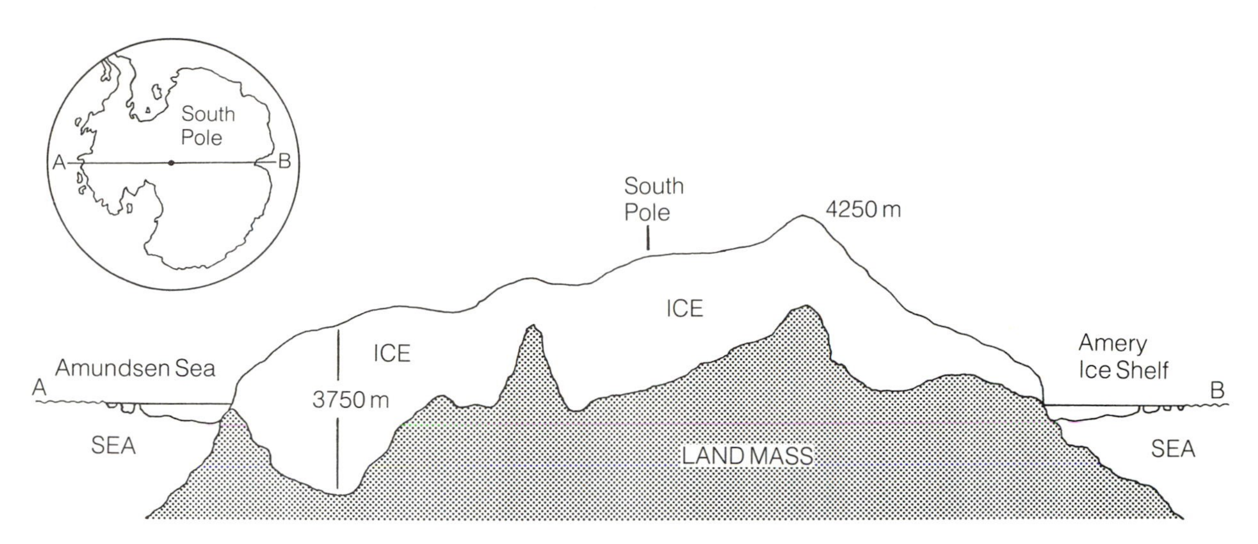

A cross-section through Antarctica to show the height of the continent and the thickness of the ice-cap. There are known to be lakes buried under the ice, lakes around the edge of the continent which are permanently frozen and lakes which are free of ice for a short period each year.

the continental interior there are large lakes permanently covered in 3000–4000 m of ice and probably fed by water melting from the base of the ice sheet, but nothing more is known about them. There are also very salty lakes where the glacier ice has withdrawn and streams now feed lakes with meltwater in summer, but very strong winds off the ice cause high evaporation rates and concentration of salts in these closed basins.

Apart from these saline lakes, most Antarctic lakes have very low concentrations of salts, except where colonies of birds or seals live nearby and their excrement (guano) drains into the water. Without such sources of enrichment, most lakes are very unproductive and have many characteristics shared with Arctic sites such as Char Lake.

There are no higher plants growing in the lakes and the principal primary producers are blue-green algae which form a thick felt-like mat on the bottom. Green algae and diatoms grow among the blue-green filaments. Mosses are also important in some Antarctic lakes and may be found growing down to depths of 20–30 m. Within the algal mats live microscopic benthic animals such as rotifers, tardigrades, protozoans and nematode worms. Some of them are able to withstand being frozen every night and revive again each morning. There are not many species but some of them can reach high numbers: the sedentary rotifers *Philodina gregaria* often form blood-red patches up to 15 cm in diameter as they stand densely packed side by side on the bottom of pools. This species can survive in the dormant state as an adult and withstand repeated freezing and thawing over many months even down to temperatures as low as −78°C.

As in Arctic lakes the phytoplankton tends to be dominated by small, motile species whose production is less than that of the benthic algae. Of particular interest is the fact that they can continue to fix carbon in darkness under the ice, but it is not at all clear how they do so. There is, however, no doubt that many of these algae can photosynthesise at much lower light levels than are usually regarded as the lowest possible elsewhere. Indeed photosynthesis is inhibited by the brief periods of bright light at the water surface during summer, and

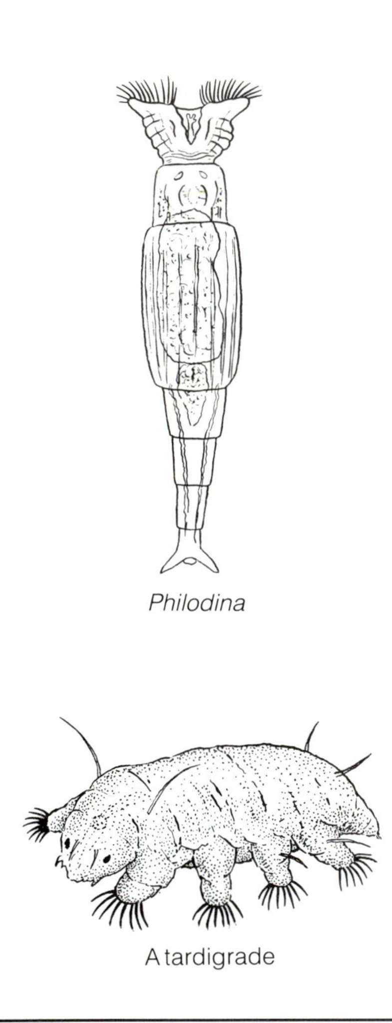
Philodina

A tardigrade

Microscopic members of the lake community

Most of the larger plants and animals that are found in temperate zone lakes cannot survive the rigours of life in very cold, unproductive waters that are ice-covered for long periods of the year. In these waters very small organisms, which do occur in other lakes but are often overlooked, assume greater significance.

Flagellate algae such as *Cryptomonas, Rhodomonas, Gymnodinium* and *Peridinium* frequently dominate the phytoplankton of polar and alpine lakes. They belong to different groups of algae (the first two to the Cryptophyta and the second two to the Pyrrhophyta or dinoflagellates) but have in common their small size, which helps them to take up nutrients from very dilute solutions, and their ability to migrate through the water to regions where light conditions are most favourable.

Philodina is a genus of **rotifers** which also have this extraordinary power to suspend life for long periods. They too are less than one millimetre long, are soft bodied and move across the substrate in a looping fashion, reminiscent of leeches. The hind end is a 'foot' bearing two 'toes', the whole of which can be retracted into the body when necessary, or extended so that the elongated body, attached by the toes to the substrate, reaches out into the water. The head end bears a double (figure-of-eight shaped) crown of cilia which, as in all rotifers, creates a current of water that brings minute particles towards the mouth at the centre. This too can be retracted into the body for protection. They can swim but spend most of their lives attached to, and creeping about on, surfaces such as plants.

Tardigrades are minute animals sometimes known as 'water bears'. They are less than a millimetre long, rather flattened and have four pairs of stumpy legs, each of which ends in one to eight claws. Many species live among moss in damp places; in polar lakes, where mosses frequently grow in the littoral zone, they may form very dense populations. Most species feed by sucking plant juices but some are predatory, living on other tardigrades, rotifers and other minute animals. Their principle adaptation to life in polar conditions is their remarkable ability to survive adverse conditions, such as freezing and drying, and to take up life again when it has apparently ceased.

the spring peak of primary production in the phytoplankton occurs under the ice before the open-water period.

The algal blanket on the bottom is better adapted to utilise strong light but is always partially shaded by the phytoplankton in the water above. These benthic algae also have the advantage that they can absorb warmth from the sun, and inside the algal mat the temperature may be one or two degrees higher than that of the water, which aids algal growth and metabolism. The algal mats also sometimes absorb heat from the sun while still covered in ice and this may be sufficient to melt the ice immediately over the mat, allowing the algae to commence photosynthesis. Sometimes the bubbles of oxygen produced lift the algal mass off the underlying rock.

Most of the larger animals in these lakes are Crustacea, both planktonic and benthic. One of the fairy shrimps, *Branchinecta gaini*, hatches from resting eggs at the end of winter, probably stimulated by the dilution of the water as the ice begins to melt. The larvae are able to take advantage of the peak period of phytoplankton growth and become sexually mature adults in about 3 months. The adults die as the lake begins to freeze over. In shallow pools where the water is presumably warmer, the life cycle is slightly shorter. Ironically these animals are close relatives of those adapted to living in temporary pools in hot desert climates, but in both cases the need for resistant eggs, which can withstand very adverse conditions, is the same.

There are two copepods in many Antarctic lakes. *Pseudoboeckella poppei* is related to species found in Australia and New Zealand; while its early larvae are found in the plankton, the later larvae and adults feed on algae and detritus from the surface of the blue-green algal mat. *Parabroteus sarsi* is much larger and preys on *Pseudoboeckella poppei* and on the larvae of *Branchinecta*. It also eats the Cladocera (three species) which feed on the surface of the algal mat alongside a couple of species of ostracods. The cladocerans and copepods all overwinter as adults in lakes and, in addition, the Cladocera produce resting eggs (ephippia). The copepods only seem to produce resting eggs in pools which freeze to the bottom.

Antarctic lakes have been incompletely studied but their fauna seems to be even more restricted than in the Arctic. Since conditions are rather similar, this must be mainly due to the isolation of the Antarctic from other continental land masses and to the fact that the lakes have been freed from the ice sheet more recently than many of the Arctic lakes, allowing less time for colonisation.

Alpine lakes

The extent to which lakes at high altitude resemble those at high latitude depends on where they are, but even at low latitudes lakes behave like those in the polar regions if they are high enough above sea level. At 47° latitude and above the tree line, lakes are only free of ice for about 20 weeks of the year.

There are many small lakes in the mountains of the Austrian Tyrol where the rocks are of limestone and the water contains quite a high concentration of calcium. Natural phosphate levels are usually low but where there are houses in the catchment area the lake may receive sewage and, consequently, be relatively productive. Typically such lakes are less than 5 ha in area, but are often as deep as 20 m, or more, because the steep mountain slopes continue beneath the water. The lake stratifies very strongly soon after ice break, so it divides into a warm (often more than 22°C at the surface in summer), relatively productive upper layer and bottom water from which the dissolved oxygen is rapidly used up by decomposition of the organic matter produced. Although the water column is mixed right down to the bottom in autumn, before the lake freezes over, a lot of its oxygen is gradually used up during the winter darkness under the ice. When it mixes again in the spring, there is sometimes a mass death of fish (a fish kill) as the deoxygenated bottom water is mixed throughout the lake. The likelihood of this is greatly increased if the lake receives sewage and a number of methods have been tried to improve the oxygen content of the lake during winter. Sewage has now been diverted away from Lanser See and in Kleiner Montiggler See oxygen is artificially pumped into the lake under the ice. For the

Piburger See, which is surrounded by high mountains whose shadow completely prevents sunlight from reaching part of the lake during December, snow is removed from the ice to increase the light reaching the phytoplankton whose photosynthesis adds oxygen to the water. As in the Arctic lakes already discussed, many of the algae which are most abundant in these lakes are able to photosynthesise at very low light intensities.

Unfortunately one of the alpine lakes which has been most intensively studied, Vorderer Finstertaler See, has now been drowned by a hydroelectric scheme. Nevertheless, the studies reveal details of ecological patterns which are typical of many alpine lakes. The lake was almost circular, only 15.7 ha in area and at an altitude of 2237 m. It contained very soft water with a conductivity of only 20–30 μS/cm and even when there was sufficient light, the level of primary production was limited by the low availability of phosphorus. This ensured that it was very unproductive, like many Arctic and alpine lakes. Also in common with Arctic lakes its phytoplankton consisted of very small forms, particularly desmids and certain dinoflagellates such as *Peridinium*. In the early winter gloom, phytoplankton activity was very low because the algae were still adjusted to brighter light, but as the winter wore on they became adapted to lower light levels and became more active. The population was still very low but during this period activity was controlled by the amount of light and not by the nutrient supply. When the snow melted more light became available, but the now dark-adapted algae migrated downwards in the water column to avoid the brighter light near the surface. In consequence there was still little growth. During the ice-free period both biomass and activity of the phytoplankton rose but was never very high because by then it was limited by shortage of nutrients.

It is hardly surprising that, as in the Arctic, this low level of primary production supported only a limited zooplankton fauna. One of the two rotifers present, *Polyarthra dolichoptera*, used bacteria as well as algae for food; the other, *Keratella heimalis*, ate only algae. These two species plus the copepod *Cyclops abyssorum tatricus* formed 90% of the zooplankton biomass. They were abundant under the ice but then declined to a minimum in May and June. Their numbers then built up again and during winter they contributed to the diet of the fish.

Only two species of fish occurred in the lake – char and trout; both reproduced under the ice from November to February and fed on the bottom of the lake, except in summer when the trout concentrated on surface organisms.

The lake basin consisted of a steep rocky littoral zone with almost no macrophytes. The rocks were covered with silt wherever there was shelter from waves and small crustaceans and midge larvae buried themselves in it and under the stones. At about 20 m depth there was an intermediate zone of highly organic mud but with relatively few living organisms, then below this, from about 20–29 m, was the relatively flat bottom of fine deep silt, full of nematode and oligochaete worms. As soon as ice cover was complete the benthic

The Arctic char

The Arctic char (*Salvelinus alpinus*) is the most widespread species of fish in cold northern and alpine lakes. It is closely related to the salmon and trout and like many species of that family there are both migratory and landlocked populations. Those that migrate enter the northern rivers in September–October, assume their full breeding colours and move upstream to spawn. The spent fish return to the sea in June. Lake populations are found in large, clear lakes all round the Northern Hemisphere as far south as the Alps. They are very variable in form but the species is distinguished by the brilliant white front edge of the fins which, in the spawning season, are blood red as is the belly. Lake forms tend to be smaller than the migratory fish which feed in the sea. In lakes they feed on small planktonic animals as well as insect larvae and molluscs. When large they may be cannibalistic on their own young. They tend to grow slowly and live a long time but landlocked populations living in isolated lakes, particularly those in Europe where their occurence is very localised, are very vulnerable to human influence such as the introduction of alien competing species, and pollution.

Some lake populations spawn in the spring and some in the autumn, while occasionally there are two populations with different spawning times in one lake. This is the case in Windermere in the English Lake District. One group spawns in shallow (1–4 m) water during November –December, while the others spawn in much deeper water (15–20 m) in the spring on a narrow tongue of gravel near the mouth of an inflowing stream which presumably keeps it free of mud. In both cases the eggs are laid in a redd in the gravel and are covered over after fertilisation. There are small structural differences between the fish of the two populations and there has been some speculation as to whether they constitute evidence that one species might be evolving into two; artificial cross fertilisation still produces fertile progeny, however, showing that they are one species at present.

Arctic char are locally important as food fish in the far north of countries such as Canada where they are netted from rivers during the migration. In most lakes they are primarily caught for sport. In Windermere the char spend most of their time in deep water where fishermen catch them using a method known as 'plumbline' angling which is peculiar to the Lake District. A short stout rod with a bell on top is projected over each side of the boat. Each carries about 25 m of line weighted at the bottom and to which six side lines are attached at 4 m intervals. The hook on each line has a small shiny metal spinner which whirls in the water as the boat is rowed slowly along. When a fish is caught the bell rings and the line is lifted in. The char used to be made into a paste and sold in shallow, round china pots with colourful pictures of the fish on the lid. These pots are now collectors' items.

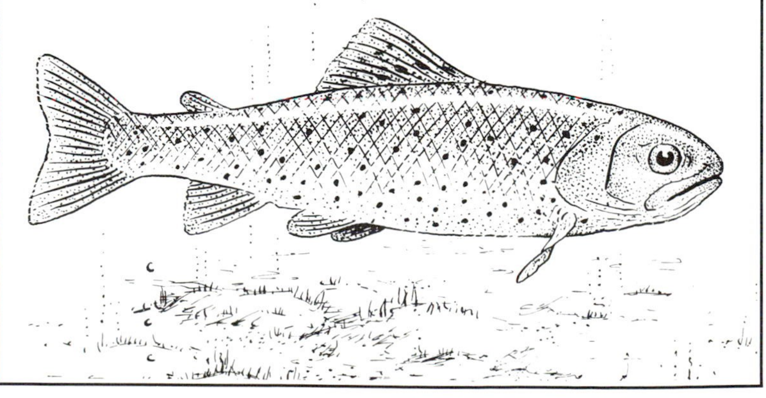

populations began to increase and reached a maximum in December –January. This apparent anomaly was due to the fact that the benthic organisms were dependent on food produced in the open water, and this only started to sink down within reach of things living on the bottom when the water was calm under the ice. Gradually this resource was used up and benthic populations dwindled to reach a minimum by the time of ice break. Those insects which survived the winter developed to their final pupal stage, and emerged from the water about 9 days after ice break to lay their eggs and start their cycle again.

In these sharply seasonal environments animals have either to complete their life history within one year or wait till the brief 'time window' the following year. There can be little leeway and mis-timing is likely to result in disasters such as young hatching when food is unobtainable or adults ready to emerge when ice still covers the lake.

The cooling effects of altitude are most clearly seen in the tropics and the lakes on the high mountains of East Africa illustrate this very clearly. Almost on the Equator, there are three isolated mountain massifs which rise to more than 5000 m above sea level and have permanently snow-capped summits. Kilimanjaro, at 5965 m is the highest mountain in Africa, followed closely by Mount Kenya (5195 m) and the Ruwenzori Mountains (5119 m). Many of the lakes above 3000 m on these mountains were formed by the glaciers which are now confined to their highest peaks, but a few were formed by volcanic action. The majority are less than 10 m deep but some, such as Lake Mahoma on the Ruwenzori and Enchanted Lake on Mount Kenya, are over 20 m deep. Despite the burning heat on the plains below, there may be frost and snow at these altitudes though only the very smallest and very highest lakes freeze over. Because water takes so much longer than air to warm up and cool down, plants and animals in these lakes are not subject to the violent fluctuations in temperature that those in their terrestrial surroundings have to endure. The water is cool, usually between 3 and 10°C at the surface, depending on the altitude. This compares with surface temperatures of 30–35°C in lakes at the same latitude but 2000–3000 m below. So, here on the Equator, one can find cool temperate conditions at high altitude.

Lake Titicaca

Lake Titicaca is the largest lake in South America and offers an opportunity to study the effects of altitude on a large tropical lake. It is situated at 3803 m above sea level on the Andean Altiplano between Bolivia and Peru. The mountains rise even higher all round and almost create a closed drainage basin. The lake is fed by five rivers and many minor streams. It drains out via the Rio Desaguardero to Lake Poopó which is shallow, salty and probably overflows intermittently into the Coipasa salt pans.

The total area of Lake Titicaca is 8100 km^2, three quarters of which (6315 km^2) is a deep, steep-sided basin with a maximum depth of

281 m and mean depth of 107 m. Puno Bay on the western shore and the Lago Pequeño to the south, are shallow areas containing a rather different ecosystem, including extensive areas of *Scirpus* marsh. Both these areas are separated from the main lake by constrictions in the shoreline, particularly marked in the case of Lago Pequeño. From the very southern end of this bay, at Guaqui, the railway runs to La Paz, the capital of Bolivia, and a boat service runs across the lake to Puna, which is the railhead for the line that climbs the western slope of the Andes from the coast of Peru. The natural vegetation of the area immediately surrounding the lake is a form of grassland known as puna but most of the lower ground is heavily cultivated. Sheep, cattle, llamas and alpaca graze any uncultivated land and range to higher elevations. Local people harvest fish from the lake and use the *Scirpus* for construction and fodder.

Less than half of the lake's water comes from the inflows, despite the fact that these drain an area of 58 000 km^2, or about seven times the total area of the lake. About 53% of the annual water input arrives as rain falling directly on the lake itself. The level of the lake fluctuates by between 0.5 and 1 m each year but also undergoes longer-term variations in level which range from +1.2 to −3.7 m above and below the arbitary zero at Puno. Such changes are typical of lakes in closed basins and although Lake Titicaca does have an outflow such a small proportion of its water loss (about 1.5%) is via this route that it hardly counts. About 90% of the water lost each year from the lake is due to evaporation and the rest seeps away. Since the rate of evaporation is more or less constant from year to year, it must be variations in the relatively low rainfall which cause the changes in lake level.

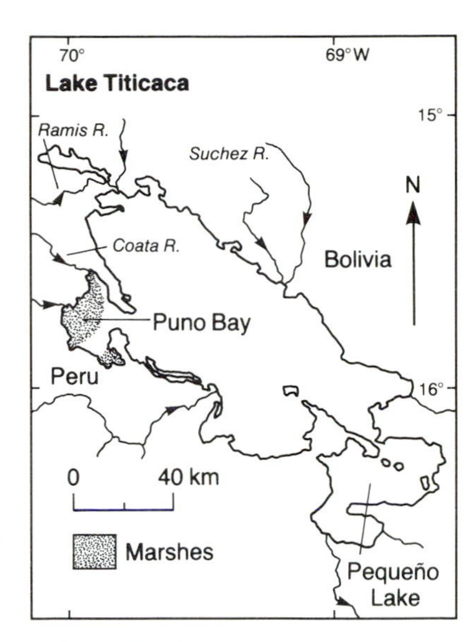

Altitude: 3812 m
Catchment area: 58 000 km^2
Principal inflows: Ramis, Coata, and Suchez Rivers
Lake area: 8100 km^2
Lake volume: 893 km^3
Maximum depth: 281 m
Mean depth: 107 m
Outflow: Desaguadera River

The Uru people utilised bulrushes (*Scirpus tatora*) from the north end of Lake Titicaca to make their boats, known as 'balsas de totoras'. These were used for fishing and for transport to and from the floating platforms of rushes on which the people lived.

At only 16° south of the Equator, Lake Titicaca lies well within the tropics and at a lower elevation could be expected to have water temperatures similar to those of Lake Kariba (p. 166) at a similar latitude in Africa. When completely mixed, Lake Kariba is usually about 22°C and when stratified the surface water temperatures reach 31°C. But Titicaca is 3500 m higher than Kariba, with a consequent marked reduction in its air and water temperatures. Monthly mean air temperatures only range up to about 11°C so it is hardly surprising that the lake water is only about 11°C when completely mixed; when it is stratified temperatures in the epilimnion range from 12 to 15.7°C. Although temperatures are reduced by the altitude they are never as low as in the temperate zones and, as elsewhere in the tropics, the seasonal variation is very slight but the daily changes are considerable. Air temperature can be 14°C colder at night than in the daytime, though in the water daily changes are only noticeable at the surface. In general the temperature of the water changes only slightly throughout the year. So in this high-altitude, low-latitude lake we have temperatures similar to those of temperate regions, but with a lack of variation more typical of the tropics.

There is nevertheless a marked seasonality in the way the lake behaves. For much of the year (October–July) the lake is stratified, with an epilimnion about 40 m deep in March, deepening to 70 m by June. The temperature difference across the thermocline is not great but, although there is frequently secondary stratification within the epilimnion, the two main layers do not mix until late July. Minor mixing and stratification within the epilimnion occurs quite frequently and irregularly depending on the weather. It seems likely that in most years the lake mixes completely to at least 100 m but often it does not mix to the bottom. This sort of regime is fairly typical of many deep lakes, particularly at low latitudes.

Although the high rate of evaporation (resulting from high altitude and bright sunshine) leads to a high concentration of salts in the water, this is not so great as to have a marked influence on the lake community. The plants and animals which live in the lake are not as peculiar as its remarkable location might suggest. The open water contains a range of algae similar to that found in other tropical lakes and most of the species are widespread ones. In summer the blue-green algae *Anabaena sphaerica* and *Ulothrix subtilissima* dominate the phytoplankton but in winter the diversity increases and the diatoms, particularly *Stephanodiscus astraea*, and green algae become important. Levels of both nitrogen and phosphorus are low in the lake water and experiments have shown that algal growth is limited by the lack of nitrogen. The ability of species such as *Anabaena* to fix atmospheric nitrogen, which might help them to overcome the nitrogen shortage, is limited by the low levels of phosphorus. The addition of both nitrogen and phosphorus to samples from the lake greatly increases algal growth and photosynthesis.

The zooplankton seems to be the same now as that reported by the first biologists to do a detailed study of the lake in 1937. Two copepods are numerically dominant: a calanoid, *Boeckella titicacae*, and a cyclopoid, *Microcyclops leptopus*. In addition there are one other calanoid, three cladocerans and a few rotifer species. This limited variety is similar to that found in other tropical lakes and probably has more to do with the absence of seasonal succession than anything else. Conditions are favourable to these species throughout the year, so once they are established there is no reason for them to be seasonally replaced by others as happens in temperate lakes. The full species list is rather short compared with that for similar lakes at higher latitudes and this is contrary to the general rule that species diversity increases towards the Equator. One interesting point is that the two calanoids are of a genus typical of the Southern Hemisphere and found in greater variety in the south of South America. This suggests that Lake Titicaca was colonised from the south rather than the north.

The natural fish community of the lake is also very restricted. It consists of one catfish, *Trichomycterus rivulatus*, and a group of about nineteen closely related species of the genus *Orestias*, of which fourteen are found nowhere else. This paucity of species does provide a contrast with the variety found in other tropical lakes but may perhaps be accounted for by the isolation of the lake basin. The genus *Orestias* is thought to have been present before the mountains were formed and to have been lifted up with them. It has now evolved in the absence of other specialist fish to fill a wide variety of niches, from open-water carnivores to shallow-water mollusc or crustacean eaters.

Rainbow trout (*Salmo gairdneri*) and lake trout (*Salvelinus namaycush*) were introduced to the lake in 1940–41; the former has thrived and for many years given good yields to the local fishermen. There are signs that reductions in the catch were due to overfishing, when fishing effort was reduced in the 1960s catches improved. The benefits of these introduced fish have been offset by the fact that they brought with them a parasite that has contributed to the disappearance of some of the unique native species.

6
The deepest LAKES

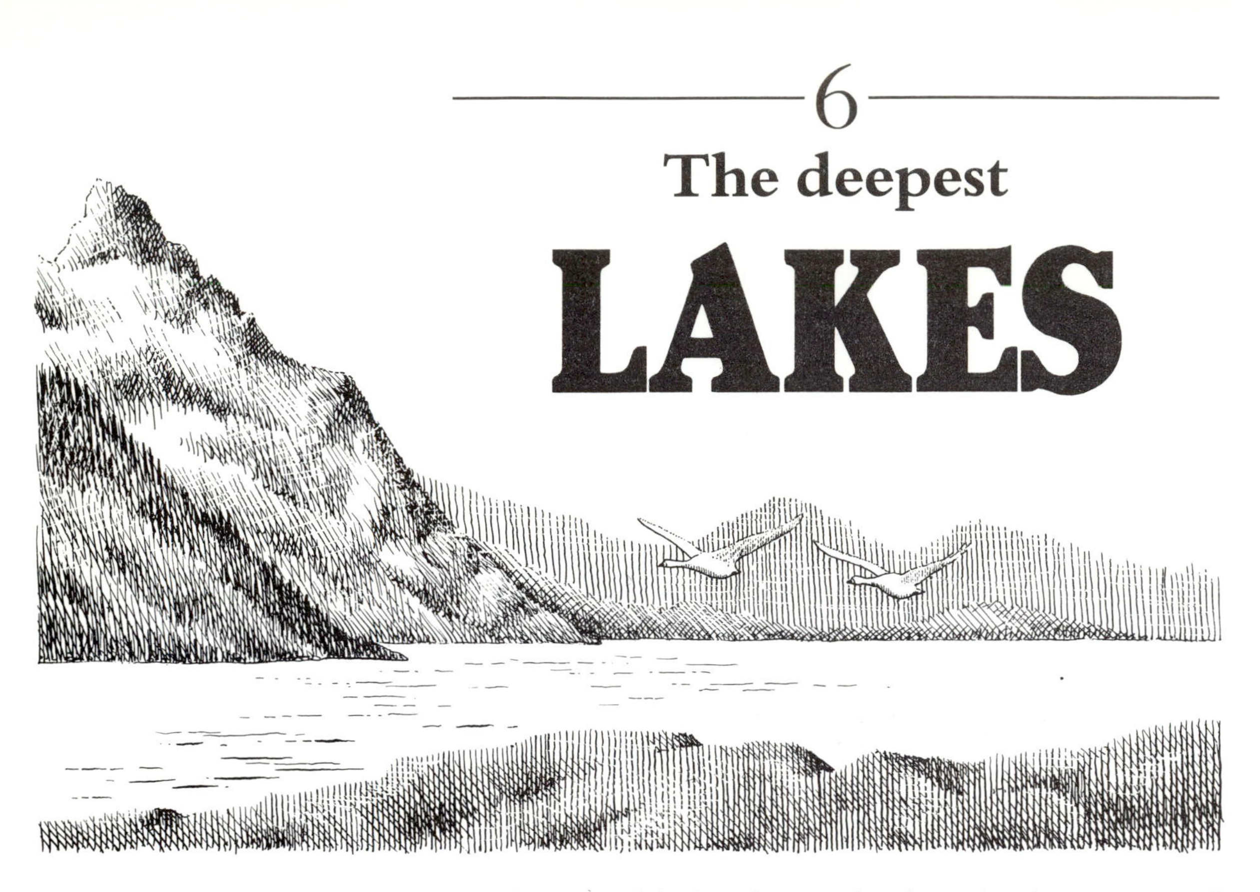

Depth is one of the key features that determine the structure and functioning of lake ecosystems. We have already seen how light, temperature, oxygen and other important factors vary with depth depending on the time of year. The actual depth of a lake depends very much on how its basin was formed and most of the deepest lakes owe their origin to movements of the earth's crust, such as those which formed the Rift Valleys of Africa and Crater Lake in Oregon (see Chapter 1). Very few lakes exceed 500 m in depth and only two are known to exceed 1000 m and have a mean depth greater than 500 m – Lake Baikal, in Siberia, and Lake Tanganyika, in East Africa. Both have the additional distinction of being very ancient, which has allowed time for evolution of species to occur within the lake basins, and the communities of both these lakes contain many endemic species.

The huge basins of very deep lakes contain immense volumes of water, most of which remains within the lakes for very long periods of time. This makes them particularly vulnerable to pollutants because there is little chance of their being flushed out of the lake once they arrive. Some, such as the Caspian Sea and Crater Lake, have no outflow and all incoming substances are destined to remain within the basin, except the water which evaporates from the surface.

The combination of large numbers of endemic species and the vulnerability to pollution make it extremely important that these lakes are treated with the utmost care by those who use their waters and drainage basins. Unlike the oceans that they seem to resemble, their size makes them no less vulnerable, because each lake is enclosed within the finite boundaries of its basin.

Lake Baikal: the deepest lake in the world

Not only is Lake Baikal the deepest lake on earth, but it is also one of the largest in area, ranking seventh in the world. Its enormous volume of water (23 000 km^3: one fifth of the world's total reserves of fresh water) has a surface area of 31 300 km^2 and is contained within a steep-sided trough 650 km long. There is evidence, from terraces high on its steep sides, that it has, at some time in the past, been even deeper than its present maximum of 1620 m. The precipitous mountains along most of the shoreline continue just as steeply down below the surface, with the result that over 31% of the area of the lake is more than 250 m deep and there is little shallow water, even at the edges. More than 300 rivers flow into Lake Baikal, but there is only one river flowing out, the Angara, which drains from the south-west corner. Thus there is considerable risk of pollutants being brought into the lake and accumulating there with little chance of being removed. To prevent trouble of this kind, a major local source of pollution (a large pulp mill) has been recently modified at great cost and new treatment works are now said to be controlling most of the major sources of pollution.

On the west side, where the mountains are steepest, inflowing rivers tend to be short and drain little more than the immediate escarpment. At the northern end and along the eastern shore the rivers drain an enormous catchment. The River Selenga, for example, is 1480 km long and comes from northern Mongolia to supply 50% of the total inflow to the lake. The volume of the inflows is greatly increased when ice and snow melt on the mountains at the end of the very severe winters. At this time the rivers carry large amounts of silt and debris into the lake, which makes the water quite cloudy.

The lake freezes over as winter proceeds, first in the north and, by the first half of January, in the south also. Ice cover lasts 4–5 months, depending on the severity of the winter, and by February or March can be 80–120 cm thick. The winter climate is intensely cold and dry, with a good deal of sunshine which penetrates through the ice to warm the water immediately below it. The bottom water of the lake

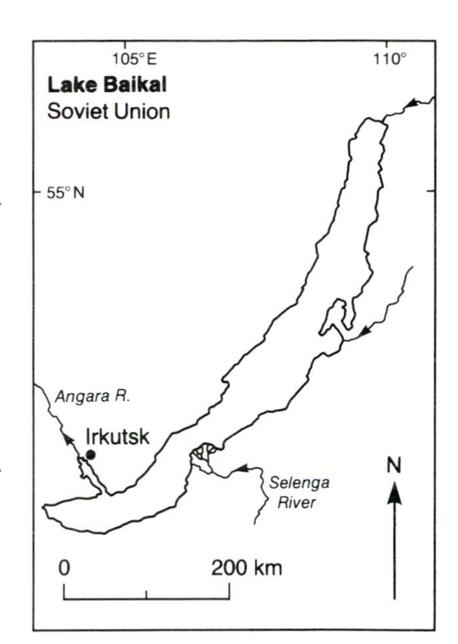

Altitude: 455 m
Catchment area: 540 000 km^2
Principal inflow: Selenga River
Lake area: 31 500 km^2
Lake volume: 23 000 km^3
Maximum depth: 1620 m
Mean depth: 740 m
Outflow: Angara River

Table 6.1 *Some of the deepest lakes in the world, with their approximate area and volume.*

	Max. depth (m)	Mean depth (m)	Area (km^2)	Volume (km^3)
Lake Baikal, Siberia	1620	740	31 500	23 000
Lake Tanganyika	1470	572	33 000	19 000
Caspian Sea	946	182	371 000	67 500
Lake Malawi	704	273	30 000	8 400
Great Slave, Canada	614	62	27 000	1 600
Crater Lake, Oregon	608	364	55	20
Lake Tahoe, California	501	313	499	156

remains at 3–3.6°C throughout the year but after the ice begins to break up in May the surface water gradually warms up until, by August, it can reach 19–20°C. So enormous is this water mass that it warms and cools only slowly, influencing the climate of the land around. It is always hotter in summer and colder in winter away from the lake. The water is well mixed so that oxygen is always plentiful, even at the greatest depths. This enables animals to live in all regions of the lake.

It is hardly surprising that such an outstanding body of water, which is also very ancient, should contain a very extraordinary collection of animals. Two main points are of particular interest: several of its freshwater groups closely resemble marine forms and many groups have evolved numerous species which occur nowhere else in the world. One species which combines both these features is the Baikal seal (*Phoca sibirica*). This is a completely freshwater species, closely related to the Caspian seal. It probably evolved from an Arctic species and it carries parasites similar to those of the Arctic seals. Baikal seals are silvery grey on the back and yellowish-white on the belly. They are small, reaching a length of 1.65 m and weighing 50–130 kg. The population numbers about 70 000 and the seals are most abundant in the northern and central areas of the lake, particularly on the four, steep-sided islands called Ushkany. The pups are born on the ice in February and March near to ice holes. These breathing holes are kept open throughout the winter by the seals gnawing at them from below.

How then did the seal and other marine-related species come to be in Lake Baikal? There is no evidence of the sea ever having been in the Baikal area of Siberia so it is supposed that they arrived by migration during the glacial period when the Yenisey–Angara River systems connected Lake Baikal to the Arctic Ocean. On the other hand, the molluscs which occur in Lake Baikal have forms resembling those of the oceans, rather than those of most freshwater lakes, and are thought to have evolved similar shapes independently because conditions in such a huge lake so closely resemble those found in the sea.

The Baikal seal, *Pusa (=Phoca) sibirica*, is found only in Lake Baikal but is closely related to another unique seal which lives isolated in the Caspian Sea.

Thirty-five percent of the plants and 65% of the animal species found in Lake Baikal are endemic. Altogether more than 1500 animals have been described from Lake Baikal and of those which live in the open waters more than 80% are endemic, comprising more than eighty-seven endemic genera and at least eleven endemic families or sub-families. For differences to be so great as to justify putting the species in separate families implies that the fauna of the lake has been isolated for a very long time, during which the animals have evolved to become very different from those found in other waters. The two groups of animals with the lowest proportion of endemic species are the protozoans and the rotifers, which have resting stages in their life cycles and are easily dispersed from one water body to another. This is also true of the diatoms which are represented by nearly equal numbers of endemic (310) and non-endemic (369) species. It is also notable that the fauna contains no amphibians and that the endemics include no insects other than the seven species of caddisfly present. The proportion of endemics is higher in the open lake than inshore.

The freshwater shrimps (gammarids) are the most spectacular example of a group evolving in an isolated place to form species found nowhere else. There are 255 species in Lake Baikal, grouped into 35 genera, thirty-four of which are endemic. One third of all the species of gammarid in the world come from Lake Baikal. To achieve such a variety of species must have taken a very long time but there must also be quite a variety of different places for them to live and things for them to feed on. Even then, as with the cichlid fish of Lake Victoria (see p. 17), there seems to be a good deal of overlap between species adapted to similar conditions. The vast majority of the Baikal gammarids are benthic; only one species has adapted to a fully planktonic existence.

Despite the principle that diversity of habitat and conditions encourages the evolution of species, the greatest assortment of endemic species in Baikal has evolved in the deep water of the lake, where almost constant conditions must have prevailed for many millions of years. The inshore areas, by contrast, contain hundreds of species which are also widespread in the waters of Siberia and the Eurasian land mass.

In the great depths of the lake the gammarids are very pale pinkish-white and have no pigment in their eyes. They have very long antennae for feeling in the darkness. Fish adapted to the abyss also often have pale colours and reduced eyes. It must be remembered that in such deep lakes everything is ultimately dependent on photosynthesis occurring in the upper layers (the top 25–50 m) of the lake where algae can grow. Below this, animals must either be detritus feeders, depending on organic matter sinking from above, or predators living on other animals.

The lake contains 50 species of fish, which range from the familiar pike (*Esox lucius*) and perch (*Perca fluviatilis*) to the family Comephoridae which contains one genus (*Comephorus*) and only two species (*C. baicalensis* and *C. dybowskii*), both endemic. These are about 18–20 cm long and completely transparent. They live in the

The fish family Comephoridae is endemic to Lake Baikal. There are only two species; *Comephorus dybowskii* is shown here. It grows to about 15 cm in length and lives in the deep open water down to a depth of more than 500 m. It feeds on gammarids and young fish, including its own.

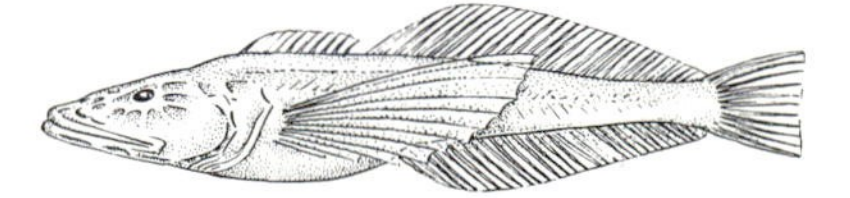

Table 6.3 *The vertical distribution of gammarids in the benthos of Lake Baikal (based on data in Brooks, 1950).*

Benthic depth zone	No. of genera	No. of species	Light & temperature	Substrate diversity	Water movement
Littoral 0–5 m	6	8+	fluctuating	great	waves
Sub-littoral 5–50 m	15	29+	damped	great	calm
Transitional 50–300 m	13	29	constant	little	calm
Abyssal >300 m	14	33	constant	none	still

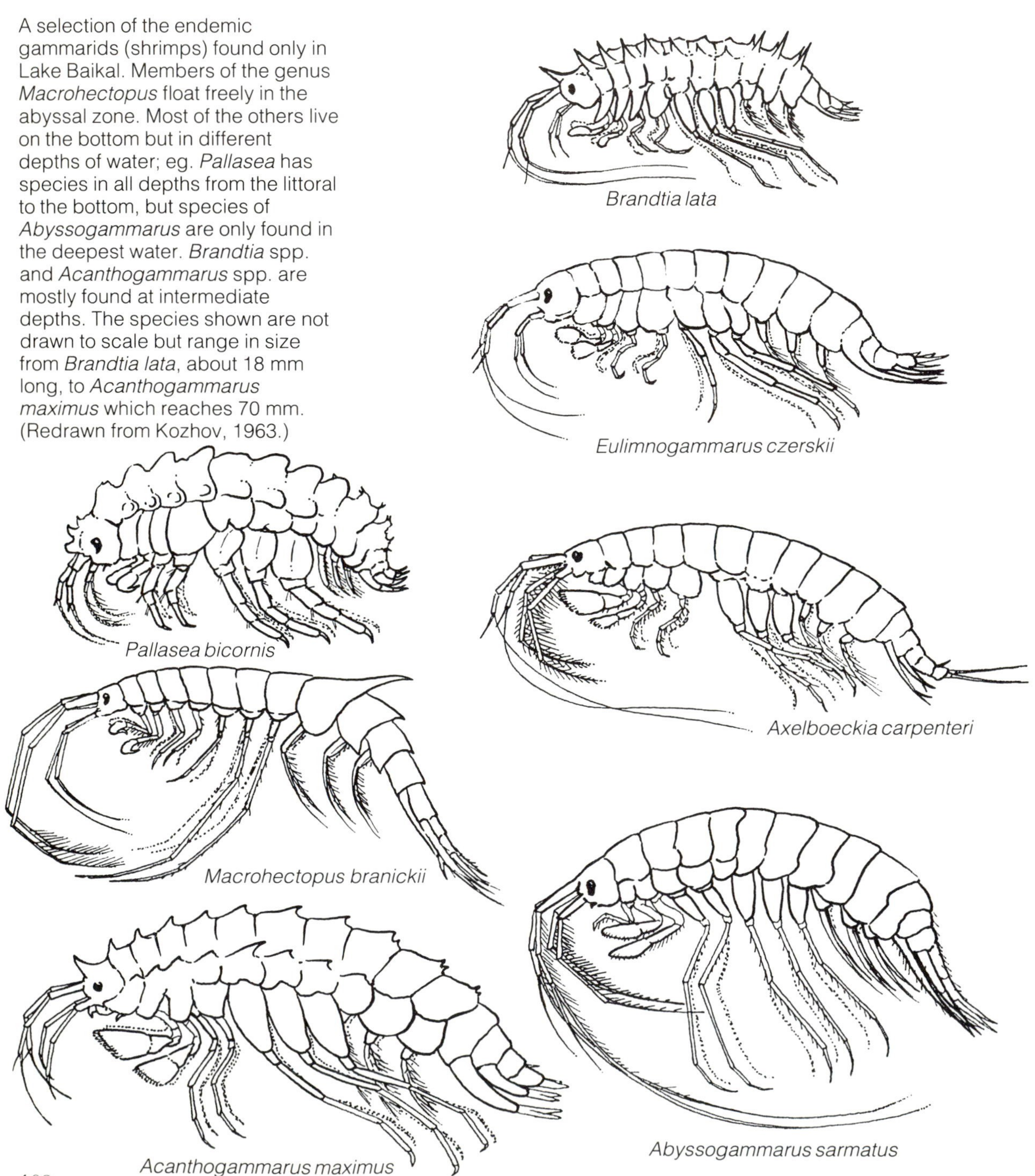

A selection of the endemic gammarids (shrimps) found only in Lake Baikal. Members of the genus *Macrohectopus* float freely in the abyssal zone. Most of the others live on the bottom but in different depths of water; eg. *Pallasea* has species in all depths from the littoral to the bottom, but species of *Abyssogammarus* are only found in the deepest water. *Brandtia* spp. and *Acanthogammarus* spp. are mostly found at intermediate depths. The species shown are not drawn to scale but range in size from *Brandtia lata*, about 18 mm long, to *Acanthogammarus maximus* which reaches 70 mm. (Redrawn from Kozhov, 1963.)

open water down to more than 500 m deep and give birth to 2000–3000 live young per female. The species of *Cottocomephorus*, which resemble bullheads, are also endemic and while they too live in the pelagic zone they tend to be more common in the littoral regions than *Comephorus*. The adults reach 11–18 cm and spawn on rocky bottoms near the shore. They are eaten by the Baikal seal and the juveniles are an important component of the diet of another fish, the omul (*Coregonus autumnalis migratorius*). This is another species, like the seal, whose ancestors must have migrated to the lake from the Arctic Ocean via the river systems of the glacial period. They occur in the shallow areas during spring and summer, where they feed avidly on young fish and zooplankton, and spend the winter in deeper water. They reach 30 cm in length and 300–450 g in weight and spawn after 5–7 years, in the nearby rivers. They form up to 70% of the commercial fish catch from Lake Baikal, so are of considerable economic importance.

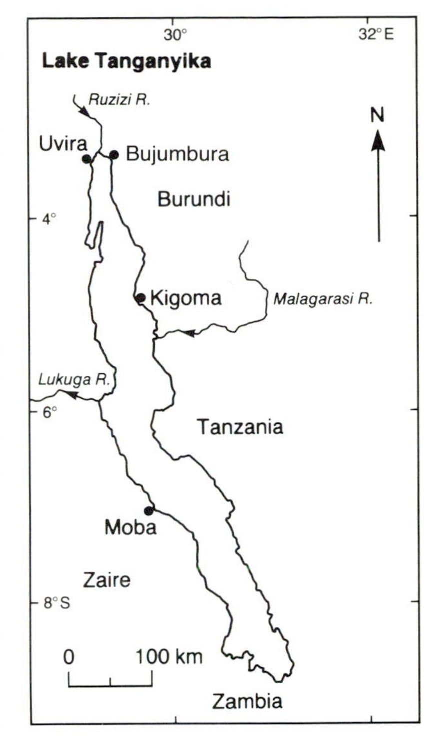

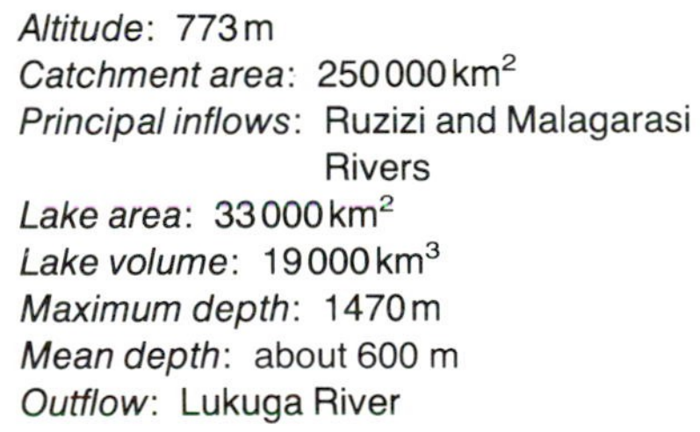
Altitude: 773 m
Catchment area: 250 000 km^2
Principal inflows: Ruzizi and Malagarasi Rivers
Lake area: 33 000 km^2
Lake volume: 19 000 km^3
Maximum depth: 1470 m
Mean depth: about 600 m
Outflow: Lukuga River

Lake Tanganyika

Lake Tanganyika, in tropical Africa, contains much warmer water than Lake Baikal and has almost certainly never been frozen in all its long history. This has resulted in an interesting contrast between the two lakes. In Baikal the degree of endemism is most remarkable in the benthic fauna of the abyssal region where, presumably, animals were able to survive even during the Ice Age when glaciers reached to the north end of the lake, the water must have been frozen to great depth and the littoral fauna was presumably wiped out. In Lake Tanganyika the abyssal zone is uninhabitable because it is totally and permanently devoid of oxygen; so the majority of endemic species are found in the littoral and pelagic zones of the lake. Nine tenths of the depth of water in Lake Tanganyika contains no oxygen because the lake is never completely mixed. The boundary between the upper oxygenated water and the deoxygenated zone varies during the year between about 100 and 200 m in depth. The upper layer stratifies periodically and it depends on the strength of the wind how deeply the mixed layer extends and brings oxygen to greater depths; the lowest limit to which this mixing and oxygenation can penetrate is about 200 m.

Lake Tanganyika is not so old (1.5–6 million years) as Baikal (50–75 million years) but it, too, is a rift lake. The lake is over 600 km long and has two major basins and a smaller middle one. The mountains on either side rise steeply from the water and fall as steeply below, so that the littoral region is very narrow along most of the lake, and only at the ends are there extensive areas of relatively shallow water. The Ruzizi River enters at the north end, carrying rather saline water from Lake Kivu; the other inflows, except the Malagarasi River, are rather short and precipitous, as they drain the adjacent mountain ranges. The lake water contains quite a high concentration of salts (conductivity 600 μS/cm), particularly in the deoxygenated zone, and it seems likely that for a very long period of time the lake had no outflow. The present outflow which drains to the Zaire River, has not always been open, even since the first European explorers arrived. It was blocked in 1874, open when Stanley was there in 1876, and almost closed again in the 1890s. Some fish found in the Malagarasi River are closely related to species found in the Zaire River system and quite different from the majority of fish found in the rest of eastern Africa, which are characteristic of the fauna of the Nile basin. This is evidence that the Malagarasi almost certainly drained westwards to the Zaire before the formation of the Rift Valley and its course has been interrupted by the filling of Lake Tanganyika whose waters now isolate these relatives of the Zaire fauna.

Lake Tanganyika contains many endemic species in an astonishing number of groups. The molluscs are of particular interest and they contrast with Lake Malawi which contains relatively few species of gastropod mollusc – twenty-seven as against sixty in Lake Tanganyika. Thirty of the gastropod species in Lake Tanganyika are endemic. Other endemics include two species of aquatic snakes, one a small colubrid which swims in the pelagic zone and feeds on the 'sardines', the other a cobra which lives on the shore but catches fish at night. The aquatic insects include a caddisfly whose adult is flightless and skims over the surface of the open water like a whirligig beetle. Even more interesting is the fact that although flightless caddisflies are almost unknown elsewhere, there are a similar species on Lake Baikal and on Lake Titicaca in the Andes.

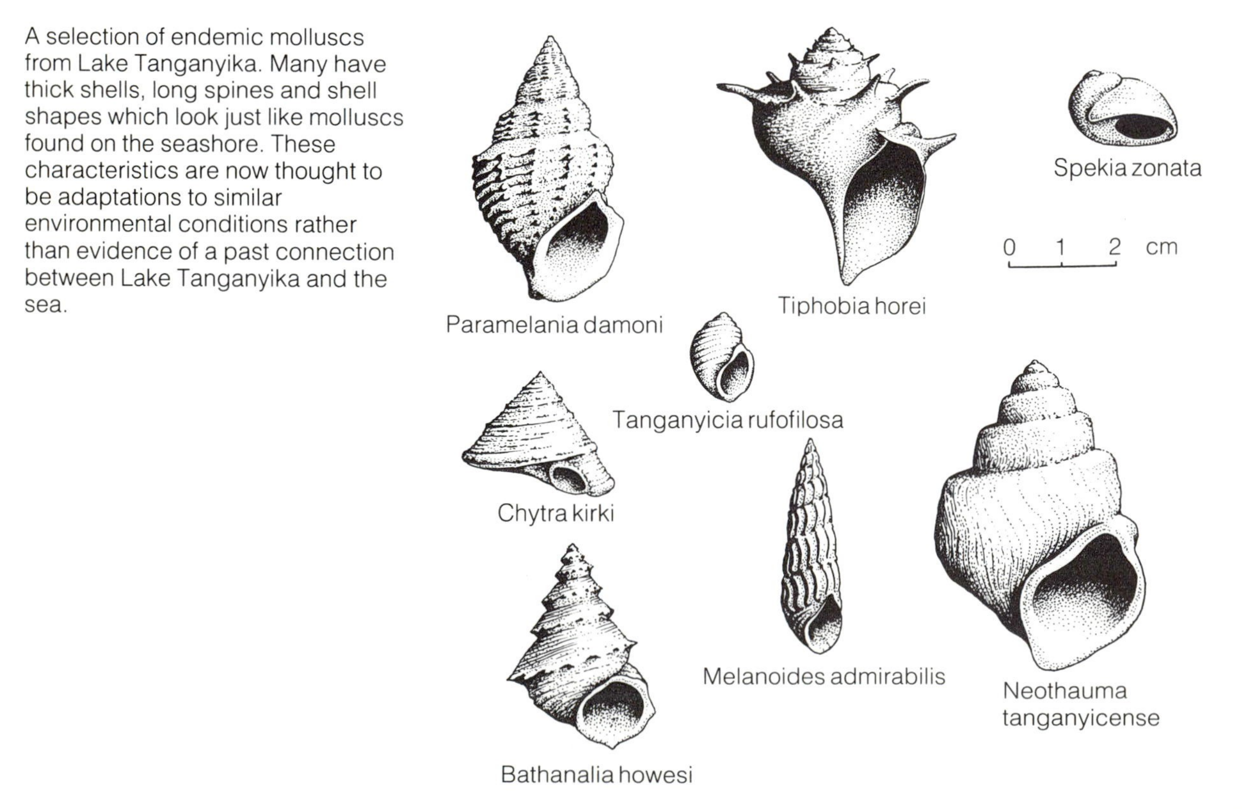

A selection of endemic molluscs from Lake Tanganyika. Many have thick shells, long spines and shell shapes which look just like molluscs found on the seashore. These characteristics are now thought to be adaptations to similar environmental conditions rather than evidence of a past connection between Lake Tanganyika and the sea.

When zoologists first began to collect animals from Lake Tanganyika they found several forms which suggested that either the lake was a remnant sea or that it must once have had a connection with the ocean. The mollusc shells collected in the inshore areas of the lake and along the beaches had shapes more reminiscent of marine molluscs than those known from lakes in Europe. When someone found a jelly-fish in the open water, the case seemed proved. The matter was hotly debated for some time but there is no geological evidence for any history of connection with the sea. The jelly-fish *Limnocnida tanganikae* has since been found in several other large African lakes and alternative explanations for the mollusc shells seem more likely, since similar species have now been found in other African lakes. Some of them have thick, globular shells and a very strong foot, which are probably adaptations to living on rocky, wave-washed shores where conditions are similar to those found on rocky sea coasts. Others have long spines on their shells and long siphons, both of which are probably adaptations to living in very soft mud whether it is at the bottom of freshwater Lake Tanganyika or on the sea bed. The Tanganyika molluscs thus appear to have independently evolved the same useful adaptations as their marine counterparts; an example of convergent evolution, not evidence of marine origin.

Those who supported the idea of a marine origin for the fauna of Lake Tanganyika would have been convinced they were right had they known at the time that the open-water fish fauna was dominated by small clupeids (herring-like fish) now known as 'Tanganyika sardines'. These too are endemic and support an important local fishery. When Dr Livingstone arrived on the shores of Lake Tanganyika he saw lights dotted all over the surface of the water as soon as it got dark: these were the fires carried on platforms in front of local canoes whose occupants were fishing for the sardines. These little fish come to the surface in large shoals at night and are attracted to the lights. They are caught in nets suspended in the water below the lights, which are raised as soon as a large number have congregated. The same method is still used today, although the fires that Dr Livingstone saw are replaced by kerosene pressure lamps.

The inshore waters yield a wide variety of species which are sold in local markets but, because the majority of the lake area contains very deep water, it is the productivity of the pelagic zone which has been of greatest interest to fisheries scientists. The principal fish caught is the 'Tanganyika sardine' which is not one species but several: the clupeids *Limnothrissa miodon* and *Stolothrissa tanganikae* and the juveniles of four species of *Lates* whose adults are predators of the sardines. *Stolothrissa* is the most abundant. They grow to a maximum length of only 95 mm and live for about one year having reached reproductive condition at 7–8 months. The predatory species grow more slowly and live longer (10 years or more). As the fishery developed and became more intensive the catch per unit effort of the predatory species declined, implying that their abundance in the population was also declining. At the same time the clupeid catch per unit effort increased. These fast-growing species

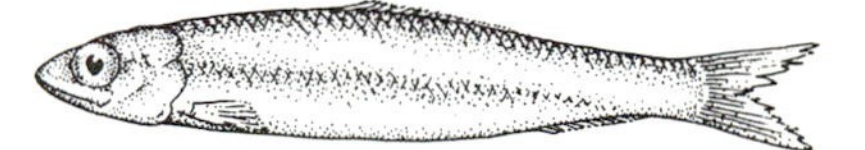

'Tanganyika sardines' are only about 10 cm long when fully mature. They are endemic to Lake Tanganyika and live entirely in the pelagic zone. *Stolothrissa tanganicae* is the most abundant species.

with their short lifespans are more easily able to replace that part of their population removed by fishermen than are the slower-growing, longer-lived species.

Looking down into the clear water of Lake Tanganyika from a boat, one can see the sunlight penetrating to great depths. The little jelly-fish are clearly visible, opening and closing their umbrellas as they surge along. The water is clear because there are not many algae to intercept the light and any sediment entering with the rivers soon sinks to a depth from which it will not be stirred up (and in any case most of the open water is a long way from the shore). This clarity led early zoologists to assume that the lake was unproductive but when more detailed studies began on the fisheries (instead of just identifying the different species), it became evident that the harvest of fish from the lake was quite high and had been so for a long time. In order to judge how much fishing the clupeid populations could withstand, it is important to know more about the rest of the food web in the pelagic zone and to discover how it is that this apparently unproductive lake supports a productive fishery.

The algae in the open water of Lake Tanganyika provide an excellent illustration of the difference between standing crop and production. The standing crop of algae (the weight per unit area, measured in dry weight or amount of carbon, present at one particular time) is very low, but estimates of primary production (the amount of *new* living material, measured as carbon fixed by photosynthesis per unit of time) made in 1975 were surprisingly high. This means that every algal cell must be fixing carbon, growing and dividing very rapidly. But most of these new algae are either quickly eaten, or they soon die and sink; they seldom increase the standing crop. Contrast this with trees where much of the new material formed is stored as wood so the standing crop in a forest gradually gets larger, or with a lake in which the water becomes green with accumulating algae because they are not disappearing as fast as they are produced. A low standing crop does not necessarily imply low production and vice versa. Zooplankton in the open water are also sparse. The dominant species is a calanoid copepod called *Diaptomus simplex* which almost certainly eats algae. It is preyed upon by the sardines whose guts are full of these copepods, leaving very few in the water. Again, as with the algae, there is a low standing crop which is probably highly productive.

The yield of fish is very high in proportion to the primary production when compared with other well-studied freshwater ecosystems and two possible explanations have been put forward. Either the transfer of food energy along the food chain (algae → zooplankton → fish) is unusually efficient, or there may be an alternative food chain based not on photosynthesis but on energy derived from the gases dissolved in the deoxygenated water which occupies nine-tenths of the lake volume. The food chain could easily be superefficient; the water temperature is high throughout the year and similar levels of efficiency have been recorded in the sea in similarly warm, clear ocean waters. The second suggestion is based on the discovery, at the boundary between the oxygenated and the deoxygenated water, of bacteria which obtain carbon from methane and also of unusually abundant ciliated protozoans, which presumably eat them. The zooplankton and the 'sardines' could feed on the protozoans when they migrate down to deep water during the day, and thus get extra food apart from that generated by algal production.

The pelagic ecosystem of Lake Tanganyika provides a good example of how basic data can be combined with ecological theory to produce new hypotheses and intriguing questions for future research. Probably both explanations of the high fish yield from Lake Tanganyika are at least partly true. The bacteria must be living off the past productivity of the lake, because methane is a product of decomposition of carbon compounds. Although they provide evidence of one route by which decomposition products can be returned from the deoxygenated zone to the water above, this still does not explain how the algae obtain sufficient nutrients to support their photosynthetic production.

Apart from the clarity of the water, one of the main reasons why the lake was assumed to be unproductive was because it seemed that, if it never mixed, dead plants and animals must sink into the deoxygenated zone and the nutrients they contained would be lost for ever from the lighted upper waters. The volume of the inflows is such a small proportion of the lake volume that, even were they rich in nutrients, there must be a deficit over the lake as a whole. But observations of physical conditions in the lake suggest further mechanisms by which some of the nutrients trapped in the deoxygenated zone are returned to the surface waters of the lake. The epilimnion is well mixed throughout the year but the strongest winds blow from the south from April/May to September and during that period, surface water is pushed northwards along the lake. This reduces the depth of the epilimnion at the south end and increases it at the north. It seems likely that nutrient-rich water from the deoxygenated zone wells up into the epilimnion at the south end and is gradually pushed northwards. Its presence can be followed as the concentration of algae increases, first in the southern basin and later further north. So the deoxygenated zone is not entirely isolated from the overlying water. Another mechanism is suggested by very precise measurements of temperature down the water column. These show that, contrary to expectation, the coldest water is not at

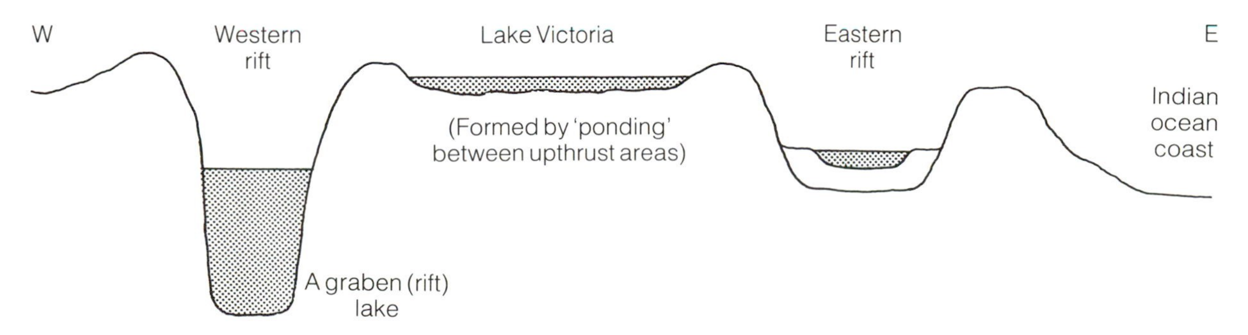

A schematic cross-section of East Africa. To the west, deep rift lakes (e.g. L. Tanganyika) have sheer escarpments, continuing below water level to give deep, steep-sided lake basins. To the east, the rift valley floor is flat, with shallower lakes lying in it. Many of these are soda lakes (see Chapter 8). On the plateau between the two rifts lies the huge, relatively shallow expanse of Lake Victoria (see Chapters 1 and 10), formed by 'ponding' between the upthrust rims of the two Rift Valleys.

the very bottom of the lake but at a depth of about 800 m. Probably the bottom water is warmed by heat from compression and from being nearer the centre of the earth. The differences are very small (less than half a Celsius degree) but sufficient to suggest that there must be convection currents stirring the depths of the lake which carry this warmer water upwards to be replaced by a sinking of the coldest water. Above the coldest zone the water is probably circulated through friction with the mixed epilimnion water above. Where these three zones of internal circulation interact, some of the nutrients might be transferred from the deoxygenated zone to the photic zone, where the algae can use them in primary production.

Lake Malawi

Although Lake Malawi (formerly Lake Nyasa) is the fourth deepest lake in the world it is less than half the depth of Lake Tanganyika. Nevertheless there are many similarities between the two. Lake Malawi lies in another rift to the south-east of Lake Tanganyika and is similarly elongate and steep-sided. It too has a large volume of deoxygenated water below the upper layers where all the activity is concentrated. Although younger than Lake Tanganyika, it has clearly been isolated from other water bodies for long enough to develop a very diverse fish fauna in which the majority of the species are endemic. In fact more species of fish have been described from Lake Malawi than from any other lake. The astonishing thing is that, despite the proximity of Lake Tanganyika, the fish are all completely different; and there are no clupeids and no *Lates* in Lake Malawi.

It was David Livingstone who made the first detailed description of Lake Malawi and one of his companions, John Kirk, sent a sample of fish to London. It is clear from Livingstone's account that about 1860 the outlet from the lake, the Shire River, was flowing strongly and the lake level was high. It gradually declined until, in 1915, the Shire ceased to flow. The exit became blocked by sand bars and vegetation so that the rising water had to reach 6 m above its lowest level before it could break through and make the river flow again. This happened in 1935 and since then the lake has not fallen below the level of that outflow. The lake level is very sensitive to local rainfall because most of the inflows in the relatively small catchment area are short streams draining the surrounding mountains. Recent changes in lake level can be seen as 'tide marks' on the rocks along the shore.

Lakes Tanganyika and Malawi provide a similar variety of habitats for fish and other animals. Along the shores there are rocks, sandy beaches, marshy bays, river inlets, and shallow open water; out in the open lake there is the surface, pelagic zone and the bottom mud. The deep abyssal zone is devoid of oxygen. Nevertheless, in Lake Malawi some species of fish have been caught below 200 m and must spend at least some of their time in water containing very little oxygen.

Altogether about 245 species of fish have been described from Lake Malawi. Of these, about 200 species are cichlids of which over 95% are endemic. Fewer of the non-cichlids (about 65%) are endemic. In Lake Tanganyika there is a similar degree of endemism among the non-cichlid species but they are grouped into more genera, which suggests greater divergence and a longer period of evolution in isolation than in Lake Malawi.

Many of the Lake Malawi fish live in small, localised areas of the lake and groups of species seem all to be exploiting the same resources. On closer examination, however, a remarkable variety of feeding habits is evident and competition between the species is probably less than appears at first sight. Bedrock and boulders of all sizes provide a structurally varied habitat, full of nooks and crannies. On the tops of the submerged rocks, and down their flanks until the light is too low, a thick mass of filamentous blue-green algae flourishes. Among the filaments grow innumerable species of other algae and a multitude of small invertebrates, mainly insect larvae and small crabs. Several species of fish eat algae (some the blue-greens, some the other kinds), many eat the invertebrates, some eat other fish, and a few have the bizarre habit of eating scales and nibbling the fins of their fellows. Even this list does not seem to provide enough variety of food to support the thirty species of fish which have been recorded in one such rocky area. But those that apparently eat the same things are further divided into those which eat in different places (on top of the rocks or in the cracks between them), those which concentrate on different species of insects, or those which feed nearer or further from shore. Altogether it can be shown that although there is overlap between the diets of many species, almost every one has some distinctive speciality of food or feeding behaviour. All of them ultimately depend on the primary production of the algal carpet. When the water level drops, light penetrates further down the rocks and algae grow for a time on previously uncolonised surfaces. At the same time the upper rocks become dry and uninhabitable until the lake level rises again and they become covered with water once more. These variations give some areas of substrate a rest and allow the algae to decay and release their nutrients to stimulate more productive growth when the right conditions return.

Quite different communities of fish inhabit the sandy bays and again different species feed on different things in different ways. Some eat the grass-like plant *Vallisneria* which grows in the sand, some scrape off the algae which grow on its leaves, some eat algae growing on the surface of the sand and between the grains. Two fish species dig in the sand and sift out small animals: different spacings between the gill rakers allow one species to trap and eat very small animals such as ostracods. This species does not eat larger prey and presumably spits them out, while the other eats the larger chironomid larvae. This community also includes at least one species which has crushing teeth for eating molluscs, found in the sand.

Lake Tanganyika also has complex fish communities in its shoreline habitats, but the species involved are quite different. This is also true of the pelagic fish. In the open water of Lake Tanganyika two species of clupeids prey upon the zooplankton. In Lake Malawi no less than seventeen species are involved in the exploitation of this

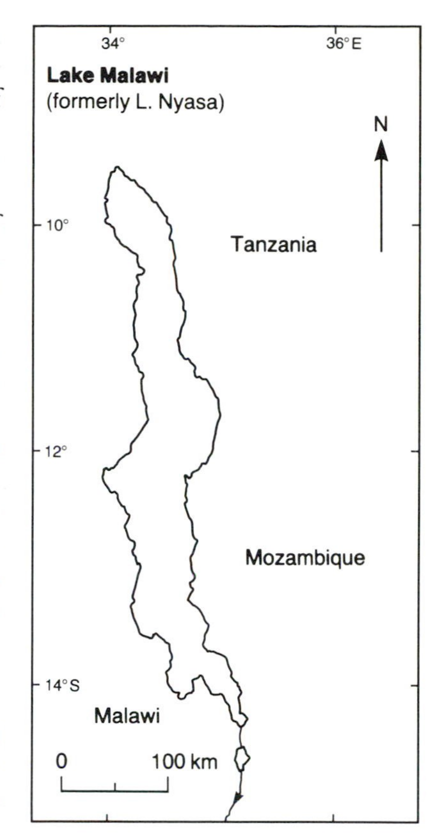

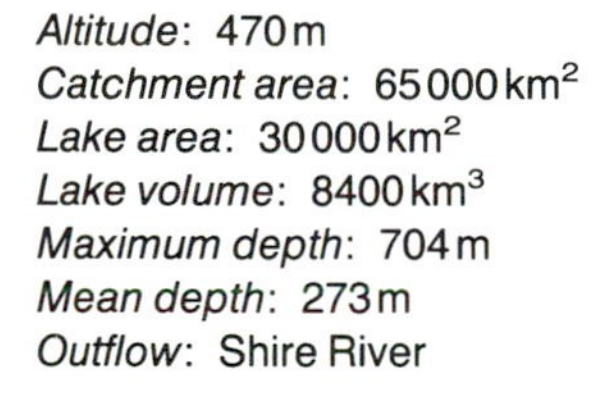
Altitude: 470 m
Catchment area: 65 000 km²
Lake area: 30 000 km²
Lake volume: 8400 km³
Maximum depth: 704 m
Mean depth: 273 m
Outflow: Shire River

food resource. The nearest equivalent to the *Stolothrissa* of Lake Tanganyika is *Engraulicypris sardella*, a cyprinid which not only feeds in but also breeds in, the pelagic zone. In addition, a group of about sixteen species of *Haplochromis* (cichlids) eat zooplankton and live most of their lives in the pelagic zone. Despite this richness of pelagic fish species, which form the basis of a local fishery, it has recently been suggested that the clupeids endemic to Lake Tanganyika should be introduced to Lake Malawi in order to increase the efficiency with which the zooplankton are utilised and the productivity of the fishery. 'Tanganyika sardines' have been successfully introduced into Lake Kariba and Lake Kivu but the former is an artificial lake whose fish fauna is derived from the Zambesi River and therefore contained no pelagic species. Lake Kivu also contained no zooplankton feeders because its fauna was severely impoverished by volcanic activity in the recent past. Even so, the zooplankton which had developed in Lake Kivu in the absence of predation has been drastically altered by the introduction. While that is regrettable it might be justified on the grounds that a productive fishery has resulted which benefits the local people. In Lake Malawi neither of these conditions applies: the lake is very old and contains a highly developed and complex community of fish which have evolved to utilise every niche available. The local people already operate a fishery and although its efficiency could probably be increased this should not be by the introduction of an alien species. Moreover, Lake Malawi cichlids are already the basis of a trade in specimens for the aquarium trade which, provided it is carefully regulated, can yield hard currency. There is also the possibility of further developing tourism for increasing numbers of scuba divers which, like the aquarist trade, depends on the interest and colourful diversity of the native fish community.

We cannot say that such an introduction would definitely be harmful but we can look at previous experiences in other ecosystems and, with few exceptions, note that the results were unsatisfactory and the possible benefits do not appear commensurate with the probable losses. Of course the introduced species might not become established but past experience of introductions does not encourage optimism on that score. To try it and see, to experiment, would be irresponsible. Again, history shows that once an introduced species becomes established, it is almost impossible to remove, even from a terrestrial environment, should it turn out to be a pest. How much more difficult it would be to eliminate an unwanted fish from one of the largest lakes in the world!

Two monster lakes: Loch Ness and Loch Morar

The deepest lake in Britain is Loch Morar with a maximum depth of 310 m, but the greatest mean depth (132 m) is found in Loch Ness which is not only better known but contains such a huge volume of water (nearly 7.5 million m^3) that it qualifies as the largest lake in Britain. Since they have many characteristics in common, apart from the myths that they contain monsters, we shall consider them together. Both fill long, narrow rocky basins in the north of Scotland. The Loch Morar basin was formed by a glacier flowing east to west and it would be a sea loch but for a narrow barrier at the seaward end of the valley. Loch Ness lies east–west in the rift called the Great Glen which is tectonic in origin but was subsequently deepened by ice. The loch drains eastwards but has also been connected to the west coast by the Caledonian Canal since 1822.

Loch Ness drains a catchment ten times larger than Loch Morar and is subject to greater human influence. The Morar catchment contains far fewer houses and less forestry: that of Loch Ness contains twenty times as many people, 100 times more roads, and eight sewage works to contribute additional nutrients to the lake. Despite these differences, both lochs contain few dissolved salts and are extremely unproductive by any standards. Although Loch Morar has a slightly higher conductivity (35 μS/cm) than Loch Ness, this is entirely due to its higher concentration of sodium (5.3 mg/l) and chloride (10.6 mg/l) ions which are derived from sea spray rather than from the land and therefore not accompanied by nutrients such as nitrates.

The water in Loch Morar is very clear and a white (Secchi) disc can be seen down to between 5.7 and 10.2 m, depending on the season. The water of Loch Ness is brown and peaty, which restricts light penetration, and the same disc is only visible down to 3.6–4.6 m. Nevertheless, Loch Ness develops rather more phytoplankton than Loch Morar. It is necessary to filter up to 3 l of water (in contrast to 0.5 l from the London reservoirs) in order to measure the algal concentrations, which reach only about 1 μg (chlorophyll)/l in Loch Morar and 1.5 μg/l in Loch Ness, even in the summer. Despite the clarity of the water few large plants grow in these lakes because long stretches of wave-washed, stony beaches shelve so steeply that there is very deep water within a few metres of the shore. Only in a few sheltered bays are there reed beds and a few submerged plants such as *Littorella*.

Both lochs are very cold (minimum about 5.5°C in the deep water) for most of the year. The maximum summer temperatures recorded in 1977–78 were only 14°C in Loch Ness and 14.7°C in Loch Morar. Loch Morar stratifies for about 3 months (July–September) but in the year of the survey no pronounced thermocline developed in Loch Ness. Whether or not a thermocline develops may well depend on year-to-year variation in spring air temperatures and wind strengths.

These features (cold temperatures, lack of nutrients, great depth) all combine to keep primary production low and ensure that few animals can live in these lakes. Insects completely dominate the invertebrate communities of the littoral zones, where an abundance of stoneflies, characteristic of cold turbulent water, is found. The most varied faunas occur in the small areas of sandy, silty substrate. The fauna of Loch Morar is particularly impoverished; a total of only seventy-one types of animal have been identified, including only one mollusc species, one leech and one flatworm.

Fish are rarely found deeper than 30–40 m, but both lakes contain populations of Atlantic salmon, brown trout, char, eel and three-

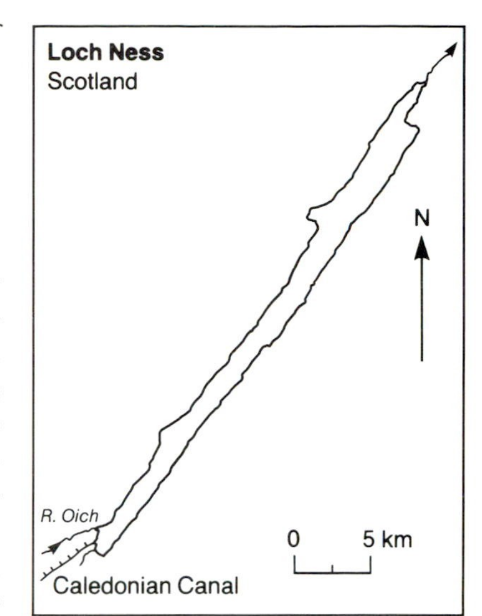

Location: 57°20′ N; 4°30′ W
Altitude: 15.8 m
Catchment area: 1775 km^2
Principal inflow: R. Oich
Lake area: 57 km^2
Lake volume: 7.45 km^3
Maximum depth: 230 m
Mean depth: 132 m
Outflow: River Ness

The Loch Ness Monster

The accounts of a monster being seen in Loch Ness go back over a thousand years but the first photograph, from which the most recent spate of interest stems, was taken in 1933. It was said to show a large beast with a small head and bulbous body. Since then many sightings have been recorded and several new photographs taken, but all were dismissed by the sceptics as unreliable or indistinct. As the necessary technology became more sophisticated, various groups of monster hunters mounted round the clock surveilleince of the entire lake using people, cameras, a submersible, sonar and, eventually, underwater television. None of this intense effort yielded any tangible evidence of a monster until, in 1975, an American group published a rather fuzzy underwater photograph which was said to establish beyond doubt that the monster existed. On the basis of this photograph, which appears to show a paddle-like fin, the 'animal' was officially described by Robert Rines and Sir Peter Scott, in the prestigious journal *Nature* and given a Latin name, *Nessiteras rhombopteryx*. However, it should be noted that this name is an anagram of 'monster hoax by Sir Peter S.', and it is not at all clear to this day who was hoaxing whom.

Does such a creature as this inhabit Loch Ness?

spined stickleback, all of which have marine affinities. Loch Ness also contains brook lamprey and pike. The latter is the only one without any marine connection and is presumably a more recent arrival than the others in the post-glacial history of these lakes.

Monsters have been reported from both of these large lakes and there are many suggestions as to what they might be. Speculation ought to take into account the food available to a large animal resident in the lake. The only sizeable prey are fish. Assuming that the low productivity of the lake could not support a monster directly, it has been suggested that they eat the salmon which feed not in the lake but in the sea. It must also be remembered that even if one monster is enough to perpetuate a myth, to perpetuate the species requires at least two of them and they must have enough food not just to stay alive, but also to produce eggs or babies. Given the unproductive nature of these lakes it seems more likely that the sightings which give rise to the stories of monsters are in fact of otters swimming far out in the lake and rolling over as they dive to fish or play. The popular idea that the monster might be a plesiosaur, extinct elsewhere for 100 million years, is undermined by the fact that the Loch was covered by thick ice only 10 000 years ago.

On a world scale Ness and Morar hardly count as large lakes, certainly not in terms of area, though in terms of depth they are not insubstantial: Loch Morar is the seventh deepest lake in Europe. They are important, if only as vast reservoirs of clean water; and Loch Morar is classed as a Grade 1 site by the Nature Conservancy Council which emphasises its importance for conservation. Whether or not they include monsters, their plant and animal communities are largely unaffected by the little human influence to which they have so far been subjected and it is important that they should remain that way. Unproductive lakes are all too easily 'enriched' by human activity.

7
Shallow LAKES

Lakes range in depth from those which are over 1000 m deep to those which periodically contain no water at all, so how can we say where deep lakes end and shallow lakes begin? It is easy to distinguish the extremes, and we have already looked at the deepest lakes, but it is much less easy to define shallow lakes. Moreover, the distinction is not just a matter of depth but more to do with function. The most useful definition of a shallow lake is one which never experiences thermal stratification for more than a few days at a time; it is shallow enough for the wind to keep all the water more or less permanently mixed. This has profound consequences for the structure of the lake community and the way it functions. Whether or not a lake is *functionally* shallow therefore depends not only on actual depth but also on its area and the degree to which the surface of the lake is exposed to the wind. Thus a small, shallow pond overhung by trees may be so sheltered that it behaves more like a deep lake, whereas a relatively deep expanse of water exposed to strong winds may be always well mixed and behave as a shallow lake, despite its greater depth. The deepest lakes discussed in Chapter 6 were also very large and one might at first assume that large lakes are inevitably also deep, but some of the largest lakes are actually very shallow: Lough Neagh in Northern Ireland is the largest expanse of fresh water in the British Isles but has an average depth of only 8.9 m; Lake Chad expands to form one of the biggest lakes in Africa but is never more than about 8 m deep; and Lake Balaton in Hungary, one of Europe's biggest lakes, is nowhere more than 11 m deep.

All deep lakes have at least some shallow areas, but since shallow lakes do not have any deep water one might wonder whether they still contain all the usual lake life. They cannot of course contain specialised abyssal communities, but they do have littoral, pelagic and benthic communities, and the balance between these depends on a number of interacting factors. If the water is clear and shallow enough for light to reach right to the bottom over a large area, then much of the lake may contain submerged and floating macrophytes and a primarily littoral community. But even in the clearest water, vascular plants cannot grow below about 10 m deep because the pressure of the water causes the spaces between their cells to collapse. If the lake water is turbid, either through the growth of algae or because the lake is shallow enough that sediments are stirred up from the bottom as the wind mixes the lake, the littoral community may be restricted and most of the lake area then contains open, turbid water. In this case there may be a truly planktonic community in the open water and a benthic community in the underlying mud. The nature of the latter will depend not only on the type of substrate but also on its stability as the water above it is mixed. If the mud is soft and constantly disturbed it is not a suitable habitat for animals that dwell in permanent tubes. The only species found may be those that can more or less swim through the surface layers of the sediment and re-bury themselves when it settles after disturbance, or those that can live in the anoxic mud below the disturbed layer.

Even if submerged and floating macrophytes are restricted by turbidity, emergent macrophytes, such as reeds, are often abundant in the shallowest areas and may form extensive, almost monospecific stands. If other macrophytes are absent there is often a sharp boundary between the reed swamp and open water. The open water of some shallow lakes may be so turbid, due to silt, that the penetration of light is obstructed, reducing phytoplankton productivity. In turn this influences the rest of the food chain. Suspended silt also interferes with the feeding mechanisms of some filter-feeding animals and may damage the gills of fish.

Shallow lakes are generally more productive (per unit area) than deeper lakes in the same regions. This is because the organic matter formed within them does not sink down to be lost for ever or recirculated only intermittently. Instead it is broken down and its constituent nutrients are frequently stirred back into the lighted zone and quickly re-used to produce more organic matter. Moreover, nutrients entering via inflow rivers are not diluted in a shallow lake to the same extent as in the large volume of a deep lake. Furthermore, a greater proportion of the water in a shallow lake is in the lighted zone, and the water of shallow lakes is frequently warmer than that of deep lakes which speeds decomposition and production. Higher primary productivity provides food for more invertebrates, which are more abundant and diverse if both algae and macrophytes are available. The plants and invertebrates are food for fish and birds. Many birds eat the fish too and shallow lakes often support large and varied bird populations. These generalisations are modified by circumstances on particular lakes, such as altitude, latitude and the

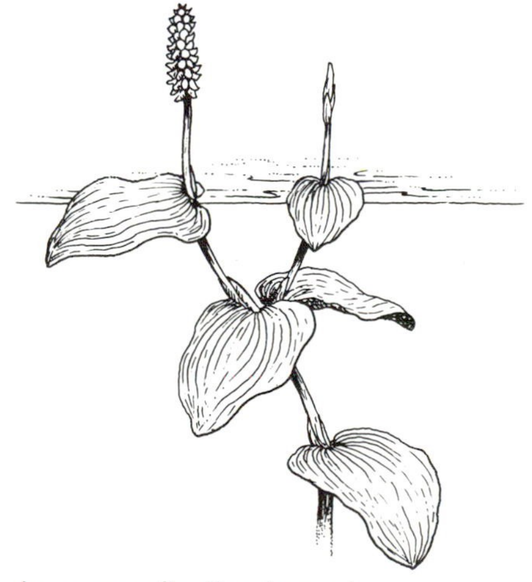

A common floating-leaved macrophyte (*Potamogeton perfoliatus*) which has its roots in the lake bed 3 m or more below the surface.

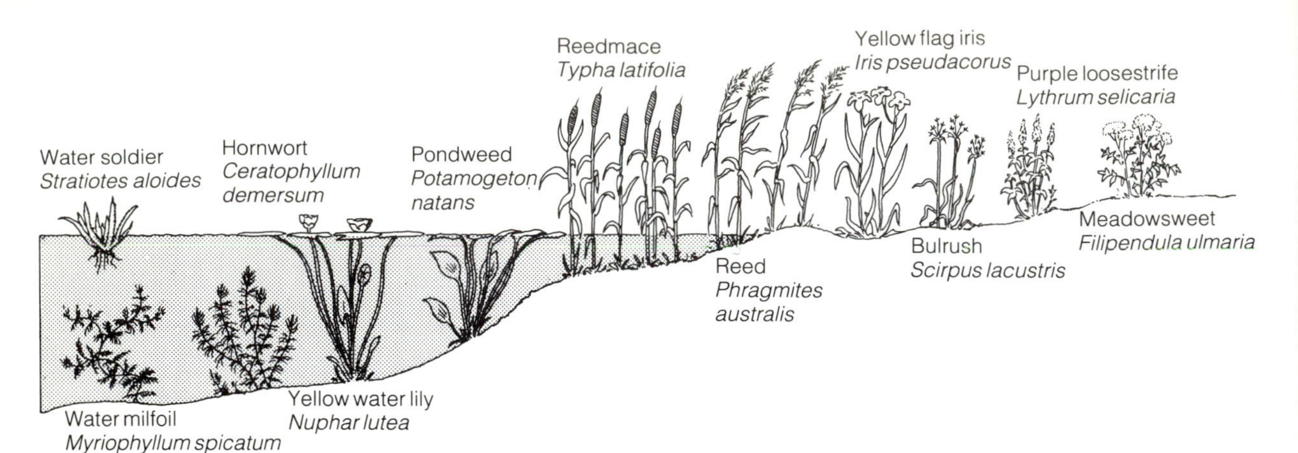

A hydrosere: the zonation of plants in the transition zone between water and land. The plants shown here are a selection of those which might be found at the edge of a shallow lake in temperate Europe.

degree of exposure to wind. Every lake, especially shallow ones, has an individual character, so that generalisations are difficult and not universally valid. We have therefore reviewed shallow lakes by reference to a series of particular examples, from which will emerge a picture of shallow lakes as a whole.

Lake Balaton

Lake Balaton, in Hungary, provides a well-studied example of an extensive, shallow lake in the temperate zone which exhibits all the features mentioned above. It is an elongated lake (77 km long, mean width 9 km) formed about 18 000 years ago as the ice retreated from central Europe. It covers 596 km^2 and has a total catchment area of 5178 km^2, rather more than half of which is drained via the Zala River which enters the lake at its western end. The main outflow is a canal which flows to the Danube. The average depth of the lake is less than 3.5 m and although it does reach a maximum of 11 m deep this is only in a small area where the width is constricted by the Tihany Peninsula. The north shore is hilly but the south shore is flat and sandy.

The lake water is alkaline (pH about 8.5) and has a conductivity of about 500 μS/cm. The calcium content is quite high and, as the algae give off carbon dioxide during photosynthesis, calcium carbonate precipitates out and flakes of it float on the water, giving the surface an opalescent white colour from a distance. The water temperature ranges from zero to 30°C through the year and most of the time is the same from top to bottom of the water column. In calm periods during summer the water may stratify and sometimes this persists for several days before the wind again mixes the lake. Frequently the wind stirs the water sufficiently for the upper layers of the sediment to be mixed into the water and this reduces the depth of light penetration and also releases phosphates from the sediment.

People have lived around the lake for thousands of years; today the lake is a very important holiday centre both in summer and in winter, when it is usually frozen for about 60 days between December and March. The very long shoreline in proportion to small volume of water makes such a lake very vulnerable to what goes on in its catchment, three quarters of which is farmland. Since the beginning of the twentieth century, increased use of agricultural fertilizers and the growing human population have led to eutrophication of the lake waters. The shoreline and water level have been increasingly regulated and marshy areas have been disconnected from the lake. These changes, and increasing pollution of the inflowing streams, have led to changes in the lake's communities over the last few decades.

The natural phosphate content of the water is quite low but the supply of organic phosphorus is artificially increased, particularly at the western end where the Zala River brings in sewage effluent. As a consequence the chlorophyll content of the water at this end of the lake has risen to more than 100 μg/l, while at the other end of the lake chlorophyll readings do not exceed 20 μg/l. During the ten years from 1966 to 1976 there was a four-fold increase in phytoplankton biomass and blooms of blue-green algae increased in frequency.

About 100 km^2 of the lake is only 1.5 m deep or less, so littoral conditions prevail over a large proportion of the lake area. Some 8 km^2 is occupied by reed beds (*Phragmites*) extending along 25% of the shoreline. The beds of macrophytes are more continuous and more varied along the north shore than the south. Beyond the *Phragmites* fringe, the floating leaves of *Potamogeton perfoliatus* and the submerged *Myriophyllum spicatum* occur in scattered patches with many less-abundant species. Canadian pond weed (*Elodea canadensis*) and water soldier (*Stratiotes aloides*) have lately spread almost throughout the lake. As in many shallow lakes, the plants not only contribute to the overall productivity by their growth but are a vital structural component of the habitat for many invertebrates and micro-organisms; more than half the invertebrate species found in the lake live among the *Potamogeton*. Dead macrophytes also provide food for detritus feeders including the zebra mussel (*Driessena polymorpha*) which is very common in the lake.

Littoral zones frequently support a much wider variety of species than open water. In shallow lakes the littoral zone forms a much greater proportion of the total area than in deeper lakes, so it is reasonable to expect a greater quantity and variety of invertebrates in shallow swamp-fringed lakes. This makes such lakes ideal for a variety of birds such as waders, herons, warblers, and aerial insect eaters. The reeds also provide shelter and support for nests, making it possible for larger numbers of birds to breed than if the lake were deeper with more open water; indeed, but for the reeds many species would not be there at all.

There are forty-seven species of fish recorded from Lake Balaton but only about one third of them are of economic significance. These include the predatory pike perch or zander (*Stizostedion lucioperca*), which forms 6–12% of the annual catch, but the most important commercial species is bream (*Abramis brama*), which constitutes 70–80% of the 1200-tonne annual haul. The bream eat mainly chironomid larvae, which make up 60% of the animals in the bottom mud of the open water; the older fish can find them even 15 cm deep in the mud. Chironomid larvae are most abundant in spring and

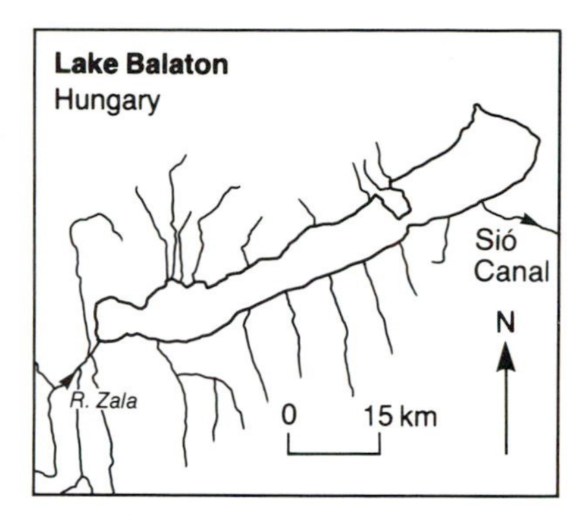

Location: 47° N; 18° E
Altitude: 104 m
Catchment area: 5178 km^2
Principal inflow: Zala River
Lake area: 596 km^2
Lake volume: 1.8 km^3
Maximum depth: 11 m
Mean depth: 3.3 m
Outflow: Sió Canal

The common reed *Phragmites australis*

The common reed *Phragmites australis* (= *P. communis*) is a large grass, widespread in the Old World temperate zones, though less common in the tropics. It does occur in North America but is less abundant there and it has been introduced to Australia. It is typically a lowland plant, but occurs at up to 510 m in Britain, at up to 1910 m in the Alps and at up to 3000 m in Tibet. It grows as far north as 70° in Norway. There are two other species, one in Asia and Polynesia, the other in tropical Africa.

The reed lives in places that are either permanently or intermittently flooded by shallow, still water. It can tolerate a water table from 1 m below to more than 1 m above the level of the soil. In warm climates it can grow in water as deep as 4 m but in Europe is not usually found in water deeper than 2 m. It is an annual plant whose shoots grow each spring from a dense mat of rhizomes. The vertical rhizomes, which branch from horizontal rhizomes, bear only one shoot in the first year but may have up to four in the second year and up to six in the third. The number of shoots per vertical rhizome then declines until it is 6 years old. The shoots are like canes and range from a few centimetres to about 4 m in height, depending on the growing conditions. The tallest reeds are found in very wet habitats where the water is warm and eutrophic. The narrow leaves grow from each node and the flower head is terminal, much branched and feathery. In Britain, shoot growth is almost complete by July and new buds form on the rhizomes. These remain dormant until the following spring unless the shoots are cut, in which case the buds grow into new shoots and thus drain the reserves in the rhizome. If cutting is delayed until September or later the buds will not develop and the stand will regenerate fully the following year. Thus the timing of cutting determines whether the reed bed is maintained as a productive resource or rapidly destroyed, as is needed to keep open water free of reeds.

The common reed *Phragmites australis* with the typical reedbed bird, the bearded tit *Panurus biarmicus*. This feeds on insects associated with the reeds in summer and on the reed seeds in winter. It also nests in the reed beds and is rarely seen anywhere else.

If the reeds are not cut, their contents are transported back to the rhizomes and only the dry structural elements remain. These eventually collapse and break up to accumulate at the base of the stand on top of the rhizome mat. In many situations the rate of accumulation exceeds that of decomposition so the layer of plant material builds up and eventually becomes so high above the water that other species are able to invade.

Regular cropping of the reeds interrupts the succession and prevents the change from reed swamp to dry land. Reeds are used for thatching and are normally cut in the winter. They are extensively exploited in the Danube delta and Scandinavia. In the Netherlands, reeds grow in enclosed areas (polders) of shallow water where they raise the level of the land and dry it out by transpiration, until they are replaced by other plants which can be grazed by cattle. Thus new farmland is formed which, in time, can be used for growing crops.

The effects of different cutting regimes on the number of reed stems in experimental plots at Wicken Fen (from Haslam, 1968)

Cutting regime	Number of stems in 12 m^2
Cut every year	73
Cut every 2 years	30
Cut every 3 years	31
Cut every 4 years	10

autumn, and one species, *Tanypus punctipennis*, reached nearly 10 000 individuals per square metre in February 1974. The biomass and species composition of the benthos has been affected by recent eutrophication of the lake. For example, certain crustaceans not previously seen there have now arrived and *Asellus aquaticus* has disappeared.

The changes resulting from increased phosphorus input have mostly been in terms of increased productivity of the dominant species already present, altering the balance between components of the ecosystem. Such changes are unnatural and should be controlled before they go too far but they are not at present harmful.

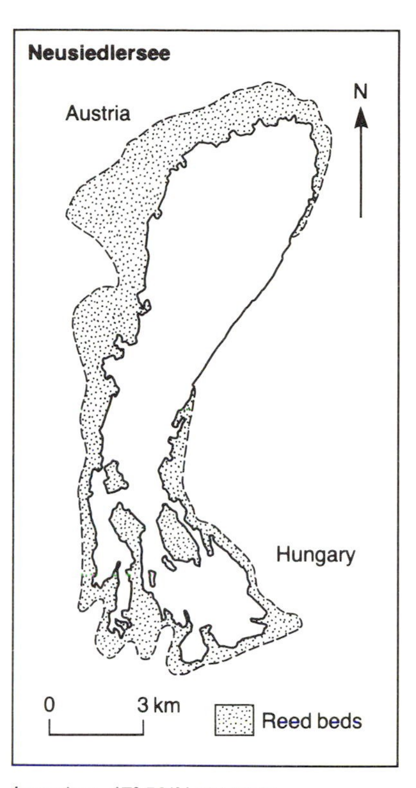

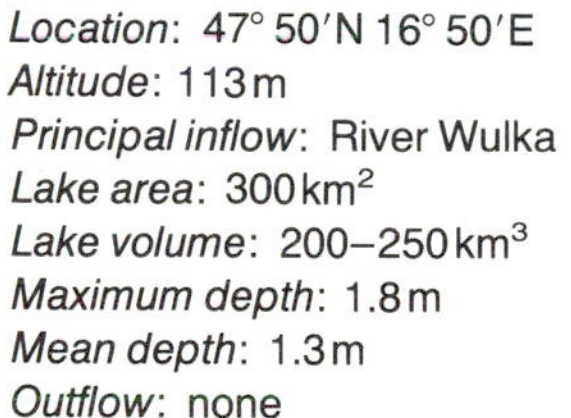
Location: 47° 50′N 16° 50′E
Altitude: 113 m
Principal inflow: River Wulka
Lake area: 300 km^2
Lake volume: 200–250 km^3
Maximum depth: 1.8 m
Mean depth: 1.3 m
Outflow: none

Neusiedlersee

Not far from Lake Balaton, on the Austro-Hungarian border, lies an even shallower lake known as Neusiedlersee in Austria and as Lake Fertö in Hungary. It has an average depth of only 1.3 m and nowhere is the water deeper than 1.8 m. It is at the same altitude (just over 100 m above sea level) as Balaton and subjected to much the same climate but, at 270 km^2, it is about half the area. The lake has an important fishery, is much used for recreation and is famous for its bird populations. There is no outflow, so the water has a fairly high conductivity of 1000–2300 μS/cm. Like Lake Balaton the nutrient input has increased markedly in recent years with more people living locally and visiting the area. In time this may become a serious problem.

The two most obvious features of Neusiedlersee are the enormous reed beds, which cover 50% of the lake, and the very high turbidity of the water, due to constant stirring of the sediment. The sediment is firm mud and sand near the eastern shore, but in the north and west the mud is very soft. When the wind blows it stirs up the mud and the water may contain as much as 800 mg/l of suspended solids. If this suspended material is filtered out of the muddy water it is found to contain more phosphorus than when it is lying on top of the sediment. This must be because it now includes the algae and zooplankton (and the phosphorus they contain). When the wind drops and the sediment settles out again the phosphorus content is again reduced so the dead algae and zooplankton must decompose without finishing up on the lake bottom. The water itself always contains sufficient inorganic phosphate throughout the year to support the phytoplankton population, so the major limitation on photosynthesis is light. At times of strong wind and high turbidity the underwater availability of light is severely reduced, so algal production in the open water depends very much on the wind.

The reed beds vary in width from 1 to 3 km and reach their greatest extent in the north and along the western shore. They are dominated by the common reed (*Phragmites australis*), one of the most widespread dominants of reed swamp in Europe. Except at the lakeward edge, where the reeds are gradually colonising open water, and at the landward side, where the soil may be dry in summer, the reeds form an almost uniform habitat over huge areas. In a way these reed beds are like forests, in that the dominant plants are not only important as primary producers but also as habitat for other species. Their presence affects such things as the light regime in the water, the nutrient condition of the substrate, and the availability of surfaces on which to live. Many species actually live within the reed plant itself, though surprisingly few actually eat it. Many more utilise the plant material once it has died and been partially broken down: thus detritivores are more numerous than herbivores. In fact the muskrat (*Ondatra zibethicus*) is the only large local animal which eats whole live reeds, though the bearded tit (*Panurus biarmicus*) eats the seeds. One of the largest invertebrates to depend upon the reeds is a moth, the reed leopard (*Phragmataecia castaneae*), whose caterpillars are stem borers. They prefer thick vigorous stems and are therefore never found at the margins of the swamp. About twelve stems in every hundred are attacked by the reed leopard larvae but almost every reed is atacked by the very abundant stem-sucking bug *Chaetococcus phragmitidis*. Neither this nor the other stem- and leaf-sucking insects seem to have much effect on the productivity of the reeds, but the gall-forming *Liara lucens* (a fly of the family Chloropidae) does inhibit the growth of infected stems and prevents them flowering. It is restricted to the landward side of the swamp where the reeds develop later in the season.

The reeds have a strongly seasonal growth cycle and the above-ground biomass reaches its peak in mid-July, by which time there will be 70–90 stalks per square metre – equivalent to about 1.8 kg of dry plant material. There is a similar weight of rhizomes under-water. Reeds are grasses, with long narrow leaves growing from nodes at intervals up the stem, the youngest at the top. By mid-July there is about 6.5 m^2 of leaf area over every square metre of water inside the swamp. This keeps the water shaded and cool but means that algae or other plants growing between the reeds must function at low light intensities. The leaves lose water by transpiration, so the reeds act like a wick and draw water up from the mud below. About 1000 l of water may be lost from each square metre of the swamp annually, but it only receives 400–600 l of rain. It is hardly surprising that reed swamps lower the water level and hasten the process of turning lake edge into dry land. Unlike some other emergent macrophytes, *Phragmites* can live where the ground is raised above water level. Their rhizomes can extend a metre down into the soil and no doubt this enables them to continue drawing up the water they need.

After the peak of growth is reached in July the above-ground biomass begins to decrease and the weight of the rhizomes increases as the plants move the products of photosynthesis down to be stored for the winter. By the autumn there may be twice as much weight of living material underground as there is above. When the remaining above-ground material dies it settles on the surface of the mud, providing food and shelter for a host of detritivores and decomposers. Usually the mud is totally devoid of oxygen, so decomposition is incomplete and some organic material accumulates each year. Eventually this gets compressed to form peat and raises the mud surface within the swamp. The process is accelerated in Neusiedlersee because suspended material in the lake water is swept into the reed belt by the wind. Each square metre at the lakeward edge of the reed swamp daily traps about 440 g of debris and mud which would otherwise have sunk to the bottom out in the open lake. This retention of sediment at the margins is what prevents the shallow lake silting up completely, but the reeds are gradually encroaching into the water. The area of open water is now about half what it was a hundred years ago. The process may be speeded up by enrichment of the inflow, particularly the Wulka River which supplies 60% of the inflow to the Austrian sector of the lake. It disperses through the reed beds which absorb part of its increasing load of phosphorus but there is now evidence that the swamps are becoming saturated in their requirement for phosphorus and more is entering the lake itself.

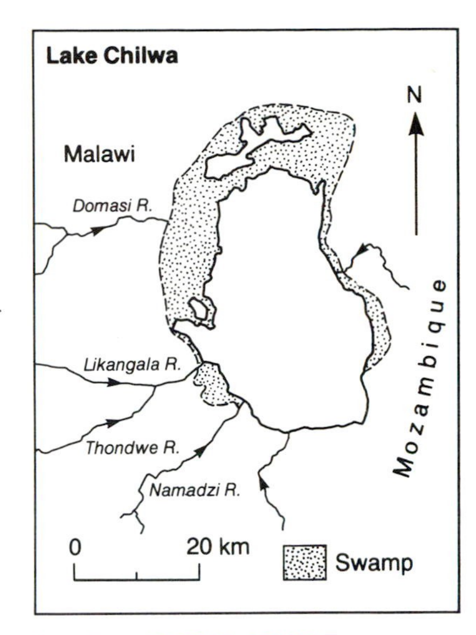

Location: 15° 30′ S; 35° 30′ E
Altitude: 622 m
Catchment area: 7500 km^2
Principal inflows: Domasi, Likangala, Thondwe, Namadzi and Phalombe Rivers
The Area, Volume and *Depth* of the lake are very variable. The maximum area of open water is about 700 km^2 (total about 1400 km^2 including marshland)
Maximum depth: about 3 m
Outflow: none

Lake Chilwa

There are interesting parallels between Neusiedlersee and Lake Chilwa in Malawi. Both are shallow, endorheic, slightly saline and surrounded by reed swamp. Unlike Neusiedlersee, however, Lake Chilwa has a tropical climate which is nevertheless quite seasonal with a cool dry period between May and October. Also, unlike Neusiedlersee, the lake level fluctuates by up to 1.5 m each year and over longer periods (20–30 years) exhibits either net downward or net upward trends. It dried up completely for a few months in 1968 and has been known to do so at least twice before in this century. The depth and area of the lake are thus very variable but even in the wettest periods the former does not exceed 3 m and the open water area is then about 700 km^2. The whole lake is surrounded by swamps which are dominated by reedmace (*Typha domingensis*) and cover a total of about 552 km^2. Each year the water level rises in the wet season and floods another 575 km^2 of marsh and grassland beyond the permanent swamp.

The swamps play a significant role in the functioning of the lake. They trap larger sediment particles as they are washed from the surrounding land and only fine silts enter the open water. These are constantly churned up by wind mixing of the shallow water, which is consequently very murky. This prevents the development of rooted submerged macrophytes outside the sheltered swamp area and the transition from swamp to open water is therefore quite abrupt. Inside the swamp there is no wind and the sediment settles out, leaving the water much clearer. But the lack of wind also leads to much greater variability (e.g. in conductivity and oxygen availability) within the swamp, whereas out in the open water its mixing effect ensures homogeneity. It follows that removal of reed swamp, here or anywhere else, would result in much more uniform conditions over the lake as a whole and a substantial loss of species diversity: hence the conservation value of reed swamps. To many people the idea of a swamp is unattractive, but when its ecology is thoroughly investigated it soon becomes evident that a swamp such as that surrounding Lake Chilwa has a vital role to play in the functioning of the ecosystem as a whole.

The swamp water is shallower, clearer and less saline but contains less oxygen than that in the main lake. This is because the sheltered conditions mean less stirring by the wind. There is also less light and therefore less oxygen from algal photosynthesis. The abundant decomposers and detritivores use up what little oxygen there is and in fact there is only oxygen in the water at the edge of the swamp. In the centre of the swamp and on its landward side the water is normally devoid of oxygen.

The basis of this detritivore and decomposer activity is of course the *Typha* reedmace, whose individual shoots live for 2 years and then die. The total biomass of *Typha* reaches its maximum between June and November, and each year 1 m^2 of the swamp produces 1.5 kg (dry weight) of new organic matter. However, even in the centre of the swamp there is seldom more than 10 cm of undecomposed organic matter on the surface of the mud, so there is little evidence of the accumulation of dead organic matter seen in papyrus swamps or the reed swamps at Neusiedlersee. Almost all the organic matter produced in this tropical swamp must be oxidised, either through metabolism of the detritivores that eat it or, occasionally, by fire. There is evidence that about one third of the swamp is burned each year during the dry season; the other two thirds feed the detritivore community.

The swamp plants represent a store of energy and also a considerable store of nutrients. Since the swamp sediment is normally deoxygenated, no barrier layer forms on top of the sediment and nutrients such as phosphates are released into the water (see Chapter 2). The plants take up nutrients through their roots and incorporate them into their tissues. While the plant is still alive nitrogen is stored in the rhizomes but other chemicals in the rest of the plant remain there and are released after the plant has died and rotted away or been converted into faecal pellets. The open water of the lake benefits from this source of food and nutrients only when the swamp water is mixed with that of the open lake. This happens when rivers passing through the swamp, especially in the rainy season, flush the swamp water into the lake, sweeping along dissolved nutrients and fine particles of plant detritus. Nutrient transfer to the open lake is also assisted by the wind which blows lake water deep into the swamp. Since the direction of the wind varies, this can occur all around the lake, causing a slopping effect which draws swamp water and nutrients back out into the open lake. Downwind there may even be a tidal effect which pushes water right through the swamp and on to the grassland beyond. As water drains back through the swamp it again carries dissolved and particulate matter with it.

Lake Chilwa has a large fish population, important in the local economy. The main populations of fish are found around the swamp margins where the largest numbers of planktonic and benthic animals occur, sustained by the productivity of the swamp. Twenty percent of the fish catch is obtained within the swamp itself. The dominant commercial fish species *Clarias mossambicus* and *Barbus paludinosus*, which make up 70% of the catch, move from the lake

Air-breathing fish

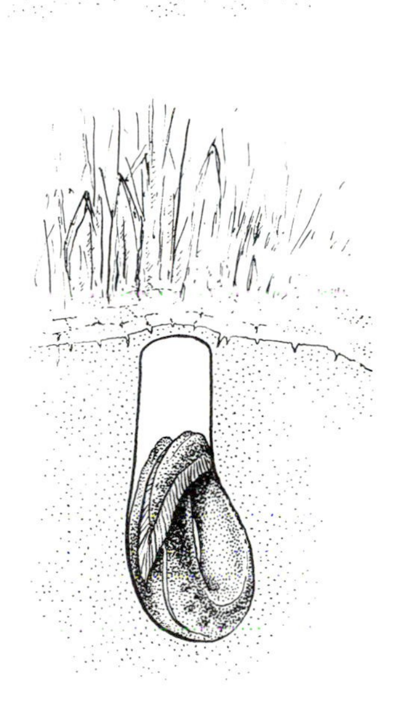

The majority of fish obtain oxygen by passing a current of water over their gills where blood vessels come very close to the surface and the blood takes up oxygen from the water and then carries it round the body. The surface of the gills is very thin and is finely divided to increase its total area. Unless the gills are supported by water they collapse and cannot function properly, so alternative means must be available if respiration is to continue in air. Many fish have developed a mechanism to breathe air as an adaptation to living either in fresh waters that are liable to dry up or in swamps where the oxygen content of the water is very low. Air breathing evolved not as the pinnacle of the evolution of fish but in some of the most primitive groups of fish known. The lungfishes are good examples, in that the species found today in Africa, South America and Australia are very similar to the fossil lungfish of the Devonian epoch, 300 million years ago.

In the lungfish the swim bladder, which opens into the gut, serves as a lung. The fish rises to the water surface and gulps air which passes into the lung. The Australian lungfish *Neoceratodus* still has four pairs of gills and uses them to obtain 90% of its oxygen requirements, but the South American lungfish *Lepidosiren* has only three pairs and the four African *Protopterus* species have only two pairs. *Lepidosiren* and *Protopterus* are almost entirely dependent on their lungs and die without access to air, even if the water in which they live is well aerated.

Lepidosiren and three of the *Protopterus* species have a further advantage for life in intermittent waters in that they can enter a state of torpor known as aestivation. They make a burrow in the mud and form a mucus-lined cocoon within it which dries to become tough and leathery; there remains an opening to the mouth from which a small tunnel leads to the surface of the mud. The head is always pointing upwards and the body curved so that the tail comes forward to cover the eyes. Instead of living off body fat, as most animals do that hibernate or aestivate, the lungfish gets its energy by absorbing muscle tissue. This creates a problem with the disposal of waste products. The kidneys separate the urea from the water and the latter is re-used while the former accumulates in the blood until aestivation ends. Even a low concentration of urea in the blood kills many animals but not the lungfish, which can tolerate extraordinary amounts until the lake fills again and allows the fish to escape from its burrow.

African lungfish *Protopterus aethiopicus* in water and in the burrow in which it aestivates as the water dries up.

Catfish of the genus *Clarias* have a structure known from its shape as the 'gill tree' developed on their second and fourth gills, which helps to obtain oxygen from air. They are known as the 'walking catfish' because they can travel overland, using their fins, in search of food and alternative water when their original home dries out. They belong to a large family of catfish (the Clariidae) which is found all over Africa and South-East Asia. They are long, cylindrical fish with a broad flattened head and horizontal mouth. Their name comes from the four pairs of long sensitive barbels around the mouth which they probably use to feel for food in dark muddy waters. They are almost all predators living on a variety of other animals, including fish.

The equivalent catfish of North America are the Ictaluridae which include the blue catfish (*Ictalurus furcatus*) and the channel catfish (*I. punctatus*). The latter is of considerable commercial importance, since its flesh is very tasty, and is grown in ponds and raceways. It can grow up to more than 20 kg and the largest specimens are highly prized by anglers.

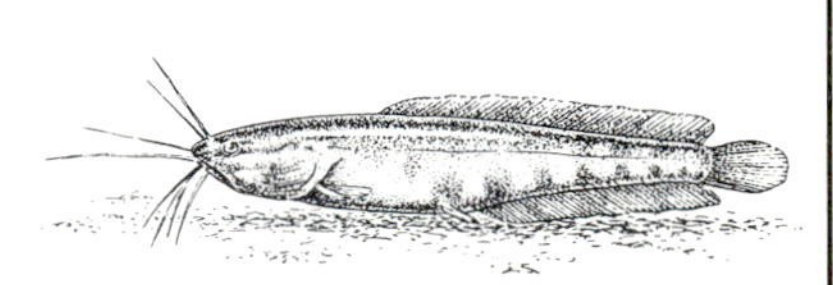

A catfish of the genus *Clarias*.

into the rivers and swamps during the flood season to spawn, then return to the lake as the waters recede.

Wide seasonal and long-term fluctuations in water level seem to be characteristic of many African lakes but are particularly evident in shallow ones and those with no outflow. As the level of Lake Chilwa falls, the water becomes more saline and many animals and plants die. Other animals produce resting eggs or take refuge in the swamps and rivers. Only those species with some mechanism for enduring these low-water periods can survive in such a lake; the rest have to recolonise from outside. Fish that can breathe air, such as the catfish (*Clarias*) and lungfish (*Protopterus*) are particularly well adapted to these situations.

Lake Chad

As might be expected, shallow lakes in the tropics are very prone to drying up, being refilled by seasonal rainstorms on the catchment area, so Lake Chilwa is not unusual among shallow tropical lakes in undergoing wide fluctuations in water level. The shallower the lake, the bigger the changes in lake area that result from more or less water being present. In endorheic lakes these changes are invariably accompanied by changes in the salinity of the water and, in those where evaporation constantly exceeds inflow, the water eventually becomes too saline for most forms of life (see Chapter 8). One of the tropical African lakes which shows spectacular changes in the area and depth of its water is Lake Chad, but this enormous lake also has an interesting mechanism by which the salinity is regulated and the water kept fresh over a wide range of water levels.

Lake Chad lies 12–14° north of the Equator in West Africa, on the edge of the Sahara Desert, which means that it is subjected to an extremely arid climate. The annual mean temperature is 28°C and evaporation exceeds 2 m per year; 92% of the water entering the lake in an average year is lost through evaporation, and the rest through seepage. Gains and losses are thus roughly equal; if gains were constant (as losses are), the lake volume would not change but gains do vary from year to year. Since the Chari River, which enters the south basin, provides 87% of the incoming water (the rest comes from rain) and the lake has no outflow, the volume of the lake, and the area it covers, are closely linked to the volume of river inflow. This in turn reflects changes in rainfall on the catchment. All over Africa, but particularly in this region just south of the Sahara, a few years of moderate rainfall seem to alternate with periods of intense drought in a more or less cyclical manner. Three times since the beginning of this century Lake Chad has almost dried up and in between there have been periods of extensive flooding. In the early 1980s Lake Chad suffered a period of drought along with the rest of the Sahel; previous records indicate that it will almost certainly recover.

Each year the River Chari, and its main tributary the Logone, have a period of reduced flow from March to June, and flood during September to December. Then they spill over to inundate vast areas

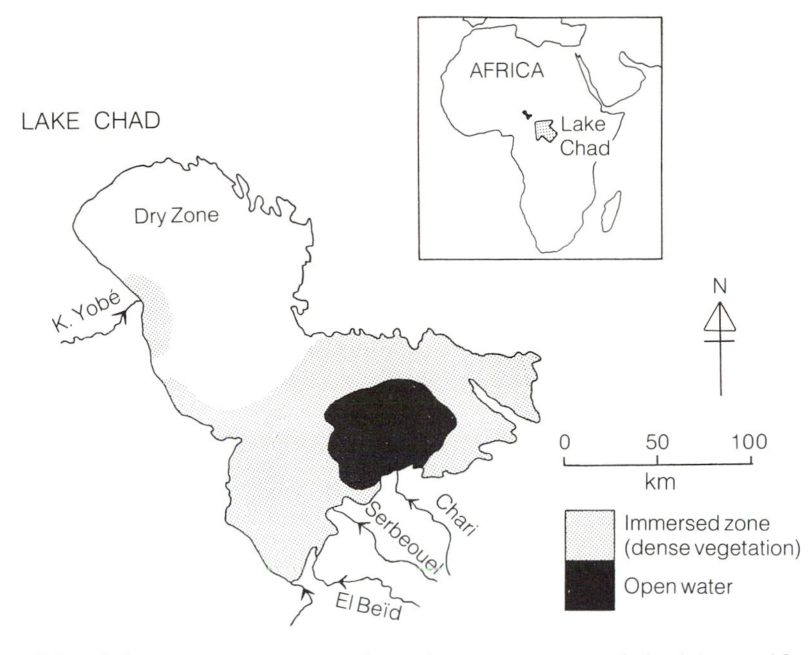

Lake Chad: a large shallow lake with no outlet whose area and depth fluctuate widely as rainfall varies in the catchment seasonally and from year to year. The outline shows the extent of the lake in the 'Normal Chad' phase. From 1967 to 1975 the level of the lake fell each year until open water remained only in the small area shaded black; this is known as the 'Lesser Chad'. The North Basin dried up almost completely and the rest of the lake basin was choked with dense aquatic vegetation, increasingly colonised by more terrestrial species.

of floodplain in Cameroon and northern Nigeria, and the lake itself increases in depth and area. When the inflow decreases, evaporation inexorably continues, and the waters recede until the next flood. Because of the fine balance between average inflow and losses to evaporation and seepage, it is only in years of high rainfall and high river inflow that the lake contains enough water to compensate for a subsequent dry year. If inflow is below average for several successive years then the level of the lake progressively drops; if there are several wetter years in a row, the level gradually rises.

At the turn of the century the lake was already shrinking compared with levels observed by earlier explorers. By 1908 it had fallen to half its 1903 level and many feared that it would dry up altogether. This led to a detailed study of fluctuations in water level as the lake expanded between 1912 and 1919. The fact that the lake had filled again put paid to the worst fears of it drying up, at least for the time being, and suggested that the lake has three phases: 'Greater Chad', when it occupies an area of about 25 000 km^2 as in the nineteenth century; 'Normal Chad', with an area of about 21 000 km^2 ranging down to 15 000 km^2; and 'Lesser Chad', when the lake shrinks below this. When a group of French hydrobiologists started to study the lake intensively in 1964 it was in the 'Normal Chad' phase but by 1972 a 'Lesser Chad' phase had clearly started and by 1974 the north basin, despite its greater depth, had dried up. This phase was still evident in 1981 and almost certainly continued through the drought suffered by the whole of the Sahel region in 1984.

Over a period of about 40 years one can visualise the lake expanding and contracting annually, but also gradually getting larger from year to year then progressively shrinking again. The process is of course uneven and the length of the cycle rather variable; such a cycle is typical of many African lakes, although not as marked as in Lake Chad.

The density of culms and total biomass of papyrus estimated for the interior of several swamps (from Thompson, 1979)

Lake Upemba Island, Zaire River
6.4 culms per m^2
1.11 kg/m^2 dry weight of papyrus
Lake Mulenda Island, Zaire River
7.7 culms per m^2
1.16 kg/m^2 dry weight of papyrus
Lake George, Akika Island
10.2 culms per m^2
2.95 kg/m^2 dry weight of papyrus
Valley Swamp, Kampala
7.9 culms per m^2
2.90 kg/m^2 dry weight of papyrus
Lake Naivasha, north swamp
3.4 kg/m^2 dry weight of papyrus

This giant sedge *Cyperus papyrus* is the plant from which the ancient Egyptians made their paper and on which they recorded their history and customs thousands of years ago. The triangular stems were peeled, sliced into strips, soaked to remove starch, and then laid in a number of vertical and horizontal layers before being pressed into a wafer-thin sheet. When dried this formed a crisp paper.

Papyrus

The catchment area of the River Nile still contains huge areas of swamp in which papyrus is the dominant plant but it no longer grows along the Nile itself north of Malakal. It is confined to Africa except for a northern extension as far as Israel.

Within this range its distribution is restricted to places where the flow is slow or almost non-existent, and where the water level changes only slowly, with an amplitude of less than 3–4 m. Consequently the largest papyrus swamps are found in large, shallow lake systems such as those of the Sudd in southern Sudan, and in the swamps associated with Lakes Kyoga, Kwania and Bisina in Uganda and with Lake Upemba in Zaire. The inflows to both Lakes George and Naivasha enter through papyrus swamps and that of the latter has been studied in some detail. There are areas of papyrus swamp at the edge of many other lakes, including Tana in Ethiopia and Lake Victoria.

In a papyrus swamp this species forms more than 95% of the plant biomass: there are very few associated species. A few shade-tolerant ferns may form an understorey and climbers such as *Ipomoea* species ascend the papyrus stalks to reach the light at the top of the canopy. Approximately 75% of the weight of the papyrus plant is in the stalks and flower heads. The rest is in the rhizomes which form a tangled mat up to 2 m thick which is not very flexible, so wave action or abrupt changes in water level soon break it up, liberating floating islands of papyrus. Such islands are independent of water depth, moving freely during floods and settling on the bottom when the water is shallow.

Each living rhizome at the surface of the mat bears four to eight stalks, known as culms, which normally grow to 3 or 4 m high. Each culm is crowned by a dome-like umbel of up to 1000 fronds bearing small brown flowers. The leaves are reduced to scale-like sheaths around the bottom of the culm so that photosynthesis takes place primarily in the stalks and flower fronds, although the leaves may expand and photosynthesise if the plant is under stress. Each culm lives about 100 days and when it dies its carbon and nitrogen content is transferred to the rhizome and used again to form the next culm. This recycling of nutrients enables papyrus to grow continuously and thus to achieve very high levels of productivity (up to 120 tonnes of carbon per hectare per year) which compare favourably with crops such as maize, which might be grown on drained and reclaimed swamp land but which cannot sustain growth throughout the year. The problem is that papyrus is inedible to humans and it has proved very difficult to find a modern, economic use for it.

Cutting papyrus soon reduces its productivity because nutrients are removed rather than recycled into new growth. However, swamps do usefully mop up excess nutrients draining from the land and also filter out silt. This is amply demonstrated by the fact that the sewage from Kampala entirely disappears within a kilometre as it drains under a swamp along one of the inflows to Lake Victoria. This effect does depend on the size of the swamp in relation to the quantity of nutrients available and also on the rate of water flow. Nutrients accumulated in the swamp during the dry season may be flushed into the lake early in the rainy season. At times of high inflow, the rivers entering Lake Naivasha (and Lake George) flow under the swamps with little alteration to their chemical composition and the boundary between river and lake water can be seen far out in the lake. This is particularly evident in Lake George where the clear water from the swamp contrasts with the pea-soup water of the lake.

In the 'Normal Chad' phase the lake contains varied habitats and large numbers of species. One of the reasons why so many species are able to survive the rigours associated with the fluctuations of this endorheic lake is because the water remains quite fresh, despite the changes. This is partly due to the waters of the Chari being very dilute (conductivity only about 50μS/cm); but even so it would be expected that in a lake with no outflow salts would gradually accumulate, particularly in one subjected to such very high rates of evaporation. There are two principal mechanisms by which salts brought into the lake basin by the river are permanently removed from the water. One is the precipitation of salts in the lake basin itself by both geochemical and biological processes; mineral particles form on the bed of the lake, particularly near the delta of the Chari. Calcium carbonate and calcium bicarbonate are also taken out of the water by plants and molluscs which either form insoluble encrustations or incorporate them into their shells. These processes result in a horizontal variation in chemical composition from the south basin (average conductivity about 150 μS/cm) to the north basin. The water in the north basin has a higher conductivity (average 625 μS/cm) and greater predominance of sodium carbonate than the water entering the south basin. In addition to this removal of salts within the lake, the salt content of the water remains fairly constant during the annual cycle of flood and recession. Much of the land inundated during the flood is gently undulating so that, when the flood waters recede, pools remain behind which become more saline as they dry up. Most of the minerals they contain are precipitated here and are only partially redissolved when the flood waters return. Salts are thus deposited on the floodplain and not returned to the lake itself, so the lake does not become more concentrated and the conductivity of the lake water remains remarkably stable.

Both north and south basins have three main types of habitat: open water, reed islands – areas of rooted plants such as papyrus and reeds growing in shallow water, and archipelagos of real islands which are the tops of drowned sand dunes. The two basins are separated by a huge area of reed islands known as the Grande Barrière. In the 'Normal Chad' phase, towards the end of June, when the water is low, monsoon winds from the south push water from the south basin across the Barrière into the north basin. This northward flow of water is accentuated by the beginning of the flood in August. The flood peaks in October and November. At the same time the level of water in the south basin rises in the south-eastern archipelago.

From 1965–72 the lake level fell by 2 m to almost half its original volume and in 1972 and 1973 the Chari floods were exceptionally low. The Grande Barrière dried up, and the two basins were completely separated. The next flood was bigger, although still below average, but, although it restored almost 'normal' conditions in the south basin, the water was hardly able to penetrate into the now much-reduced north basin because of the dense growth of vegetation which had developed on the Grande Barrière during the dry period. The north basin then completely dried out. A similar situation continued with subsequent rather small floods, none large enough to overcome the barrier and do more than seep into the north basin through the vegetation. The north basin was completely dry by 1975 and the lake entered the 'Lesser Chad' phase, which consists of only the south basin. This received sufficient river water to maintain it in a more or less normal condition but undergoing rather wider fluctuations each year than before its isolation from the north basin.

A lake like Chad is thus a much more dynamic geographical entity than deeper and more stable lakes: sometimes it's there, sometimes it isn't. Similarly, its plant and animal communities will tend to wax and wane following the fortunes of the lake itself. This favours opportunist species and denies the opportunity for the evolution of new ones provided by long-term stability. Thus, even though it is an isolated system with a great variety of habitats, Chad has no counterpart to the rich endemic fish fauna of the deeper lakes in eastern Africa.

Lake Naivasha

Lake Naivasha, in the Eastern Rift Valley of Kenya is also unusual in that it is an apparently endorheic basin which contains fresh water. It was once part of a much larger basin that included the present lakes Nakuru and Elementeita and had an outlet through the Njorowa Gorge. The level of this outlet is now high above the present lakes; all three have no outlets. Both Nakuru and Elementeita now contain highly saline water, yet Lake Naivasha has remained fresh. This is probably because it loses water and salts by seepage through the highly porous volcanic rock which underlies it.

Since the period of generally higher water levels in this part of the Rift Valley there have been several dry phases alternating with wetter periods. The basin now occupied by Lake Naivasha has certainly been completely dry at least once in the relatively recent past, maybe only 300 years ago. Probably because of this it had a very impoverished fauna at the beginning of the twentieth century and its present fauna is almost entirely composed of introductions. These include the North American black bass (*Micropterus salmoides*) and several species of tilapia. The introductions of fish have brought numerous species of fish-eating birds to the lake to add to the variety of other species already associated with the fringing vegetation and the mud exposed by seasonal changes in water level.

There is a narrow fringe of swamp around much of the lakeshore and, with the seasonal fall in water level, large areas of the lake bottom may be exposed and the fringe of papyrus left stranded. The swamp vegetation survives the dry period thanks to the water held in the rhizome mat and the associated sediment. Large areas of bare mud are exposed on the lakeward side of the papyrus and these are rapidly colonised by swamp plants and also by a number of ephemeral semi-terrestrial species. These newly germinated plants provide food for terrestrial animals, both wild and domestic, whose feet churn up the drying mud, encouraging its oxidation and the decomposition of the organic matter it contains. When the water returns, seedlings of plants such as water lilies grow but most of the rest are

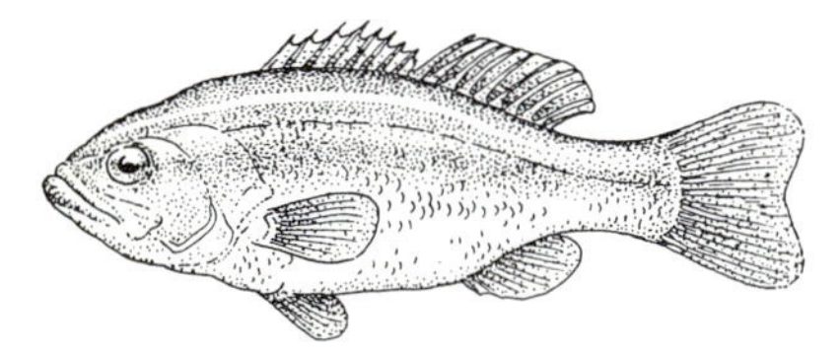

The black bass *Micropterus salmoides* is a common sport fish in North America which has been widely introduced to lakes elsewhere, including Lake Naivasha in Kenya.

The density of culms and total biomass of papyrus estimated for the interior of several swamps (from Thompson, 1979)

Location	Number of culms per m^2	Dry weight of papyrus (kg/m^2)
Lake Upemba Island } Zaire River	6.4	1.11
Lake Mulenda Island } Zaire River	7.7	1.16
Lake George, Akika Island	10.2	2.95
Valley Swamp, Kampala	7.9	2.90
Lake Naivasha, north swamp		3.4

drowned and contribute, along with the dung of the grazers, to a flush of nutrients into the shallow lake water.

Loch Leven

Loch Leven lies just north of Edinburgh (Scotland) and is 13.3 km^2 in area, with several large islands providing ideal nesting grounds for ducks. Although its maximum depth is 23 m, the mean depth is only 3.9 m and much of its area is very shallow. It is surrounded by productive farmland from which run-off loaded with fertilizers has increased the nutrient supply to the lake. The fields also provide grazing for geese in winter and one farm is now a bird reserve. The lake was intensively studied during the International Biological Programme of 1966–72. The studies have continued and demonstrate that, as in many shallow, well-mixed lakes subjected to a very variable climate, events in the lake vary greatly from year to year and are rather unpredictable.

Despite the fact that the loch is privately owned and has been a nature reserve for many years (and a famous trout fishery for much longer), it is still not immune from the effects of twentieth-century changes in its catchment area. Nutrient input has increased markedly, not only from agricultural run-off but also from the sewage of increasing urban development. Blooms of blue-green algae have increased in frequency since 1960 and the number of species of macrophyte, plus the area of macrophyte growth, have declined. Over the same period the nature of the zooplankton changed, from dominance by *Daphnia hyalina* to dominance by the copepod *Cyclops strenuus*. This change was reversed in 1970 and at the same time the phytoplankton changed from being dominated by small species (which can be eaten by *Daphnia*) to predominantly larger species. This may have affected the macrophytes, which have begun to spread again.

Another change was the input of the insecticide dieldrin, used as a moth-proofer by a woollen mill situated on one of the inflowing rivers. *Daphnia* are particularly susceptible to this poison and one might suspect that it was implicated in the changes in the zooplankton, except that the mill stopped using it in 1964 and the *Daphnia* did not reappear until 6 years later. Maybe it took that long for the poison to disappear from the system and for the populations to readjust. Unfortunately, as so often happens, these major changes took place either before, or at the very beginning of, the period of intensive study, so it is impossible to be precise about cause and effects.

The more varied the subdivisions of the habitat, the greater the variety of birds and other animals that can be supported. A lake with extensive areas of swamp, swamp fringes, floating and submerged macrophytes, a variety of substrates on the bottom and varied depths of water will support the greatest number of species. Still more are associated with the interface between the lake water and the land. In addition terrestrial species such as swallows and martins are frequently seen feeding on insects emerging from the lake water.

A study of birds associated with British lakes has shown that the mean number of fish-eating species declines with increasing altitude and it is assumed that lakes at lower altitude tend to be more productive. Only gulls, terns and waders seemed to have more species associated with unproductive lakes. In Britain the communities of birds associated with lakes of all types change markedly between the seasons as many species arrive from further north to overwinter. They come because it is rare for many lakes to freeze over completely for any length of time in our relatively mild maritime climate.

A lake such as Loch Leven thus plays a dual role as habitat for birds: it offers facilities for nesting and feeding in summer, plus feeding and roosting for overwintering species such as geese. Loch Leven is of major international importance as a site for wildfowl and has the largest concentration of breeding ducks in Britain. Mute swans (*Cygnus olor*) and seven species of duck (tufted: *Aythya fuligula*; mallard: *Anas platyrhynchos*; wigeon: *Anas penelope*; gadwall: *Anas strepera*; shelduck: *Tadorna tadorna*; shoveler: *Anas clypeata*; and teal: *Anas crecca*) nest there regularly and some are present all year. Three additional species of duck arrive for the winter, along with whooper swans (*Cygnus cygnus*) and greylag (*Anser anser*), pink-footed (*A. fabalis brachyrhynchus*) and barnacle (*Branta leucopsis*) geese. A total of thirty species of waterfowl was recorded at Loch Leven between 1966 and 1972, plus twenty-three waders (of which six bred), eleven species of gulls and terns (three bred), and sixteen other species of water birds.

Lake Mývatn

Lake Mývatn in Iceland is a good example of a shallow lake which is more productive than might be expected from its northerly location. It is also very famous for the number and variety of its waterfowl populations, particularly nesting ducks. The lake is about 38 km^2, divided into two main basins with a very irregular shoreline and many islands. It is surrounded by extensive areas of boggy ground and many smaller ponds and lakelets which provide ideal breeding grounds for a variety of birds, while the main lake supplies rich food resources during the breeding season. It lies in a very volcanic area which is still highly active and numerous springs, both hot and cold, increase nutrient supplies to the lake. The lake is nowhere more than 4.2 m deep and has a very short retention time, so nutrients are constantly replaced. Although covered by ice for about 190 days each winter, the lake lies in the rain-shadow of the Icelandic ice cap and gets a lot of bright sunshine during the long summer days. These factors result in a high level of primary production by both planktonic algae and the blanket weed (*Cladophora*), which covers large areas of the bottom. This provides a substrate for small invertebrates and thereby takes over the role of macrophytes which are uncommon in Mývatn.

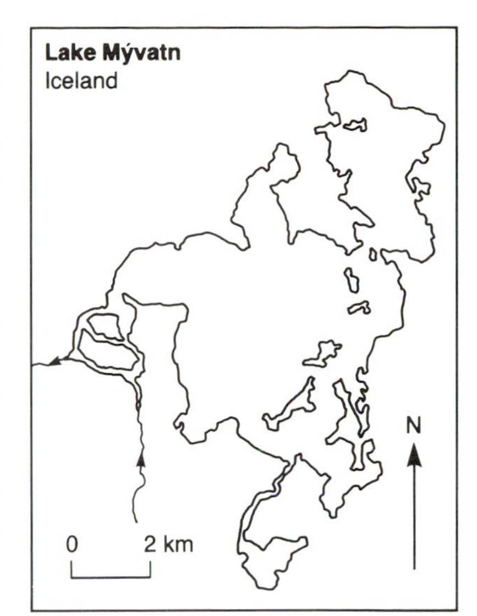

Location: 65° 30′ N; 17° W
Altitude: 277 m
Inflow: groundwater
Lake area: 37 km²
Lake volume: 76.5 million m³
Maximum depth: 4.2 m
Mean depth: 2.0 m
Outflow: River Laxa

As usual in high-latitude lakes (see Chapter 5), the community is not very diverse. Because of the rapidity with which water passes through the lake, the benthic plants (such as *Cladophora*) and animals develop more significant populations than the planktonic species, which tend to be swept away. Important components of the benthos are the small bivalves *Pisidium* and several species of worms. The most spectacular of the invertebrates are the midge larvae (of which there are at least eight important species, dominated by *Tanytarsus gracilentus*) which develop in the bottom mud and then emerge from the water as adults, in vast swarms. From a distance these resemble smoke but at close quarters one sees the air filled with a blizzard of tiny flying insects. Their moulted pupal skins (exuviae) float on the lake surface and are blown by the wind into a scum 10 cm thick along the shore. The zooplankton is dominated by rotifers but benthic Cladocera, such as *Eurycercus lamellatus*, are more significant in terms of biomass and, together with the benthic insect larvae, provide food not only for birds but also for trout, char and sticklebacks. Surplus food from the lake is washed out into the River Laxa, where it is eaten by some of the biggest salmon in Iceland. This river is also a prime site for the harlequin duck (*Histrionicus histrionicus*), an American species that does not regularly occur anywhere else in Europe. It is a diving duck which nests only near flowing turbulent water. It feeds primarily on larvae of the black fly (*Simulium*) which cling to the rocks of the river bed and filter tiny particles from the water flowing out of the lake.

A further twelve species of duck regularly nest at Lake Mývatn. Most of them are migratory and spend about 5 months (from April to October) in the area. Some, such as the common scoter (*Melanitta nigra*) and goosander (*Mergus merganser*), leave even before they have moulted. The mallard (*Anas platyrhynchos*) are partly resident and Barrow's goldeneye (*Bucephala clangula*) almost entirely so. Four of the species, gadwall (*Anas strepera*), shoveler (*Spatula clypeata*), pochard (*Aythya ferina*) and tufted (*A. fuligula*) ducks, are relatively recent immigrants to Iceland: the shoveler and the pochard are still rather rare but the numbers of tufted ducks have increased greatly since the beginning of this century.

In addition to the ducks, many waders breed and feed in the wetland areas surrounding the lake and other aquatic birds such as red-necked phalarope (*Phalaropus lobatus*), Slavonian grebe (*Podiceps auritus), black-headed gull (Larus ridibundus*) and Arctic tern (*Sterna paradisaea*) are abundant on and around Mývatn. Whooper swans (*Cygnus cygnus*) are year-round residents but very few of them breed; they mostly use the lake as a safe feeding and moulting site. They and the greylag geese (*Anser anser*) are herbivores, the former eating aquatic plants and the latter grazing the wetlands adjacent to the lake. A few pairs of black-throated (*Gavia arctica*) and red-throated (*G. stellata*) divers breed on the lake and feed on the fish.

The birds of Mývatn have been protected for centuries and their predators, such as Arctic foxes (*Alopex lagopus*), ravens (*Corvus corax*), gyr falcons (*Falco rusticolus*) and skuas (*Catharacta skua*), have traditionally been killed. This is because the farms around the lake have used duck eggs as a supplementary source of income since earliest times. Written records of the egg harvest go back to 1712. The harvest has always been strictly regulated so that only 'surplus' eggs are taken, that is those laid over and above the average number that usually survive to fledging. Thus four eggs are left in each nest, except in the case of Barrow's goldeneye where the number is five. When breeding is finished the down nest linings are also collected. Since 1900 the annual total of eggs taken has ranged from just under 5000 to nearly 16 000 on one farm alone and suggests that some 40 000 eggs were collected per year from the Mývatn area as a whole. The farms which harvested the greater proportion of the total lie on the shore of the north basin and the eastern shore of the south basin – an indication of where the most favourable and abundant nesting grounds are situated.

On one farm, egg records have been kept for each species for almost 60 years and show that less than 10% came from dabbling ducks while nearly 95% came from diving ducks. Among the latter the records also show the increasing contribution from species of

Changes in the relative proportions of different types of duck found on Lake Mývatn in Iceland. Most of the ducks present are diving species, whose species composition has altered. The proportion of dabbling ducks (which feed in the shallows) has not changed much.

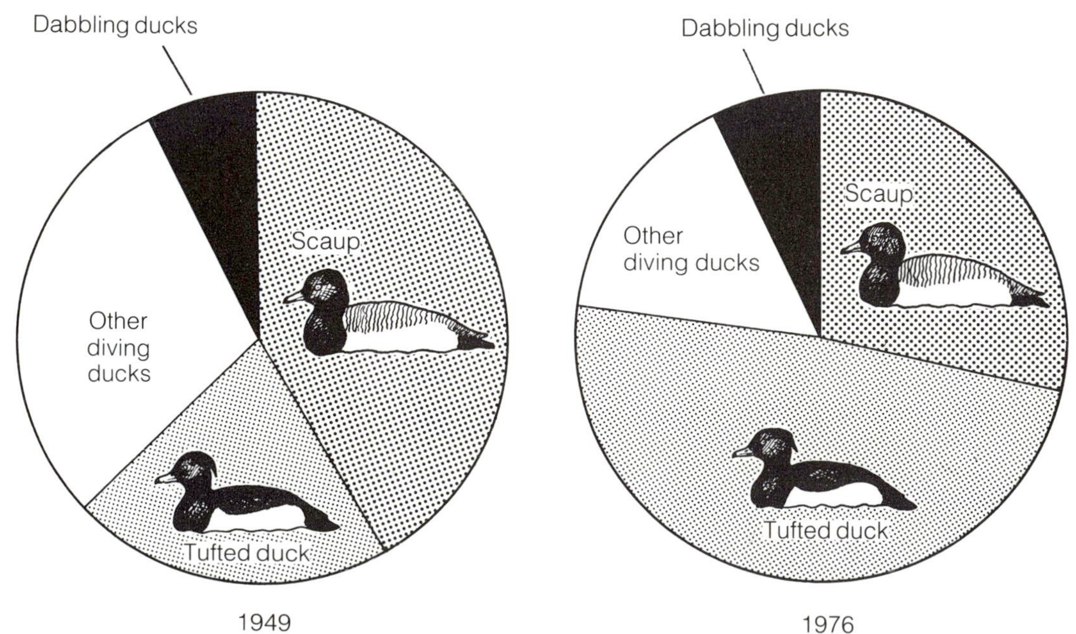

Above *An aerial view of English gravel pits showing lakes of different ages; the varying colours of the lake water indicate differences in depth and in amounts of algae or suspended silt present.*

Right *The mute swan* Cygnus olor *is one of the many species of birds that breeds in the vegetation fringing gravel pit lakes, a substitute habitat for rivers disturbed by boats and dredging.*

Facing Page Top *Gravel excavation in progress in the Thames Valley; the high water table ensures that pits fill with water immediately, but the new lake supports relatively few plants and animals.*

Facing Page Bottom *After a few years the banks and water of a gravel-pit lake are colonised by a rich diversity of wildlife.*

Two contemporary uses of lakes; fishing with a cast net in South-east Asia, and water sports in the United States.

The desolate and blistering hot margin of the Dead Sea; the lowest point on Earth.

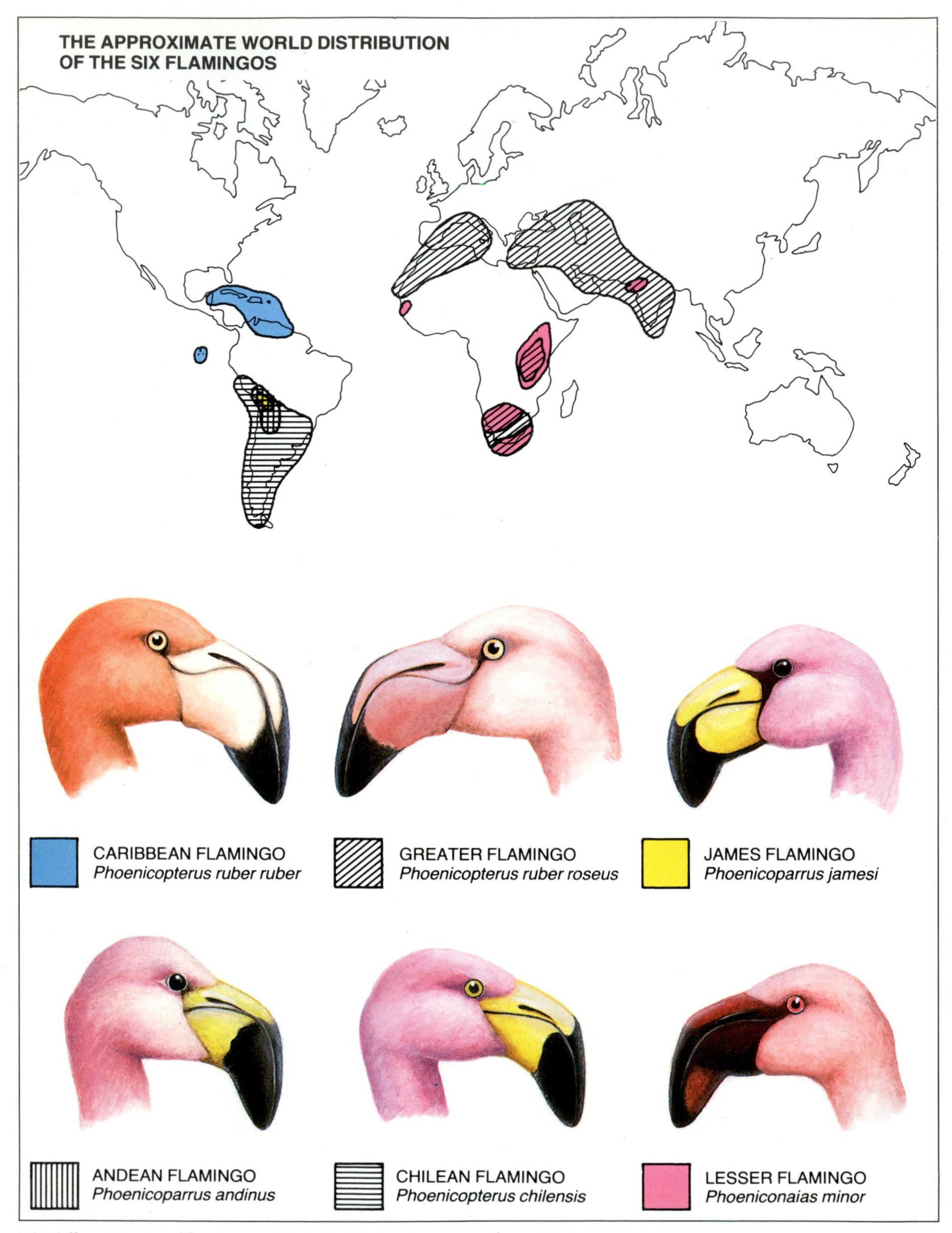

The different species of flamingo and their distribution between various salt lakes around the world.

Top *The construction of dams generates lakes in desert areas where there is otherwise no standing water. This is the Roosevelt Dam in Arizona which controls the supply of irrigation water to the city of Phoenix.*

Above Left *The water hyacinth* Eichhornia crassipes *floats high out of the water on its bulbous stems. Originally from South America it is now widely distributed on tropical waters.*

Above Right *The floating fern* Salvinia *reproduces rapidly to choke the light and life out of ponds and lakes in the tropics.*

Above *The thousands of tourists, both local and international, who visit Lake Nakuru National Park in Kenya make it an economic asset even though nothing is actually harvested from the lake itself.*

Right *Parakrama Samudra is a man-made lake in the dry zone of Sri Lanka built centuries ago to provide irrigation water. It now also supports a fishery.*

Facing Page Top *Chicago is one of many large cities on the Great Lakes; its waterfront on Lake Michigan resembles a seaside resort.*

Facing Page Bottom *If effluent from lakeside cities is not properly treated, fish are among the first and most obvious victims of polluted lake water.*

Tufted ducks

The tufted duck *Aythya fuligula* is a much more familiar sight in western Europe today than it was at the beginning of this century. This is primarily due to its adaptability and to the fact that most recent changes have involved organic enrichment of lakes and waterways. Except in cases of extreme pollution this has resulted in increased primary productivity and consequent increases in populations of the chironomid larvae, worms and molluscs on which the tufted ducks feed. They have also taken advantage of the proliferation of man-made water bodies, such as reservoirs and gravel-pit lakes, which has encouraged their spread.

Tufted ducks range widely through Asia and Europe, and south as far as the Nile Valley of Africa. It is a small, chubby species, 40–47 cm in length and with a wingspan of 67–73 cm. The striking black and white livery of the males makes them quite conspicuous on open water, particularly as they are highly gregarious for most of the year. Their common name refers to the small black tuft of feathers on the head. Shades of brown make the females much less obvious. They form a monogamous pair-bond for the duration of each summer. In some areas a proportion of the population may be resident, but it is primarily a migratory species with most of the population moving southwards and westwards in winter. In parts of Britain the summer population may leave in September to be replaced by different individuals from further north, thus giving the impression that more are resident than is the case.

They are principally found on shallow lakes in the lowlands and seem to avoid lakes of more than 15 m depth unless there are extensive shallow areas. They are diving ducks and must be able to reach the bottom for feeding. They normally feed in areas less than 4 m deep but can dive to 14 m. This enables them to utilise really open areas of the lake away from peripheral vegetation and thus avoid competition with the dabbling ducks.

They nest on land close to the water's edge in the fringing vegetation, preferably on an island. The eggs are greenish brown and are incubated for 25 days. At Loch Leven the average clutch size is 9.5 eggs with an average weight of 55.5 g. Eighty percent of the clutches are started just as the biomass of chironomid larvae in the mud reaches its peak. This ensures that the young ducklings hatch when the number of large chironomids rising to the lake surface is at a maximum. This timing is probably critical because the ducklings initially feed by dabbling at the surface and do not become expert divers until they are older.

Table 7.1 *Mean percentage of species in egg harvest 1901–57 at the farm Grímsstadir, Lake Mývatn (from Gudmundsson, 1979).*

Aythya spp. (tufted duck and pochard)	70.0%
Clangula hyemalis (long-tailed duck)	10.1%
Melanitta nigra (common scoter)	10.1%
Bucephala islandica (Barrow's goldeneye)	1.7%
Mergus serrator (red-breasted merganser)	3.1%
Diving ducks	95.0%
Dabbling ducks	5.0%

Aythya, probably due to an increase in tufted ducks. If the eggs are harvested in proportion to the abundance of the various species then we can assume that the percentage composition of the harvested eggs reflects that of the populations. This assumption may not be wholly accurate, but a direct census of the birds is difficult. Counting ducks on the lake with any degree of reliability is less easy than counting eggs. The best estimates of total population size are obtained by counting males in spring and multiplying by the observed sex ratio to account for females. A systematic survey was made in 1949 and detailed studies in 1975 and 1976. The areas counted were not quite comparable in the surveys so the absolute numbers are difficult to compare for the whole lake, but the percentage compositions can be compared and, as in the egg records, the increase in tufted ducks is evident. Although the total contribution of *Aythya* species has increased, that of the scaup (*A. marila*) has declined and the pochard (*A. ferina*) has remained insignificant. The surveys of 1975 and 1976 showed that in the spring of those years there were about 12 000 ducks in the whole Mývatn area and 7240 and 7986 on the lake itself. In both years more than 3500 tufted ducks represented by far the most abundant species.

The egg harvest records indicate wide fluctuations from year to year in the numbers of ducks nesting at Mývatn and suggest that numbers have reduced overall. This may be due to the fact that the egg harvest has declined in importance as a source of income to the farmers during the last few decades with the increasing modernisation of their farms. The decline is also ascribed to the arrival of mink (*Mustela vison*) in the area during the 1950s. Certainly the most closely packed breeding colonies of ducks disappeared immediately from the islands they formerly occupied and now the densest colonies are found in sedge beds and on small islets in the lake, often in association with black-headed gulls and terns, which are noisy and may succeed in driving off predatory mink. Besides the changes in proportion of the *Aythya* species, the absolute numbers of scaup have declined, as have those of long-tailed duck and scoter. It may well be that in an ecosystem whose productivity is very sensitive to climatic changes, fluctuations in populations are normal and that during the 1970s they were at a particularly low point. It is also possible that these long-term fluctuations in relative abundance are one of the mechanisms which enable so many species to coexist on such a narrow range of food resources.

8

Saline and soda LAKES

All the lakes we have discussed so far contain what is generally called fresh water: true 'fresh' water consists only of hydrogen combined with oxygen but even the purest highland stream contains more than this. Many chemicals dissolve in rainwater as it passes through the atmosphere, and more are picked up as the water washes over plants and rocks and percolates through the ground before running into rivers and lakes. The amount of chemicals dissolved in the water of a lake therefore depends very much on the nature of the rocks through which its inflows pass. It also depends on the balance between the amount of chemicals entering the lake and the amount leaving it.

In most freshwater lakes, water, with its dissolved chemicals, leaves the lake through the outflowing river and the total quantity of outgoing water and chemicals more or less balances what comes in. Some water (a good deal in tropical countries) is also lost through evaporation from the surface of the lake but in this case the dissolved chemicals do not go with it: evaporation removes only pure water and the chemicals are left behind. Where a lake has no outlet, the lake basin is the final destination of all water and chemicals from its catchment area and is called a closed or endorheic basin. If it receives water very rich in chemicals and, moreover, lies in a very warm part of the world, a lot of water will be lost through evaporation and the chemicals will be trapped within the lake, forming a very strong salt or soda solution.

Perhaps the best known examples of this phenomenon are the Dead Sea and the Great Salt Lake (United States), but the largest is the Caspian Sea into which the Volga River flows. The water of the Volga is 'fresh' but the Caspian has no outlet and thus salts accumulate in it. In all these examples the predominant chemical is the chloride ion, either in the form of common salt (sodium chloride), which dominates in sea water, or combined with another common element such as magnesium. A few lakes contain similarly strong solutions of chemicals that are dominated not by chloride but by carbonates. They are usually associated with volcanic regions where inflowing waters contain a high proportion of carbonates and frequently come directly from hot springs. The most spectacular series of these soda lakes is found on the floor of the Rift Valley in East Africa where specialised microscopic plants and animals form the basis of food chains supporting some of the most diverse and abundant communities of birds in the world.

The concentration of salts in these lakes, whether saline or soda, changes with the water supply and this is frequently very variable from year to year, but because they have been accumulating salts over thousands of years they never actually become freshwater lakes, even when heavy rains dilute them. Some scientists draw the line between fresh and saline or soda lakes at a concentration of salts of 3 g/l, others draw it at 5 g/l by comparison, sea water contains a total of 35 g/l. The Caspian Sea, and many other lakes, are quite salty but still support a range of fairly normal plants and animals. However, this chapter is mainly concerned with those lakes in which so much salt has accumulated that the water supports only a few highly specialised species (often not found anywhere else) which sometimes occur in great numbers.

In some situations the concentration of salts becomes so high that the water cannot hold them all in solution at once and the salts begin to crystallise out. In some lakes in dry years the water evaporates completely and only a large expanse of blindingly white crystals remains. Where this happens the salts are often mined or, as at the Dead Sea, the water is evaporated in shallow lagoons and the resulting solid salts collected for export.

Saline and soda lakes occur in many parts of the world. In addition to those already mentioned, there are large groups of them in the

A comparison between the water and chemical balances in a flow-through lake and a closed (endorheic) lake. In a flow-through lake water and chemicals, brought in by the inflow, rain and ground water, are carried away by the outflow. When a lake has no outflow, water is only lost by evaporation; this removes water but leaves behind chemicals brought in from the catchment and these accumulate in the lake. The majority of endorheic lakes therefore contain high concentrations of chemicals such as salt.

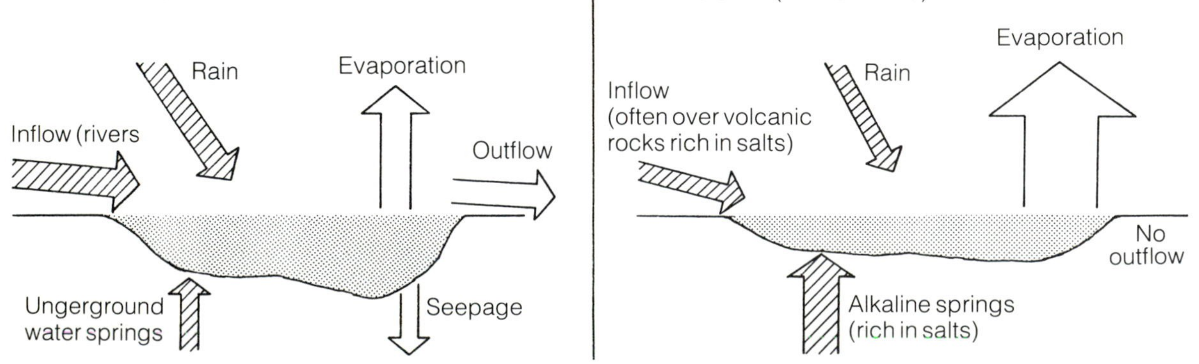

western United States, in the Andes of South America, in Central Asia and in Australia. Most occur in hot, dry places but not all; those of the Andes are very cold. Most but not all are relatively shallow. They are all in closed basins but not all lakes in closed basins have high concentrations of chemicals; we have seen some examples in the previous chapter.

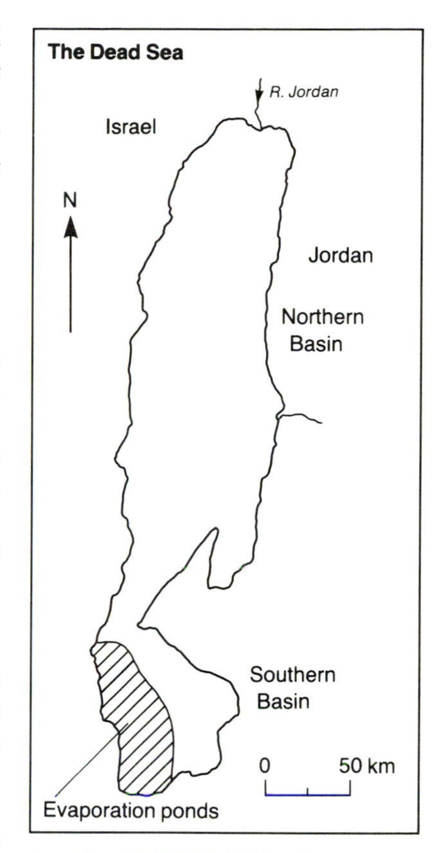

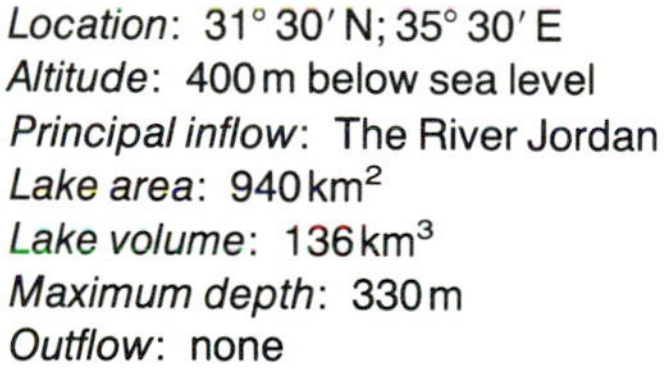

Location: 31° 30′ N; 35° 30′ E
Altitude: 400 m below sea level
Principal inflow: The River Jordan
Lake area: 940 km²
Lake volume: 136 km³
Maximum depth: 330 m
Outflow: none

The Dead Sea

The Dead Sea is the most concentrated natural salt lake in the world. It lies at the lowest point on the earth's land surface, 392 m below sea level, and the bottom of the lake bed is 748 m below sea level. It lies in part of the Syrio-African Rift, which extends for 6500 km from Syria, through the Red Sea, and into Africa as far south as Lake Malawi. The Dead Sea occupies about 1000 km^2 towards the north end of this Rift and steep escarpments rise on either side of the water. The very deep north basin is almost completely divided from the much shallower south basin by a narrow spit.

During the Pliocene (5–2 million years ago) this area was connected to what is now the Mediterranean, whose level was much higher than at present. At that time huge deposits of rock salts were laid down and much of the drainage to the present Dead Sea derives its salt content from these rocks. Some of the rocks remain, like the craggy column known as Lot's wife, the Biblical (as opposed to geological) origin for which is described in *Genesis* **19**.

The main inlet is the River Jordan from the north, whose water is already enriched with chloride but whose discharge into the Dead Sea is now much reduced by upstream abstraction for irrigation. There are some freshwater springs around the shore, such as that at the oasis of Ein Gedi, and there are also some hot saline springs which directly feed the lake. Rainfall is very low (less than 100 mm per year on average) and evaporation removes 1900–2000 mm of water per year. Thus the lake has a decreasing inflow, very low rainfall, no outflow and very high rates of evaporation. It is hardly surprising that the salt water is ten times more concentrated than sea water. The addition of any salts to water increases its density and Dead Sea water contains so much salt that it is also very dense: bathers find that sinking or diving are almost impossible, their body

Table 8.1 *The chemical composition of some saline and soda lakes in comparison with sea water and a freshwater lake (n.d. = no data).*

	Sodium (mg/l)	Potassium (mg/l)	Calcium (mg/l)	Magnesium (mg/l)	Chloride (mg/l)	Carbonate/ bicarbonate (mg/l)
Windermere (= fresh)	5	0.6	7	1	8.6	14.4
Ocean water	10 713	385	406	1 291	19 264	14.3
Great Salt Lake (Utah, USA)	100 667	3 378	346	5 616	112 896	183
Lake Eyre (Australia)	43 780	10	910	300	67 960	40
Lake Bogoria (East Africa)	28 400	387	n.d.	2.2	6 390	57 900
Dead Sea (Middle East)	25 176	5 469	9 876	30 781	149 996	trace
Lake Nakuru (East Africa)	3 300	237	n.d.	0.9	1 020	7 296

is so buoyant. Although the most abundant chemical in this water is chloride, the majority of it is combined not with sodium, as in the sea, but with magnesium as magnesium chloride. Thus whilst ocean water has a pleasant salty tang, the waters of the Dead Sea taste absolutely disgusting.

The concentration of salts in the Dead Sea is about 350 g/l; the water is saturated and simply won't hold any more. Add some salt crystals and they do not dissolve. The accumulation of various salts here has led to a vast natural chemical source, consisting of an estimated 22 million tonnes of magnesium chloride, 2 million tonnes of potassium chloride (potash), 215 million tonnes of calcium sulphate, and plenty of calcium and sodium chloride. There is even some gold, though not enough to be worth extracting.

Amounts like this, and concentrations expressed as grams per litre, are hard to visualise; so, on a visit to the Dead Sea, we tried an experiment. We took a familiar pint milk bottle from home, filled it from the Dead Sea and boiled the water dry on a camping stove. Instead of delicate salt crystals, we got a solid white slab in the bottom of the billycan that required recourse to our hire car's tool kit to break it free. Our pint of water had yielded over 200 g (7 oz) of cement-like salts; put back in the bottle they filled nearly one third of it. So if you have problems visualising 350 g/l, reflect that the Dead Sea's water is actually one third solids. Small wonder that things float in it so well.

Before the inflow of the River Jordan was so reduced, its less-saline water passing into the Dead Sea used to lie over the top of a layer of denser 'fossil water' in the deepest (80–320 m) part of the lake. This fossil water was very much more saline and more dense than the overlying layer and the lake was rarely, if ever, completely mixed. There was a steep chemical gradient at a depth of 30–40 m, forming a transition zone between the two layers. The Dead Sea was thus chemically stratified and the considerable difference in density between these two layers prevented mixing of the whole lake. Each summer this was reinforced by a temperature stratification. With the reduction in the volume of inflow water, particularly since 1964, the level of the lake was lowered by 3.5–4 m and the surface layer of water became progressively more concentrated. This reduced the difference in salinity, and thus density, between the two layers and by 1975 they were very nearly the same. Each winter the temperature of the surface layer dropped from 37°C to at least the 21°C of the deep water and sometimes as low as 18°C. This further reduced the density difference between the two layers and each year allowed mixing of the upper layer to greater and greater depths. This process gradually mixed more and more of the fossil water into the overlying water and in February 1979 the lake was mixed throughout its whole depth for the first time on record. The resulting salinity and density were slightly higher than those of the formerly fossil water and were the same over the whole depth of the lake.

It is hardly surprising that almost nothing lives in this water. The only exceptions are a few extraordinary micro-organisms, including the bacterium *Halobacterium halobium*. Indeed, these bacteria die if the

A halophilic bacterium – *Halobacterium halobium*

Halobacterium halobium is probably the best known species of halophilic (i.e. salt-loving) bacterium. Unlike the alga *Dunaliella*, not only does it tolerate high salt concentrations in the external medium but it positively requires them. Its cells disintegrate if the salt concentration falls below 150 g/l and it grows best at 200–300 g/l. This restriction to high salt concentrations in the environment is due to the fact that the cell enzymes are modified to function at high concentrations of sodium, potassium and chloride which would inactivate the enzymes of most other organisms. Moreover, the lipids and proteins of the cell envelope are quite distinct from those of other bacteria and the primary structure of the cell proteins is unique and coded for in the genes. These protein adaptations are irreversible and are what makes the cells unable to function in lower concentrations of salt, so *Halobacterium* is more restricted in its range of tolerance than *Dunaliella*. The viruses that infect these *Halobacterium* are similarly restricted by their need for high salt concentrations.

water is diluted even to only three times the concentration of sea water. They are distinguished by possessing in their cell membranes patches of a purple pigment, bacteriorhodopsin, which they can use to trap energy from the sun; biotechnologists are seeking ways to exploit these bacteria commercially for desalination and even for generating electricity. Their adaptation to life in highly saline water is described in the box above.

Exploitation of the mineral salts dissolved in the Dead Sea has been going on for many years. By building closed lagoons, the dilution effect of the inflow is cut off and evaporation is even more rapid than in the main lake. Potash, for fertilizers, and bromides are the two main products extracted at present. Each litre of water yields 12 g potash and 4.5 g bromides. Sodium and magnesium are among many other substances that it may be economic to exploit in the future.

Because it is so shallow, the south basin may dry out completely before long but it will be many years before the north basin begins to show the effects of extraction. Nevertheless, it has been suggested by the Israelis that to dig a canal from the Mediterranean to the top of the escarpment above the Dead Sea would not only supply a head of water to generate hydroelectric power but also return the waters of the Dead Sea to their former higher levels. They claim that adding salt water to the Dead Sea could hardly be called pollution.

The Great Salt Lake

The Great Salt Lake in Utah (United States), provides examples of the biological aspects of a salt lake and also of the geographical and biological implications of fluctuating water levels in a shallow, closed lake basin. The lake we see now is the remains of Lake Bonneville which, 14 000–16 000 years ago, covered an area of more than 50 000 km^2 from Idaho in the north to eastern Nevada in the west. This lake was over 300 m deep and drained out to the north via

Changes in the water level of the Great Salt Lake since 1875 when the first gauge was installed at a time of very high water level. Changes in level also indicate major changes in lake area.

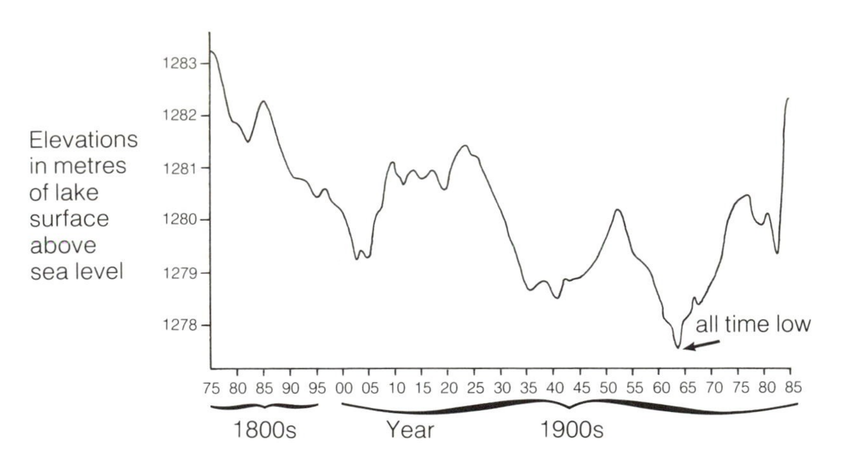

the Snake River. From the end of the last Ice Age the level gradually declined but did so in a series of fluctuating stages which have left very distinctive raised beaches on the surrounding mountains. There is evidence that during the last 10 000 years the lake has been both extremely low, but never completely dry, and also rather high, but never again as high as the level of Lake Bonneville. Throughout this period it has been a completely closed basin. As the volume of Lake Bonneville declined, large deposits of salt were left on the surface of the land and these are still being washed into the remaining lake. The hard, flat Bonneville salt pans were made world famous in the 1930s and again in the 1960s and 1970s by their use as a smooth open 'road' for setting land speed records. The composition of the lake water is rather similar to that of sea water, with sodium chloride the dominant salt, but the actual concentration of salt in the lake at any one time depends entirely on the amount of water in the basin.

When the Mormons (members of the Church of Latter-Day Saints) arrived in the valley of the Great Salt Lake they thought they had found their promised land so they named the main inflow the Jordan River and built their city along its banks. Two other major inflows enter the lake, one in the south basin and the other, the Bear River, in the north-east arm. These rivers drain high mountains to the east which are snow-covered in winter. They provide about two thirds of the total water supply; the rest comes from rainfall on the lake itself and from groundwater. Since there is no outflow, the lake level depends on the balance between inflow and evaporation. During winter and spring, when evaporation is low and inflow high, the lake level rises; during summer intense evaporation causes a decline in level. The average seasonal rise and fall is about 0.6 m. Nevertheless, the water budget is seldom exactly balanced and the lake shows longer-term trends of falling or rising water level. The lake was first surveyed by Fremont in 1843, when he estimated the level of the water surface to be 1280 m above sea level. A great rise in the level prompted the establishment of the first gauge in 1875 but the highest level in historic times (1283 m) had been reached two years previously.

In 1963 the lake level had fallen to an all-time low of 1277 m, the water was extremely concentrated (about ten times the concentration of sea water), and there were fears that the lake would dry up altogether. Indeed in Salt Lake City Museum there is a display entitled 'Can the lake be preserved?'. This outlines a 'Master Plan' drawn up by engineers to ensure the survival of the lake. Several aspects of the rescue plan implicitly recognised that human activity was part of the problem, but further interference was suggested as a solution. Fresh river water would be stored behind dykes, instead of being 'wasted' on mudflats, and would be used to dilute the concentrated lake water. The building of a solid railway causeway in 1959 had cut off the lake's northern arm and, except for two culverts, prevented the natural flow into the north arm of more dilute water from the southern part of the lake (which receives about 95% of the inflow). Consequently, the water in the north arm became more and more concentrated. Further dyking would increase the area thus isolated and this could be exploited for chemicals while the main body of the lake would be reduced to an area whose volume and concentration could be controlled and made suitable for recreation. When the water is very concentrated swimming is a novel experience rather than a pleasure and the salt corrodes boats and other structures as well as damaging human eyes and skin.

If, instead of trying to control every aspect of their natural surroundings, people had paid more attention to the lessons of history, they might have anticipated that the lake would make a comeback without their aid or interference. They would have recognised that the 'death' of the lake was merely the drying phase of a natural long-term cycle. By September 1982 the lake had risen again to 1280 m, the same level as when Fremont first surveyed it. During the following winter exceptional rainfall and snow on the mountains caused the lake to rise by an unprecedented 1.6 m and the following summer it fell by only 15 cm. It rose again the following winter and was still very high in the summer of 1984 (1282.6 m) but was still below the historically highest level, let alone the previously much higher Bonneville levels marked so obviously on the nearby mountains and even within Salt Lake City, where some streets run along the terraces which were former lakeside beaches. It did drop back during summer 1985 and everyone we spoke to that summer seemed to have heaved a sigh of relief that the worst was over, but heavy winter snow in the Wasatch Mountains may ensure that it rises again in 1986. If it reaches 1284.7 m, the local topography dictates that the lake will suddenly expand its area by almost one third, into the Great Salt Lake Desert to the west.

Failure to recognise the long-term nature of lake level fluctuations cost the local population dearly. Because the lake lies in a very shallow basin every small rise and fall in water level causes great changes in the area inundated. As the lake level fell, people bought the land that was exposed, industries were set up to extract chemicals from the lake water, and embankments were built to carry the railway securely across the lake and to carry the Interstate Highway across the salt flats at the south end of the lake. Many of these

Map showing Lake Bonneville and its present day remnant, the Great Salt Lake of Utah. Changes in lake area during recent times are shown within the outline of the ancient Lake Bonneville.

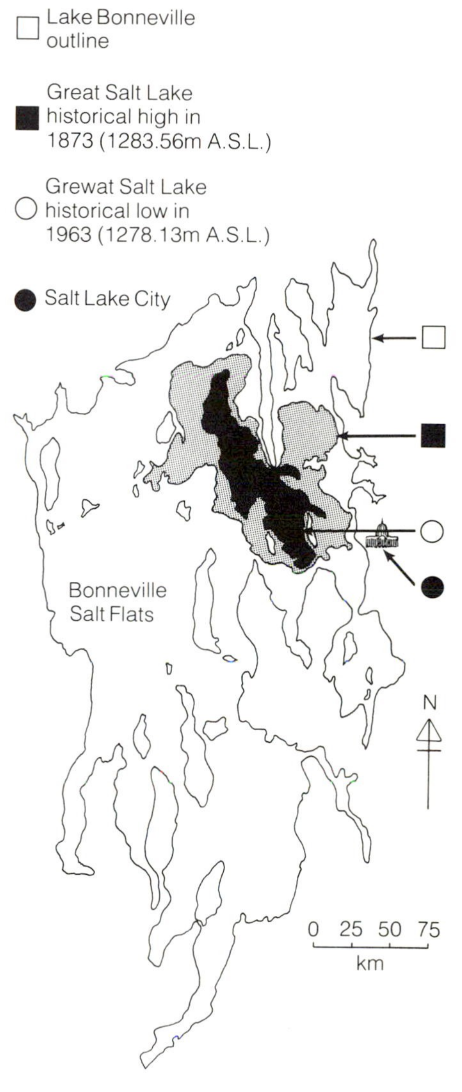

developments have been inundated by the rising water and expensive works have had to be undertaken in order to raise the level of the railway and build protective dykes along the road. It is estimated that the rise from 1281.4 to 1281.7 m (a rise of only a foot) alone cost US$55 million. If the level should reach 1285.6 m, the international airport at the southern end of the lake would also be threatened. The comings and goings of Lake Chad (see Chapter 7) illustrate how the process continues over many years.

There have also been major biological consequences of the dilution of the lake. The brine shrimps (*Artemia*) have all but disappeared, not because they cannot live in the less salty water, but because the dilution appears to have allowed the development of huge populations of predatory waterboatmen (corixids). Large numbers of these insects could be seen in the water of the Great Salt Lake in the summer of 1985 and are known to feed on brine shrimps. When the water is very concentrated the waterboatmen breed in nearby pools of fresh water and fly to the lake to feed but they may possibly breed in the lake when the water is diluted and then greatly reduce the brine shrimp population. Along the shores of Lake Sevier, which is marked on the maps as a dry salt lake bed in southern Utah, we found drifts of dried corixids left behind as the lake water receded due to summer evaporation. It looked as if the population had expanded explosively, with a rich supply of brine shrimp food, and had then died in millions, perhaps of starvation. Such instability is typical of communities composed of only a few species and living in sharply fluctuating environments. Some of the brine shrimp eggs will no doubt continue to hatch and provide food for some corixids but others will remain on the bottom of the lake or dried up along the shore until the water again becomes so concentrated that their predators cannot survive. Then the brine shrimps will stage a comeback.

The mudflats and marshlands at the mouth of the Bear River, in the north-east corner of the Great Salt Lake, provide food, resting and breeding grounds for large numbers of birds, particularly waders and ducks. Dykes have been built to form shallow freshwater lagoons and to prevent the encroachment of saline water from the lake. An enormous area was set aside as the Bear River Migratory Bird Refuge, one of the world's largest bird reserves and a major stopover point for migrants travelling north or south through the desert states of the western United States. With the rise in lake level, some of the increased flow of river water has backed up and greatly increased the depth of water in the lagoons. In many places it is now too deep for waders, and dabbling ducks cannot reach the bottom. On the lakeward side, saline water has inundated the freshwater marshes and killed the dominant *Typha*, thus spoiling the feeding grounds and habitat of innumerable birds.

Serious interference with the lake began with the building of the railway crossing. The first trans-continental railway was completed in 1869 and the final spike, which joined the Central Pacific and Union Pacific Railways, was driven in at Promontory, north of the lake. This route involved a considerable and expensive detour to the north and the railway companies very soon began to examine the possibility of crossing the lake instead of going round it. Using Promontory Point as an anchor, there were 21 km of water to be crossed to the east and 33 km to the west. They started to build solid causeways but on the western stretch, although the water was only 7–10 m deep, the rubble sank deep into the mud. The western causeway was therefore finished off with 19 km of wooden trestle built on piles, each more than 36 m long, driven down through the mud to a firm foundation. This was completed in 1904 and for the next 50 years carried single-line traffic across the Great Salt Lake. Despite constant maintenance and strengthening it became unsafe and was replaced by a solid causeway in 1959.

The causeway which now carries the railway across the lake divides it in two with only two culverts allowing exchange of water between the north arm and main lake. Because 90–95% of the freshwater supply enters the south basin, for most of the time water passes from south to north and thus the salinity of the water in the south is about one third that in the north. Evaporation is intense in summer over the whole lake, since relative humidity is often less than 8%. When water does flow back from north to south, it is such a concentrated solution of salts that it is heavier than the water in the south basin and sinks below it to form a separate layer which cannot mix with the overlying water because of the density difference.

There is no oxygen in this lower layer but there is not very much in the rest of the water either, particularly when the lake level is low: the high temperatures and altitude plus the salt concentration combine to reduce the ability of the water to dissolve oxygen. Its average oxygen content is only about 0.6 mg/l and it rarely reaches 1 mg/l. Normal fresh water, even as warm as 25–30°C can hold as much as 7.5–8.0 mg O_2/l. Around the shore of the north arm in summer a red scum of algae and bacteria gets trapped under the crystallising bank of salt. The algae continue to photosynthesise and produce oxygen which cannot dissolve in the saturated water, so it accumulates as a gas. Gradually the pressure builds up and lifts the crust of salt into a dome which eventually cracks to release several litres of almost pure oxygen.

The summer air temperatures are very high, but the altitude (1280 m above sea level) and mid-continental situation ensure extremely cold winters which result in wide annual fluctuations in the water temperature (−5°C to +30°C, except in the shallow margins where it gets even hotter); these are quite different circumstances from those of the Dead Sea. In any saline lake the salt content of the water lowers the freezing point which is why at −5°C the water of the Great Salt Lake is still liquid and not ice. (The same principle applies when salt is used to stop ice forming on roads.) At low lake levels and with cold winter temperatures below 4°C, sodium sulphate crystallises out of the water to form a jelly-like layer on the bottom of the lake which is pushed by the wind on to the shore. There it dehydrates and forms crystals, but they are redissolved in spring when the temperature begins to rise and rain falls. When lake levels are low, salt crystals form on the surface of the water in the north basin on calm days in summer and are then pushed to the shore by the

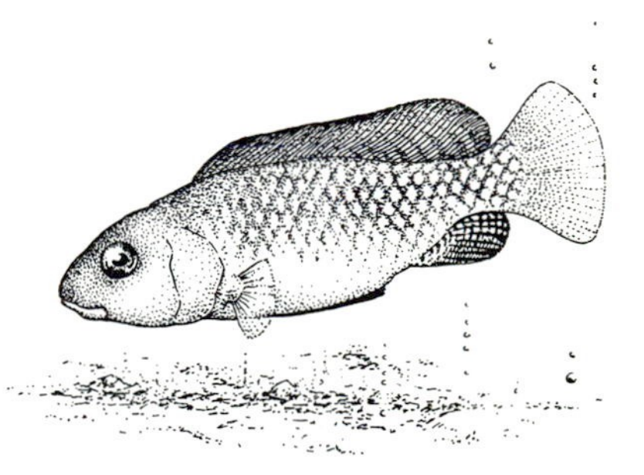

The desert pupfish *Cyprinodon* lives in small pools in the deserts of the western United States. It can live and breed at higher water temperatures than other bony fish and is also tolerant of saline conditions.

A halophilic alga – *Dunaliella salina*

Dunaliella salina has the widest range of salt tolerance known and can grow in solutions from 20 g of salt per litre (dilute sea water) to saturated sodium chloride at 350 g/l; it grows best at 120 g/l. This is possible because it keeps the osmotic pressure of the solution in its cells the same as that outside, not by gaining or losing water, but by altering the concentration of glycerol in the cell. The glycerol does not affect the functioning of the cell; indeed the enzyme function of the alga is thus rendered entirely independent of the salt concentration of its environment. It takes about one hour for the cells to adjust to a new outside concentration, by either synthesising or breaking down glycerol. Healthy cells do not leak glycerol even when the external medium is diluted, but when the cells die their contents provide a substrate for the associated populations of bacteria in the lake.

The glycerol also acts as an anti-freeze and enables *Dunaliella* to function at very low temperatures. The cells are still moving at −3°C and only cease to do so at −18°C. Since most salt lakes occur in warm climates these algae also have to tolerate very high temperatures and this too may be aided by the glycerol.

breeze. They pile up into loose banks of salt which are, at least partially, redissolved when it rains. Salt also crystallises out in the open water and sinks to the bottom, where it becomes compacted into a hard concrete-like layer which slowly gets thicker. It's a wonder that anything at all can live in such conditions.

In fact those species that do live there occur in unbelievable numbers. As with all extreme environments the community is relatively simple, because most things find the conditions intolerable. The food chain is based on bacteria and algae, which are eaten by the brine shrimp *Artemia* and the larvae of the brine fly *Ephydra*. There are a few other species including a fungus, a ciliate protozoan and some viruses.

In fact the waters of both the Great Salt Lake and the Dead Sea contain some of the most specialised living organisms. All living cells contain salts, similar to those in sea water, dissolved in their watery cytoplasm. If the cells are placed in distilled water they absorb water, by osmosis, until they burst, but if they are put into a solution stronger than that inside the cells, then water is drawn out and they shrink. This means that many cells can only live in solutions that are in osmotic equilibrium with their cell contents; but there are some which can regulate the movement of water and salts across the cell membrane, provided that the difference in concentration between the inside and outside is not too great. Even so, there are very few organisms that can live in salt solutions as strong as the waters of the Great Salt Lake, and fewer still in the Dead Sea.

Single-celled algae of the genus *Dunaliella* and bacteria known as *Halobacterium* not only survive but thrive in highly saline lakes all over the world and they are adapted to life in strong brine by two quite different methods (see boxes). Both *Dunaliella salina* and *Halobacterium halobium* are coloured red due to the accumulation of carotenoid pigments. They impart a pinkish tinge to the salt crust at the drying edge of a salt lake. In less concentrated salt solutions *Dunaliella* are green but in the Great Salt Lake these green forms are only found in shaded places such as under debris on the lakeshore.

The special characteristic animal of the salt lake habitat is the brine shrimp *Artemia salina*. It belongs to the group of Crustacea known as fairy shrimps (Anostraca). Most of these are adapted to live in temporary freshwater pools that either dry up or freeze soon after their formation. *Artemia* is found all over the world in waters containing a high concentration of salts in which sodium chloride dominates. Brine shrimps are totally defenceless so they can only live where potential predators cannot survive because of the salt. *Artemia* live quite happily in freshwater in the laboratory but are not found in natural freshwaters where they would be eaten by fish and other predators. Brine shrimps filter the algae and bacteria from the water with their limbs as they swim through the water on their backs. They are sometimes coloured pink, in this case due to the presence of the red blood pigment haemoglobin which helps them to extract oxygen from water in which its concentration is very much reduced by a high salt concentration and warm temperatures.

The most remarkable feature of *Artemia*, and one it shares with the Anostraca of temporary freshwater pools, is the ability of its eggs to hatch after long periods of dryness. People who keep fish in home aquaria sometimes buy *Artemia* eggs and hatch out the larvae as a convenient food for small fish. They are also used for commercial rearing of larval fish. You may even see packets of 'instant life' for sale in novelty shops. This mysterious powder consists of dried *Artemia* eggs in which life will remain suspended for years: you just add water and brine shrimps hatch out.

In fact female brine shrimps have the option of producing two sorts of eggs. Some batches are thin skinned and are held in the egg sac until they are ready to hatch, which takes 5–6 days. At other times the eggs are covered in a hard brown shell and are released to fall to the bottom of the lake. Here they either hatch after a few days into nauplius larvae or they enter a resting state known as diapause. If this happens they will only hatch if they are first dehydrated either by drying or by extremely high salinity in the water. The advantage of such eggs is obvious for animals inhabiting temporary pools or saline waters in either very hot and dry or extremely cold environments. While they are dehydrated the eggs are very resistant to extreme conditions; they can survive heating to more than 100°C and cooling below −190°C: life on the sun-blistered dry mud of lakeshores is easy by comparison. When they are put in water, development of the egg is completed in 24–38 hours and the larvae hatch out.

The brine shrimp industry

In the 1950s an enthusiastic keeper of tropical fish, C.C. Sanders, discovered that brine shrimps made very good food for fish and published an article in *The Aquarium* magazine. This led to the founding of an industry which initially sent frozen adult brine shrimps, harvested from the Great Salt Lake, all over the United States and abroad. The firm also collected the resistant eggs of the brine

A halophilic animal – the brine shrimp *Artemia salina*

To live in temporary or highly saline waters an animal requires some very special adaptations: *Artemia* has them all. It can live in salt solutions ranging from 10 g/l up to 220 g/l (sea water contains 35 g/l). In order to survive *Artemia* must have efficient control over the amount of water and salts in its blood. The skin over most of the body is impermeable to both water and salts and their transfer is confined to limited areas; the skin on specialised parts of the limbs can control the passage of salts and excrete them to the outside in order to keep down the amount in the blood, and the intestines regulate the water content of the blood. In most circumstances the concentration in the blood is less than in the environment and there will be a strong tendency for salts to pass into the body via the gut, and for water to be drawn out by osmosis. Only at quite dilute environmental salinities (about 10 g/l) is the blood more concentrated than the outside solution. Even at salt concentrations as high as 150g/l the blood is still only slightly more concentrated than sea water.

Adult *Artemia* are about 12–15 mm long and clearly divided into segments. Eleven segments immediately behind the head each carry a pair of legs which are flat and leaf-like. Behind these is the long narrow abdomen, the first two segments of which are fused together and in the female carry an egg sac; in the male there is a copulatory organ.

The larvae, like those of many Crustacea, have only two or three body segments and only three pairs of legs. They undergo six moults, increasing the number of segments and legs each time until they look more like the adults. They then have three or four further moults while sexual developments take place. They grow and reach maturity very rapidly, which is another adaptation to life in a precarious environment.

The Anostraca (to which *Artemia* belongs) are sometimes grouped with other Crustacea such as the Cladocera and called branchiopods, which means 'gill feet', because they absorb oxygen from the water through the surface of their legs. Fairy shrimps swim on their back, beating their legs synchronously to move them along. As they do so a continuous stream of water moves over the legs from which oxygen can be extracted; the water also contains the tiny plants and bacteria on which the shrimps feed. So the legs trap food particles with a fringe of bristles known as setae. These form a sieve which is swept through the water with the beating of the legs. The food drops into a groove at the base of the legs and is passed forwards to the mouth. Someone once called *Artemia* the 'beats as it eats as it breathes' animal.

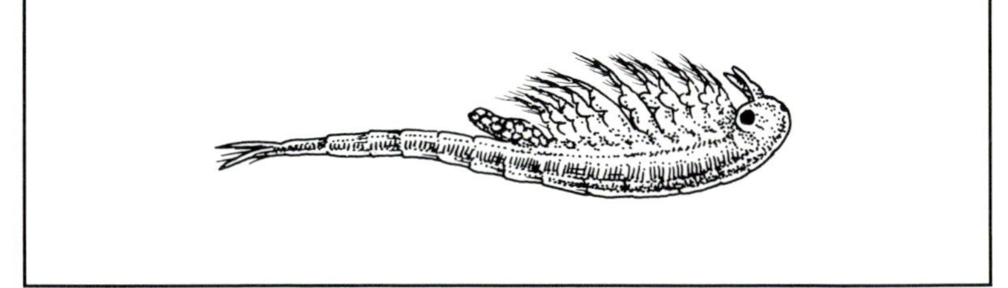

shrimps from the lakeshore, dried them and packed them for shipment. Aquarists could then hatch the eggs and feed the larvae to their fish. The frozen adults soon became unprofitable but the egg-based industry continued.

When the industry started, the lake level was quite high and shrimps and eggs were gathered from the east shore but this location became impractical as the water level sank and the operation moved to harvest eggs along the western shore. Several firms have licences from the Utah Division of Wildlife Resources to collect brine shrimp eggs and individual people can take up to 4.5 kg of animals or eggs per week for non-commercial purposes. This figure gives some idea of the enormous numbers of this one species which live in the lake when conditions are favourable. The commercial harvest, on which royalty was paid, reached 77.2 tonnes in 1965. The eggs are raked and shovelled into piles on the shore and then bagged and hauled to the warehouse in four-wheel drive vehicles. They are dried and stored for a year before being marketed.

This industry is highly dependent on the salinity of the lake and its variations in level. After completion of the railway causeway and isolation of the North Arm, the salinity rose beyond the tolerance of the brine shimps and so the industry had to move. Now the lake level is so high and the water so dilute that the number of brine shrimps is greatly reduced and the brine shrimp business is having a thin time.

Soda lakes of the African Rift Valleys

Eastern Africa is one of the major lakeland areas of the world and contains many lake types. They range from some of the largest and deepest lakes on earth to tiny lakes in volcanic craters, from pure freshwater lakes to concentrated soda, and from ice-covered mountain tarns to waters fed by hot springs that are never cool. There are also a number of small soda lakes, particularly in association with the volcanic explosion craters to the south of the Ruwenzori Mountains. Most of the lakes in the Eastern Rift are very different from the large, deep lakes of the Western Rift, and many of them are highly saline. Unlike the Great Salt Lake, sodium carbonate rather than sodium chloride is the principal salt in most of the African saline lakes. They also contain a great variety of other salts and the total concentration of all salts is most conveniently expressed by the conductivity of the water.

To the north, in Ethiopia, there are two groups of large lakes on the floor of the Rift Valley. The first group, at altitudes ranging from 1558 to 1680 m above sea level, includes Lake Zwai, which is the largest in surface area (654 km^2), the shallowest (maximum depth 8 m) and the least saline, and Lake Shala, the deepest (maximum depth 266 m) and most saline (20 000–30 000 μS/cm). The second Ethiopian group consists of Lakes Abaya and Chamo, which are at lower altitude (1283–5 m) and contain relatively fresh water. Less is known about these larger lakes than about the Bishoftu group of small crater lakes lying further north on the edge of the Rift and more accessible for study from Addis Ababa. These are all roughly circular in outline and have no obvious inflow and no outflow, but they differ in the extent to which they are protected from the wind by the crater walls around them. They vary in depth from Lake Kilotes (maximum depth 6.4 m) to Lake Bishoftu (87 m maximum depth). This group of lakes is also distinguished by its greater altitude of 1870–2000 m, which makes the lakes somewhat cooler.

The soda lake fish *Oreochromis alcalicus grahami* (= *Tilapia grahami*)

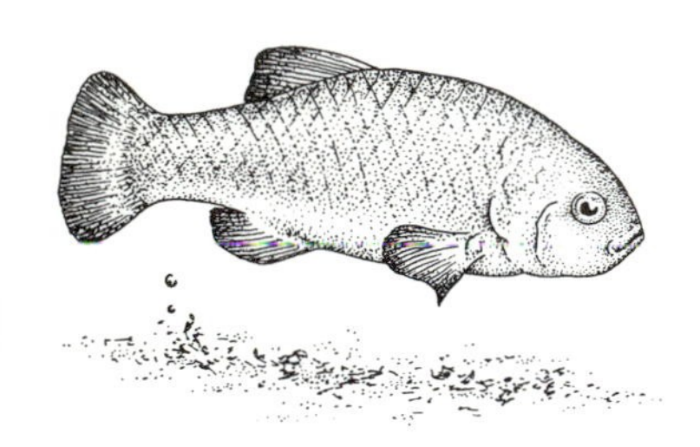

The soda lake fish *Oreochromis alcalicus grahami* formerly called *Tilapia grahami*.

This little cichlid, the smallest of the genus *Oreochromis*, lives in some of the most extreme conditions in which fish have ever been found. It was first discovered in the warm, alkaline springs (pH about 10.5) that run into Lake Magadi. The water has a temperature of 37–45°C and drains into lagoons where it cools to about 26°C. This is still a lethal temperature for many fish but these little *Oreochromis* are tolerant of temperatures up to 40°C. They feed on the blue-green algae that encrust the rocks and gravel of the lagoons and they graze as close as they can to the source of the hot water. A clear browse line marks the position of the temperature (41°C) beyond which they cannot go. The algae are tolerant of even higher temperatures and flourish luxuriantly where the fish cannot reach them. The fish also eat copepods and fly larvae (together about 10% of their diet).

Members of the genus *Oreochromis* are mouth-brooders. The female protects her eggs in her mouth, irrigated by the water that she passes over her own gills. When they hatch, they are released into the surrounding water; *O. alcalicus* is exceptional, among this group of cichlids, in not taking the babies back into the mouth once they have been released. Perhaps this is because there is no need: few if any predators can live in the extreme environment of the springs where they have evolved.

Groups of males in breeding condition each make a shallow pit in an arena on the bottom of a shallow pool. Here each defends his station against rivals until a female is attracted to him for mating. If he leaves to feed, another male takes his place and will defend the pit when he returns. The returning male is nearly always repulsed by the usurper. This may be a mechanism that ensures only the dominant males will breed and only the number for whom there is space in the arena. When the population is dense only a proportion of the males is able to breed because of a shortage of breeding sites. This helps to regulate the population density in relation to the capacity of the rather limited environment to support them.

The East African Lakes Magadi, Natron and Manyara may once have been part of one big lake during a wetter climatic phase. There are similar, closely related fish in springs and pools associated with each of these lakes: *O. alcalicus alcalicus* at Lake Natron, *O.a. grahami* in Lake Magadi and *O. amphimelas* at Lake Manyara. Fossils similar to *O. nilotica,* which is widespread in the less-saline lakes of East Africa, have been found in silt beds about 12 m above the present level of Lake Magadi. Perhaps these were the ancestors of the species and sub-species found in the present saline lakes which are now quite separate from each other. At Lake Magadi the springs and lagoons are also normally isolated from each other but when it rains heavily they are interconnected by shallow, relatively fresh water and the separate populations of fish can mingle and interbreed. Thus they are not likely to evolve separate species in each spring as has happened with other small fish in oases around the Dead Sea and with the pupfish in the western deserts of the United States.

The fact that *O. alcalicus* can survive the inundation of fresh water when it does rain implies that it is very tolerant of sudden changes in temperature and salinity. This must have been a great asset for survival when they were introduced to Lake Nakuru from Magadi. Lake Nakuru is at a higher altitude than Magadi and therefore cooler; the fish have acclimatised to temperatures between 19 and 25°C and to the wide changes in salinity which have occurred in recent years. They grow larger in these conditions too: in Lake Magadi the males may, exceptionally, reach 10 cm in length, in Lake Nakuru they grow as long as 25 cm.

Even further south is the large soda pan of Lake Chew Bahir (formerly Lake Stephanie) which lies to the east of the Omo River as it drains the Ethiopian Highlands and feeds Lake Turkana (formerly Lake Rudolf), the largest lake of the Eastern Rift. South of Turkana, in Kenya, the floor of the Rift is dotted with a string of smaller, shallower lakes, all of which are endorheic and the majority of which contain very saline water; Lakes Baringo, Naivasha (see Chapter 7) and Kitangiri are the only exceptions. The most famous is Lake Nakuru, whose huge populations of flamingos and great variety of other birds have made it the centrepiece of a National Park and one of the greatest wildlife spectacles in the world, earning significant tourist income for Kenya in return for ecological protection. A threat of pollution damage to the lake prompted detailed studies of its biology at the beginning of the 1970s.

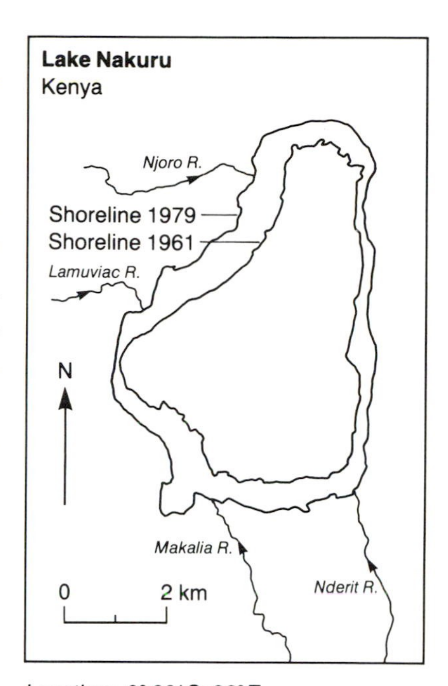

Location: 0° 20′ S; 36° E
Altitude: 1759 m
Catchment area: 1800 km^2
Principal inflows: Njoro and Nderit Rivers
Lake area: about 40 km^2 variable
Lake volume: very variable
Maximum depth: about 2.8 m
Mean depth: very variable
Outflow: none

Lake Nakuru

Lake Nakuru is fed by two perennial rivers, the Nderit and Njoro, and several which flow intermittently, but it has no outflow. The amount of water in the lake and its salinity therefore depend on the balance between four factors: the inflow from these rivers, which is influenced by rainfall on the catchment; the amount of water diverted for human use; input from the alkaline springs along the shores; and evaporative loss from the surface, which is probably of the order of 1.8 m per year. Since rainfall in East Africa shows considerable year-to-year variation, it is hardly surprising that between 1960 and 1980 the lake area varied from 36 to 49 km^2 and its mean depth from less than 0.5 m to 3.5 m; the maximum depth was never more than 5 m during this period.

The chemicals in the lake water are dominated by carbonates and bicarbonates, not chlorides. Their concentration varies depending on the amount of water present, but the lake is always more concentrated than most plants and animals can tolerate. In 1961, for example, the conductivity was 165 000 μS/cm during a very dry period, while in December 1976 it was down to 9 500 μS/cm after heavy rains earlier that year. The plants and animals that do live in Lake Nakuru can withstand very wide fluctuations in salt concentration, but occasionally the changes go beyond even their tolerance and radical changes occur. Fortunately one of these periods of change occurred while scientists were studying the lake and they were able to see what happened. We, too, were fortunate in being able to observe the magnitude of these changes. On a visit to Nakuru at Christmas 1972, enormous flocks of flamingos were massed around the shore and could be viewed closely from a small hide at the north-east corner of the lake. The following Christmas the flamingos were barely visible across a huge expanse of dried mud and there was grass growing around the hide. In 1979 the hide had almost disappeared underwater and it was still barely visible at Christmas 1983.

In its 'normal' condition, the water of Nakuru contains no large plants but a phytoplankton community almost totally dominated by one species of blue-green alga, *Spirulina platensis*, which is found in many soda lakes. This forms the basis of the food chain: it is eaten by a calanoid copepod *Lovenula africana*, a small cichlid fish *Oreochromis alcalicus grahami* and the lesser flamingo *Phoeniconaias minor*. The zooplankton also contains five species of rotifer, of which *Brachionus dimidiatus* is the most abundant. The insects are confined to four species of waterboatmen and two midge larvae which live in the mud. It is a very simple community, again typical of an extreme environment, with very few different kinds of organisms, but each present in great numbers.

This paucity of aquatic species contrasts with the tremendous diversity of birds which exploit the lake's resources. More than 100 species are dependent on the lake itself or its inflows for their food, and over 400 species have been recorded in the National Park as a whole. Besides the lesser flamingo (whose numbers can vary up to 1 500 000) there is also the greater flamingo (*Phoenicopterus ruber*). These are much less numerous than the lesser flamingos and feed mostly on the copepods and midge larvae which they stir out of the mud with their feet. The massed ranks of flamingos form a pink rim around the lake which is constantly ebbing and flowing as they move. Many of the other birds are fish eaters who have only come to

Spirulina platensis

Species of the genus *Spirulina* are blue-green algae (also called blue-green bacteria or Cyanobacteriales, because they are more nearly related to bacteria than to true algae). Like all green plants, they contain chlorophyll which enables them to photosynthesise. As implied by its name, *Spirulina* consists of a filament of cells twisted into a corkscrew shape about 0.2 mm long. It can form gas vacuoles which aid flotation and cause the development of extensive rafts at the water surface.

Spirulina contains an unusually high proportion (up to 65% of dry weight) of protein which is deficient in only one of the essential amino acids and, in at least two quite different parts of the world, Mexico and Chad, local people harvest the thick scum of *Spirulina* from saline lakes to make highly nutritious biscuits. Furthermore, unlike higher plants, the cell contents are not enclosed in cellulose cell walls which most animals cannot digest. This has led to a number of experiments aimed at culturing it in bulk to use the yield either directly or as a supplement to animal feed. Dry weight yields, per m^2, of nearly ten times those of wheat and soyabean have been obtained, and *Spirulina* cultures yield seventy times more protein than wheat and more than ten times that from soyabean. When fed to animals their growth was comparable to when conventional feedstuffs were used and there were no ill effects. *Spirulina* therefore seems a likely candidate in the search for new, high-yielding sources of protein for the future. *Spirulina* has some distinct advantages for this purpose besides its protein content. It grows best in the hottest parts of the world where protein is frequently scarce but the necessary sunshine is not. It also grows in water with a high pH and with such a high concentration of salts that few other organisms could be grown in it instead.

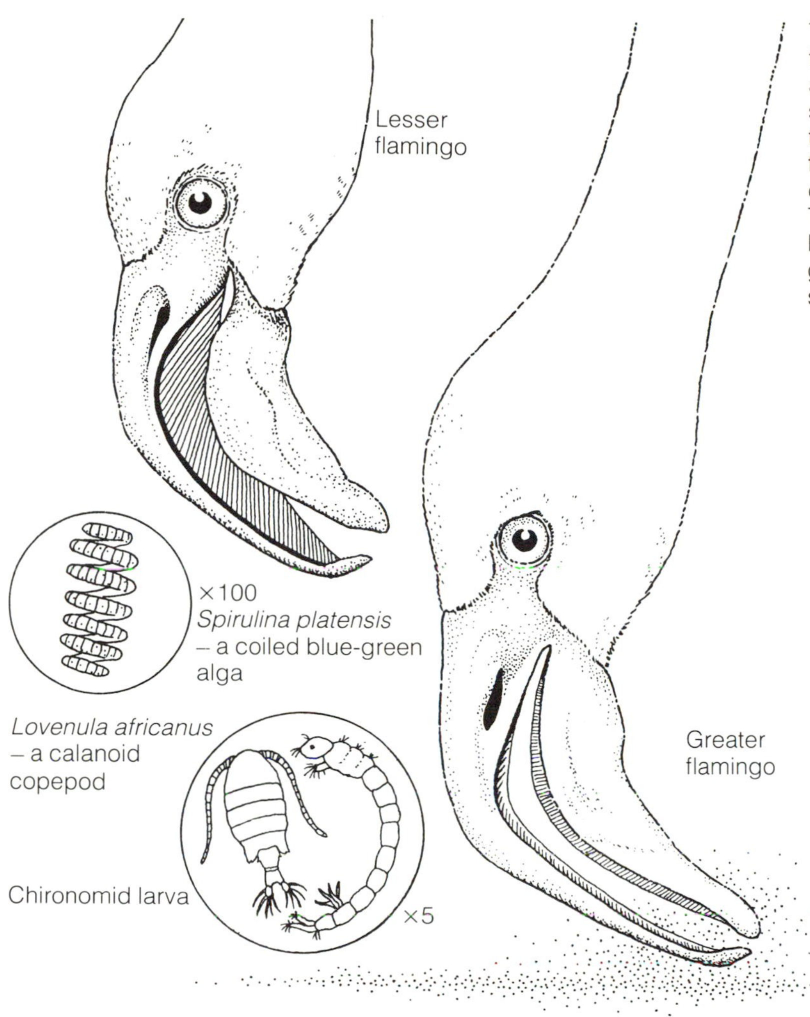

The feeding of greater and lesser flamingos (*Phoenicopterus ruber* and *Phoeniconaias minor*). The finer spacing of the filter in the beak of the lesser flamingo enables it to feed on much smaller organisms compared to the greater flamingo. The former is essentially herbivorous, eating mainly blue-green algae; the latter eats mainly small invertebrates.

Lake Nakuru recently, since the introduction of the fish. These were brought in to control mosquitos in the less-saline areas, initially in 1953 and again in 1962 after the lake had dried up and then refilled. These birds include hundreds of cormorants who nest in trees at the mouth of the Njoro River, anhingas, fish eagles and many species of heron, as well as large numbers of the great white pelican. These huge birds (adults weigh 9–12 kg) sail like majestic ships over the dark green water and rhythmically scoop their bills, often in unison, through the water to catch the fish. They do not breed at Nakuru but prefer the quiet of Lake Elementeita about 10 km away. They nest on a rocky promontory and are safe from disturbance there because Elementeita is surrounded by private rangeland. During the breeding season the parents fly to and fro between Nakuru and Elementeita to fish and feed their young. They consequently export nutrients and biomass from Nakuru to Elementeita and their droppings and those of their young must enrich the water of the latter. Like the flamingos they demonstrate the interdependence of the soda lakes in the Eastern Rift.

Flamingos

Flamingos are the only birds specifically associated with saline and soda lakes and are very specialised for living in these rather peculiar environments.

Fossil evidence shows that flamingos were once more widespread than they are now. Today there are five species, one of which has three sub-species. They all have long, thin legs, webbed feet and long neck, plus a beautiful pink coloration which makes them instantly recognisable.

The most abundant is the lesser flamingo (*Phoeniconaias minor*), whose total population is estimated at about 5 million, mostly living in the Rift Valley of East Africa. Their range overlaps with the eastern sub-species of the greater flamingo (*Phoenicopterus ruber roseus*), which is a much more widespread species. There are large breeding populations in India, Iran, Tunisia and on saline lagoons in the Camargue in Southern France. They also breed, with what is probably a separate population of lesser flamingos, at Etosha Pan in Namibia.

Probably the best-studied population of the greater flamingo is that which breeds in the south of France. This is the most northerly, and the only European, site where flamingos breed frequently, though irregularly. Breeding attempts were made in 36 out of the past 60 years and every year for a period of 17 years (1947–63). At other times there have been gaps of 4 or 5 years between breeding attempts. On average these Camargue colonies have produced about 800 young per year. The colonies are mainly based on the man-made sea-water lagoons (salines) from which water is evaporated during summer and the salt harvested in September. Artificial islands have been constructed for the nesting flamingos; they feed both in the lagoons and on the adjacent, salty marshes. Many of these flamingos fly south to Spain, Sardinia and North Africa towards the end of the year and return between February and June. Those that stay behind take a risk in cold winters such as 1984–85 when there was heavy mortality among them.

The Caribbean flamingo (*Phoenicopterus ruber ruber*) is another sub-species. It nests on the island of Great Inagua in the Bahamas, also in association with salt extraction works, and on Bonaire, off the north coast of South America. There are other smaller nesting groups on the Yucatan coast of Mexico and on the Galapagos Islands. These populations disperse widely in the Caribbean and along the coast of South America outside the breeding season. In 1973 it was estimated that there were about 60 500 of this sub-species.

The other three flamingo species are confined to South America and breed primarily on salt lakes at high altitude in the Andes. The Chilean flamingo (*Phoenicopterus ruber chilensis*) has the greatest range and is the most numerous; the Andean (*Phoenicoparrus andinus*) and James' (*P. jamesi*) flamingos are much more restricted in their occurrence and much less numerous. James' flamingo was actually thought to be extinct until it was rediscovered in 1957; in 1973 its population was estimated at about 50 000.

The conditions under which the South American species live could not present a greater contrast to those of the other two. All three occur, for example, on the Altiplano in northern Chile, most of which is more than 4000 m above sea level. The Chilean and James have nested on the Salar de Surire, the largest (250 km^2) of a series of salt lakes in this area. There are also freshwater lakes, peat bogs and some thermal springs, all of which are used by some of the flamingos. The warm springs provide a refuge when ice forms, even on the saltiest lakes, during the bitter cold at night. However, there are no warm springs near the nesting areas and parent birds must protect their young from icy winds.

At the beginning of 1974 the water level at Lake Nakuru rose so far and the water became so dilute that radical changes occurred in the community of aquatic plants and animals. *Spirulina platensis* became quite scarce and was replaced as the dominant alga by another, smaller species of *Spirulina* which was too small to be retained by the filtration apparatus of the lesser flamingo. Unable to obtain sufficient food, many of them moved away and caused consternation among conservationists all over the world. It was suspected that industrial effluents from the town of Nakuru, which touches the north-east shore of the lake, were causing the changes that had driven away the National Park's greatest asset. The copepods and waterboatmen also disappeared. Because the concentration of algae suspended in the water was much less, more algae were able to grow on the surface of the mud and some flamingos, particularly the greaters, adapted to eating these. The lake community reverted to its former structure during 1976 and it is likely that these events actually resulted from natural changes in the water balance of the lake and its chemical composition. No doubt similar changes will occur again in the natural course of events but this time they did draw attention to the importance of the lake to Kenya's tourist industry, which relies heavily on worldwide interest. International efforts were made to ensure the proper treatment of domestic and industrial effluents from Nakuru town and it must be hoped that these attempts to protect the lake from pollution are sustained because it is certainly very vulnerable: like all endorheic lakes, what goes in does not come out.

Other Rift Valley soda lakes

South along the floor of the Rift Valley from Lakes Nakuru and Elementeita the altitude gets lower and the climate increasingly severe. Lake Magadi lies just north of the Equator at 579 m above sea level and here mean monthly temperatures are over 25°C throughout the year and may reach 31°C at the height of the dry season; maximum temperatures must be a great deal higher. Rainfall only exceeds evaporation from March to the middle of May, so it is small wonder that the lake often does not contain water at all. The sediments of this lake form the second largest expanse of solid trona (sodium carbonate and its derivates) in the world and are commercially exploited for soda ash. Over one million lesser flamingos nested here in 1962 but this was unusual. They normally nest on Lake Natron (900 km^2) which is larger than Lake Magadi (95 km^2) and consists mainly of soda flats covered by a shallow layer of highly alkaline water fed by its one, large inflow river.

The numbers of flamingos at individual lakes fluctuate widely and, since they migrate at night, their movements are difficult to follow. Nevertheless, it is clear that they depend on more than one of the Rift Valley soda lakes for their long-term survival. They certainly feed at Lakes Nakuru, Elementeita and Bogoria, and the same population may even move as far north as Ethiopia in times of need. Until 1954, when the breeding ground at Lake Natron was discovered, no-one knew where the lesser flamingos nested. Smaller numbers of the East

African population of the greater flamingo also breed at Lake Natron.

The flamingos nest gregariously (up to one million pairs together) on mudflats where each pair builds a mound of mud on which to lay a single egg. Both parents participate in incubation of the egg and subsequent care of the young. The downy, grey chicks hatch after 28 days and are dependent on their parents for food until their specialised bills are fully formed. This takes about 65–70 days. They can fly at 70–75 days but until then the growing chicks herd together in large groups, many thousands strong, and trek long distances across the mudflats while their parents are away feeding. Throughout this period the chicks are unprotected from brilliant sunshine on the exposed salt, whose temperature may reach 50°C on the surface. When, in 1962, over a million pairs of this species nested at Lake Magadi, the water evaporated before the chicks could fly and thousands perished with heavy anklets of soda crystals encrusting their legs.

Both these species, like all flamingos, feed with their head inverted, filtering food from the water by means of specialised structures within the bill. Differences in the sizes of their filters lead to differences in what they can eat. The lesser flamingo feeds primarily on algae, particularly *Spirulina platensis* which occurs abundantly in tropical soda lakes. The very small gaps between their filter platelets permit them to feed on organisms which are only 40–200 μm by 20–50 μm in size. They filter about 30 l of water in an hour and spend almost all the daylight hours (12 hours per day on the Equator) feeding. When the population of *Spirulina* is very concentrated this enables the birds to obtain the 66–72 g (dry weight) per day that they need. If conditions in the lake change, as they did at Nakuru in 1974 when the lake level rose markedly and the water became more dilute smaller algae may become dominant. Under such circumstances the lesser flamingos may be unable to obtain sufficient food in a day. This is probably why they moved from Nakuru to other lakes at that time; numbers at Lake Nakuru did not return to their usual level until 1976.

The filter structure in the bill of the greater flamingo is coarser and most algae pass through it, but the copepods swimming in the water and the midge larvae are trapped. These small animals are excluded from the filter of the lesser flamingo by special structures which prevent clogging. Thus, these two flamingo species can feed side by side without competing for the same food. Because the greater flamingo needs to reach the surface of the mud it tends to be restricted to the shoreline where the water is shallow enough for it to reach the bottom, if necessary by up-ending. The lesser flamingo can feed while walking or swimming and is thus able to use the whole area of the lake. The coarse filter of the greater flamingo also enables it to feed on the insect larvae and crustaceans found in British, and other northern estuaries, so zoo escapes often manage to survive for months or even years.

Lake Manyara, like Nakuru, gives its name to a National Park which extends along much of its western shore but in this case does

Flamingos and the Rift Valley lakes

Strung out along the floor of the Rift Valley for over 500 km the soda lakes of East Africa are linked together by a thin pink line of migrating flamingos. Their only regular breeding ground is at Lake Natron and they move between there, and all the other lakes where they feed, in long overnight flights.

Lake Nakuru is protected within a National Park but neither the breeding ground nor other feeding sites enjoy any protection beyond that afforded by their remoteness and the exceedingly inhospitable climate and conditions of their surroundings. If the lesser flamingos are to maintain their spectacular numbers the whole system of soda lakes must continue undisturbed.

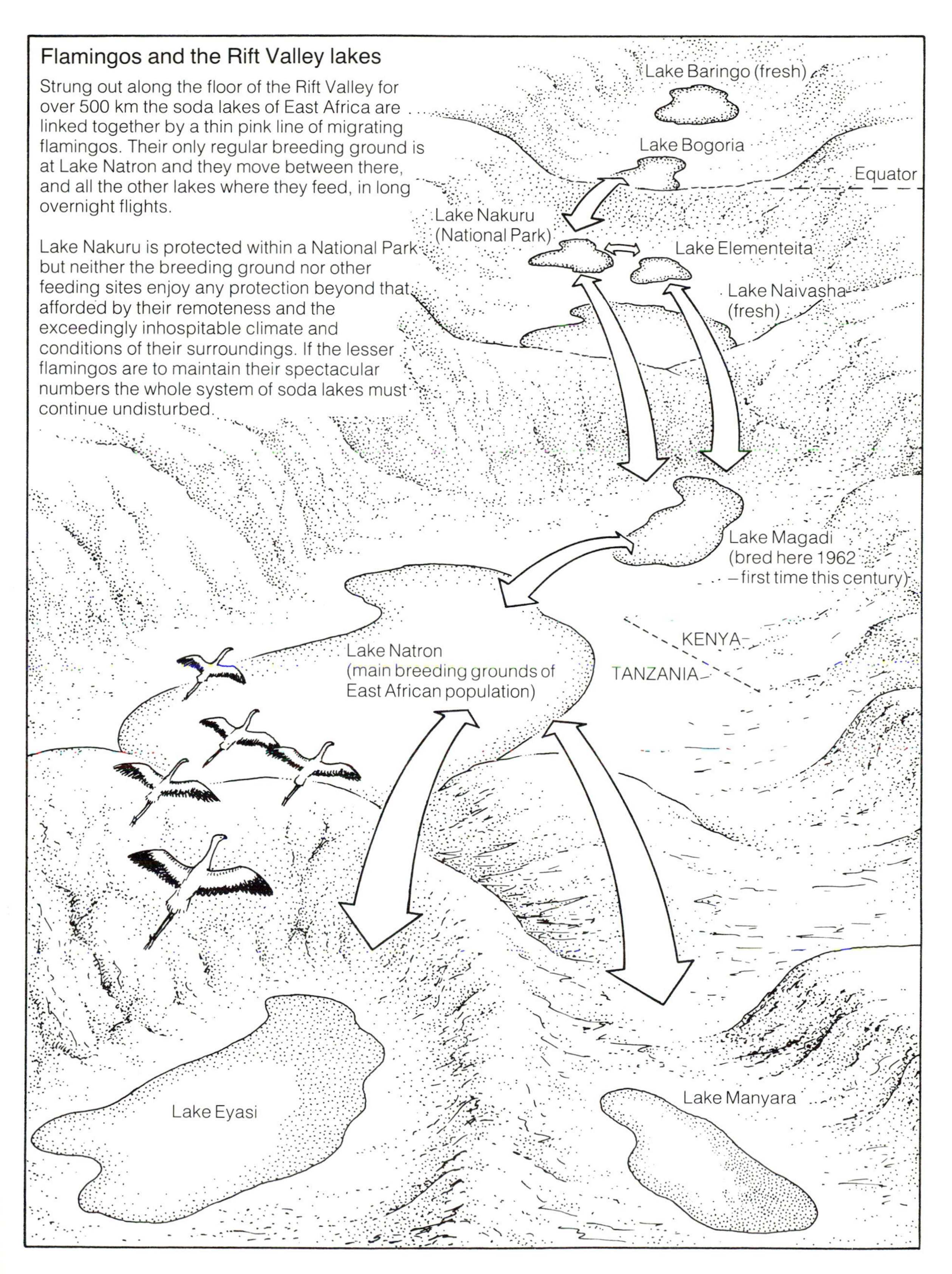

not include the whole lake. The lake water is too concentrated to contain fish except in the springs around its edge, and it does not attract the large number of fish-eating birds that add to the diversity of Lake Nakuru. The lake level varies as in other soda lakes, and influences the vegetation and animals of the Park. When the lake level is low, extensive swards of grassland develop between the water's edge and the forest. These are much favoured by the large herbivores such as buffalo which are forced to retreat into the forest and may be short of food when the grasslands are inundated by high lake levels. On top of the escarpment, above the Park, settlement and forest clearance are increasing. This has led to soil erosion, and the rivers which tumble down the escarpment carry increasing amounts of silt which is deposited as delta fans in the lake. In time this will alter the pattern of interaction between the lake, its shoreline vegetation and the animals that live there.

The most southerly in this string of shallow, saline lakes is Lake Rukwa which lies in an entirely separate rift of its own. The lake occupies two basins: the northern one is so shallow that it sometimes dries out altogether but even the southern basin is nowhere more than 6.5 m deep. As in Lake Nakuru the water is saline, but it is not so concentrated that it cannot support some normal species of plants and animals. In fact the aquatic fauna is much more diverse than that of Nakuru and another major difference is that the phytoplankton is dominated not by blue-green algae but by diatoms.

Lake Rukwa has a similar influence to Lake Manyara on the grazing game animals. Fluctuations in the level of the lake coincide with migrations of the animals into and out of the Rukwa Valley. These patterns of expansion and contraction in the range of large wild herbivores are easily accommodated by the natural ecosystem but may well be disrupted if human settlement occurs in areas that the animals only need at irregular intervals when lake levels are high. The largest breeding colony of great white pelicans (*Pelecanus onocrotalus*) is probably that at Lake Rukwa where they, too, are vulnerable to fluctuations in lake level. When the water is low the nests are left high and dry and accessible to predators; when the water level is high the nests are swamped. Despite these vicissitudes, pelicans seem to be remarkably faithful to their traditional nesting ground although this results in very variable breeding success which would be serious in a less long-lived species. The fluctuations of Lake Rukwa also influence the population of the red locust *Nomadacris septemfasciata*, since the grasslands adjacent to the lake are one of the principal breeding grounds for this species.

9

Man-made LAKES

Some artificial lakes are formed as an incidental consequence of quarrying or mining subsidence but the largest are formed by damming a river valley, creating a reservoir of water behind the dam. In some parts of the world which do not have many natural lakes (such as southern Africa, South America, the Great Plains of North America), the number and size of such reservoirs equal or exceed that of natural water bodies. Most such reservoirs are small and built for local purposes such as watering stock or supplying a small community, but large ones have been constructed to provide water and power for industry. In some cases it may be argued that the big projects were too grandiose, even unnecessary, and were built as 'aid' to a gullible nation. Certainly some of the larger projects have been of major social, economic and ecological consequence. During the last few decades the number and scale of reservoir projects have greatly increased, so that man-made lakes are now significant features of the landscape and their functioning a matter of national and international importance. Big dams are built most often to provide electricity, but frequently to store water for use during periods of low river flow, and sometimes to regulate the flow of a river in order to control flooding. In many countries the lakes held back by dams also have important secondary uses for recreation, while in less-developed countries the development of fisheries as a source of food and employment is an important consideration.

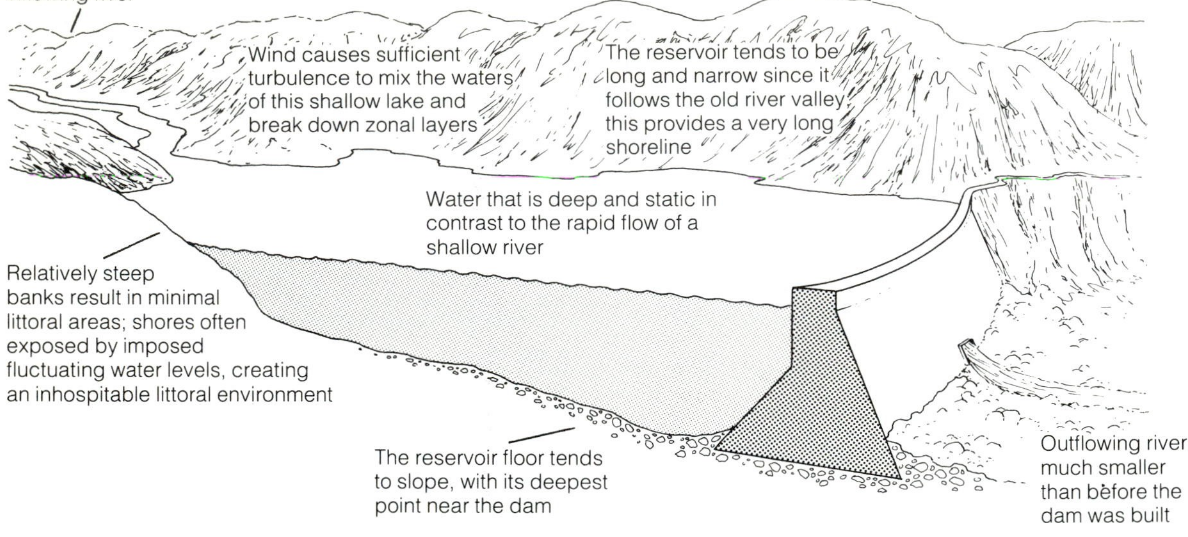

A cross-section of a reservoir lake formed by damming a valley in mountainous country.

Dammed valley reservoirs

It might be assumed that, after the initial filling period, a dammed reservoir behaves just like a natural lake. In many ways this is true, but there are some significant differences between these reservoirs and natural lakes. A reservoir fills a former river valley, and often the tributary valleys too, so it is usually long, narrow and dendritic in shape. This gives it an unusually long shoreline in relation to its area. Reservoirs are frequently fed primarily by one large inflow river and the ratio of the drainage area to the surface area of the lake is usually larger for a reservoir than for natural lakes. Events in the catchment area, then, have a great influence on the reservoir, which responds very rapidly to changes. The retention time of reservoirs is usually much shorter than that of natural lakes because the inflow is such a larger proportion of their volume.

These contrasts are well illustrated by a comparison (see table opposite) between some reservoirs on the Kansas River in the Great Plains of the United States and glacial lakes of similar size in Michigan: glacial lakes also fill valleys, so are particularly suitable for such a comparison. It is also evident from this comparison that the mean and maximum depths of the reservoirs are generally less than those of the natural lakes. In most reservoirs the deepest part is close to the dam wall and the height of the dam sets an upper limit to the depth of the lake. As with natural lakes, the shallower the water and the more exposed the surface area to wind, the more the lake water will be mixed and the less likely it is to stratify for any length of time. This is particularly important to reservoir managers because it will influence the extent to which the bottom waters are likely to become deoxygenated. Since reservoirs are most usually drained from the bottom, rather than the top, this determines the quality of the water supplied both to the users and to the river below the dam. Because the outflow is controlled for human convenience, periods of maximum inflow and outflow rarely coincide. Water is frequently

Table 9.1 *Comparison of Kansas river reservoirs with glacial lakes in Michigan (after Marzolf, 1984, in Taub).*

	Area (km^2)	Drainage area/ surface area	Retention time (days)	Mean depth (m)	Max. depth (m)
Michigan lakes					
Burt	68.5	14.8	379	9.2	22.2
Charlevoix	68.0	10.8	1168	17.0	37.2
Mullet	66.5	9.7	328	11.2	42.4
Higgins	41.7	2.1	5694	14.9	41.2
Walloon	17.3	5.3	1198	8.8	30.5
Douglas	15.1	3.7	1095	5.5	27.1
Gull	8.3	8.0	1500	12.4	33.5
Mean		7.8	1622	11.3	33.4
Kansas reservoirs					
Tuttle Creek	63.9	390	62	8.2	16.8
Milford	63.1	1021	165	5.9	21.3
Perry	49.4	58	166	3.7	11.9
Wilson	36.4	135	912	7.6	16.8
Cedar Bluff	27.8	514	1095	5.8	12.8
Kanapolis	14.3	1422	78	4.5	17.4
Norton	8.8	200	511	3.2	7.0
Mean		507	427	5.5	14.8

let out of the reservoir in copious quantities during dry periods and this leaves large areas of the shore uncovered, often when the plants and animals of the littoral zone are least able to cope with such sudden changes. This also tends to be at times of lowest inflow because that is when there is the greatest need for the stored water.

The building of a dam, and the impounding of water behind it to form a lake, causes not only a transformation of the landscape but also a major upheaval in the ecosystem of the valley. Previously there were slopes covered in terrestrial vegetation from which water drained into a river that carried silt, chemicals and organic matter downhill towards the sea. Now there is a broad expanse of water which flows, sometimes very slowly and sometimes rather faster, down towards the dam and the much diminished river below. The terrestrial plants and animals have been drowned and, less obviously, the river fauna, which was accustomed to fast-flowing, well-oxygenated water, now has to contend with a lake environment.

Some new reservoirs fill within months, others take years. Lake Kainji in Nigeria filled in only 3 months because the basin is small in relation to the enormous inflow of the River Niger. Its retention time is about 3 months. Lake Nasser/Nubia, on the River Nile at the border between Egypt and Sudan, took 10 years to fill up to its final maximum depth of 130 m. This lake is in an almost unvegetated region of the Sahara Desert but most of the big dams built in Africa have flooded large areas of savanna woodland, as at Kariba on the Zambesi, or tropical forest, as in the Volta Lake in Ghana. The drowned vegetation and soils start to decompose immediately and the activity of the bacteria not only releases nutrients into the water but also uses up oxygen. In tropical countries, where the warm water cannot hold much oxygen anyway and warmth encourages high rates of bacterial activity, the water at the bottom of the reservoir quickly becomes devoid of oxygen and hydrogen sulphide may form. Lake Kariba has a distinctly seasonal climate which leads to thermal stratification of the water column and in the first few years the hypolimnion was completely deoxygenated for several months of the year after the dam was closed. There was concern that the hydrogen sulphide formed under these conditions would damage the turbines of the hydroelectric power station, but each year thereafter the period of deoxygenation became shorter as the terrestrial organic matter finally rotted away.

After the experience of Kariba, a similar effect was anticipated when the dam was closed on the Volta River and the lake began to form. Some of the forest had been cleared to facilitate fishing on the future lake but most of it remained. The oxygen levels dropped even more rapidly than expected and within 4 weeks of the dam closing there was less than 16% saturation at the surface and almost no oxygen at 10 m depth.

The majority of reservoirs in Britain have been built in the uplands from which the tree cover has long since been cleared, and where the cool environment does not encourage rapid decomposition, so deoxygenation has not been a problem. It must have been more of a consideration when reservoirs were built in the lowlands, involving the inundation of rich soils and agricultural land. The first of these big lowland reservoirs was Grafham Water and the largest and most recent is Rutland Water.

The problem of deoxygenation was aggravated at Lake Kariba by the explosive growth of the exotic floating fern *Salvinia molesta*, which covered large areas of the lake within a month of the dam closing. Under this floating mat of vegetation the whole depth of the water column was without oxygen because the *Salvinia* blocked out the light and prevented algae producing oxygen by photosynthesis. The floating mat blanketed the surface, preventing wind mixing and

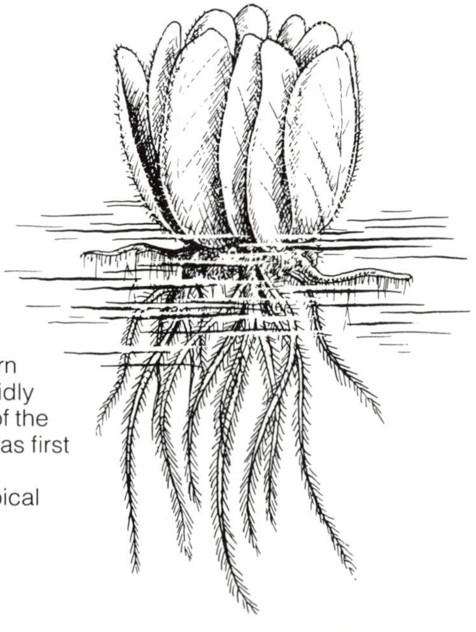

Salvinia is a small floating fern whose explosive growth rapidly covered a large proportion of the surface when Lake Kariba was first formed. It has also proved a nuisance on many other tropical lakes to which it has been introduced.

the entry of oxygen from the air. The *Salvinia* grew most strongly in sheltered bays, just where fish were most likely to feed and breed. Fortunately, the population of *Salvinia* declined naturally over the next four years as the nutrients were gradually used up. This and other species of non-native, floating plants have repeatedly caused havoc on newly created reservoirs all over the tropics. Most of them cannot survive in fast-moving water but human interference provides them with all the right conditions for growth and they are well able to exploit the opportunity.

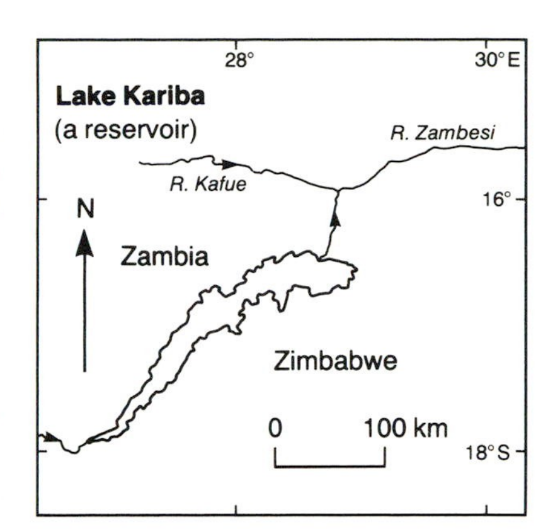

Altitude: 487 m
Catchment area: 663 820 km^2
Principal inflow: River Zambesi
Lake area: 5364 km^2
Lake volume: 156.5 km^3
Maximum depth: 93 m
Mean depth: 29 m
Outflow: River Zambesi

Another typical coloniser of newly created lakes in the tropics is the water hyacinth *Eichhornia crassipes*. This is surely one of the most beautiful of aquatic plants but also probably the world's most pestilential aquatic weed. Its pale mauve flower spike is superficially very similar to the familiar hyacinth grown as a house plant and indeed they are both members of the Order Liliales. There are seven species of *Eichhornia* found in both the Old and the New Worlds but *E. crassipes* is now far and away the most widespread. Its lovely flowers are held upright in the centre of a whorl of large, shiny dark green leaves and the whole floats on the surface while a tangle of roots descend into the water.

The plant originated in slow-flowing and stagnant waters in Brazil. Its international spread began when some specimens were sent to a horticultural exhibition in New Orleans in 1884. They were much admired and prople were prompted to take some for their own ornamental ponds. Hyacinths grow so well that surplus plants soon began to be dumped in the warm waters of the southern United States. In just 6 years it had spread widely from Florida to Texas. In 1895 it was taken to Australia and by 1902 had arrived in India. By 1907 it had already become enough of a menace in Sri Lanka that a special law was enacted in an attempt to prevent its further spread. It appeared in Malaya early this century and in Central Africa during the 1930s. It has now reached even Tahiti and the Solomon Islands.

On every continent, wherever the water was warm enough, the subsequent story has been the same: a phenomenally rapid spread until *Eichhornia* covered the surface of lakes, rivers and swamps, bringing navigation to a halt and making fishing impossible. It took only 3 years to choke 1500 km of the Zaire River. This astonishing spread owes much to the plant's adaptations for rapid reproduction and dispersal. It reproduces vegetatively by sending out horizontal branches called stolons, from which sprout new plants alongside the old. When the stolons break, the daughter plants are liberated and drift away to form new colonies. Seeds are also produced and are shed into the water where they germinate in the mud at the bottom before floating away. The seeds can also lie dormant for years in dry mud at a lakeside or if the water dries up.

The stems are full of tiny air bubbles, and the base of each leaf stem forms a large flotation bladder the size of a golf ball, so plants are very buoyant and float high out of the water. They are thus easily spread by currents, wind and clumsy attempts to destroy them. They can even sail upstream if the wind is strong enough in the right direction. In many parts of the developing world canals and artificial lakes not only provide ideal new habitat for the hyacinth, but are actually designed to prevent the seasonal flushing and scouring of water courses which might have swept away the plants and prevented them from becoming established.

Solid mats of water hyacinth 50 cm thick now blanket thousands of square kilometres of freshwater habitat on four continents. The mass of plants blocks out sunlight so that the water below is permanently dark and therefore unable to sustain the development of algae and the normal populations of plankton and fish. The water then rapidly becomes stagnant and once its oxygen content is used up the hyacinth blanket prevents any more being absorbed from the air. The hyacinths thus strangle the life out of a lake and, by offering a large surface area of leaves to the hot tropical sun, the plants accelerate evaporation of water to the atmosphere. This in turn speeds up the process of turning lakes and ponds into swamps and ultimately dry land; all this in countries which often need every drop of water they can get.

This is particularly true in the Sudan, a huge desert country heavily dependent on the River Nile. *Eichhornia* was first recorded there in 1958 and now infests over 3000 km of the Sudanese White Nile. Fortunately it has not so far spread into the Blue Nile, nor into the Gezira irrigation system, and strenuous efforts are being made to prevent it doing so. Further north on the White Nile it forms an intermittent border, 2–40 m wide, on the outside of the normal fringe of papyrus. In this region there may be 63–73 individual plants per square metre and on average they weigh more than 1000 g each. The tallest shoots can be 89 cm high and the total fresh weight of *Eichhornia* in 1 m^2 of the floating mat may be up to 70 kg. If it grows right across a river or lake, even large ships may be unable to penetrate and it can block the intake screens of hydroelectric power plants.

Apart from its beauty, *Eichhornia* apparently has no redeeming features at all, at least in an ecological context. It is too large for most herbivorous fish to eat and if it is fed to domestic animals they do not like its bitter taste and will not thrive. The plant is even unsuitable for compost because its spongy air-filled tissues mean that vast quantities need to be collected: 20 tonnes of hyacinth yield only 1 tonne of dry fertiliser. The fibre content of the plant is less than 15%, so low that it is of little use for making paper or packaging materials.

At present, efforts are being directed towards trying to control the spread of water hyacinth. In Sudan, traditional manual and mechanical methods of control are no longer adequate. Such methods tend to break up the stolons, allowing individual plants to float away and form new colonies. If the hyacinths are removed and piled on the land, they form huge rotting heaps which are almost as much nuisance as the living plants. In many countries the synthetic hormone-based weed killers (like the notorious 2, ↑ 4-D) have to be used to control water hyacinth. Up to 90% of the plants can be destroyed by spraying the mats of vegetation but such herbicides are very expensive for poor countries, especially since they have to be used repeatedly. There are also serious ecological side effects to their

Table 9.2 *Some of the biggest man-made lakes in Africa*

Lake (river)	Area when full (km^2)	Dam closed	Filling period (years)	Max. depth (m)	Max. drawdown (m)	Annual outflow volume ratio
Kariba (Zambesi)	5300	Dec. 1958	4.5	125	14	1 : 9
Volta (Volta)	8300	May 1964	5	74	3	1 : 4
Nasser/Nubia (Nile)	6216	1964	10	130	20	1 : 2
Kainji (Niger)	1270	Aug. 1968	0.25	60	10	4 : 1
Cabora Bassa (Zambesi)	2739	1974	0.33	151	36	1 : 1

use and the dead hyacinths sink into the water where they rot and use up any oxygen that might have been left: a 'cure' that's almost as bad as the disease! It would be much better if a biological means of controlling this plant could be found. Recent experiments carried out in India, in Florida and Sudan have used moth caterpillars, an aquatic grasshopper, a mite and two species of weevils, all of which attack water hyacinth.

Other experiments aim to use the water hyacinth as a means of cleaning up sewage by removing dissolved chemicals. Sewage, even when filtered and 'purified', still contains nitrates and phosphates which act as fertilisers if the effluent is released into a lake. By allowing the filtered effluent to lie in lagoons where *Eichhornia* is grown, the nutrients are mostly absorbed by the rapidly reproducing hyacinths. The plants can be skimmed off regularly or the water drained out from below. Rid of its excess dissolved nutrients, the water can then be allowed into a river or lake without causing ecological damage. Some of these large floating plants may one day be used in manned space stations as a way of removing chemicals from sewage water, allowing it to be re-used.

In a river most of the water mass is in fairly close proximity to the substrate and the water is oxygenated throughout, so there is little to prevent fish from obtaining invertebrate food from the bottom. When a deep lake is formed, and particularly if most of it contains no oxygen, much of the bottom is put out of bounds. In Lake Kariba the newly formed shallow inshore areas were quickly colonised by chironomid larvae which provided food for fish; additional feeding places were available on the trunks of the inundated trees. This habitat was studied in great detail in the Volta Lake, where it was much more abundant since high forest was flooded rather than savanna woodland. Above the level of the deoxygenated water, the trunks of the trees rapidly became smothered in attached algae which provided a substrate and rich feeding for invertebrates including mayfly nymphs (*Povilla adusta*). This species was present but not common in the original river but soon colonised tree trunks throughout the lake. Nymphs burrowed into the bark and emerged from their holes to graze on the covering of algae, at which time they fell prey to many species of fish. Many other small creatures colonised the drowned trees but the mayfly made up more than 90% of the invertebrate biomass present and formed an important link in the food web of the lake during its formative years.

This sort of colonisation of flooded trees has also been studied in some of the reservoirs on the River Volga in the Soviet Union. There, the dominant colonisers were not mayflies but chironomids, and the density of these animals varied with the extent to which the trees were affected by waves. Numbers were higher on trees exposed to full wave action than on those in more sheltered situations, but the average size of the chironomids was smaller. This was primarily due to a change in the predominant species in these different conditions. There was no equivalent on the trees in the Volga reservoirs to the large burrowing mayfly nymphs found in L. Volta. In all these reservoirs the animals were more abundant in association with the bark than the wood itself; so as the trees began to rot and the bark to disintegrate, their populations decreased as the reservoir aged.

Rotting vegetation not only provides habitat for invertebrate animals but also releases into the water nutrients which can be utilised by algae. It is therefore hardly surprising that the general pattern in newly formed reservoirs has been an initial increase in productivity, followed by a decline as the nutrient store is used up. In many reservoirs this initial increase in productivity has been evident in the form of algal blooms and increased catches of fish.

In many ways it seems surprising that fish catches should increase because species accustomed to a riverine environment have suddenly been presented with a lake habitat and might not be expected to thrive. However the species that benefit from the new environment are those (often the majority of the community in large tropical rivers) which would normally spawn and feed avidly during the flood season, when the river overflows its banks onto the floodplain, covering large areas of semi-terrestrial vegetation with shallow, slow-moving water. Such species particularly benefitted in those reservoirs which filled slowly, because each successive flood covered new areas of land to replace those which had been permanently inundated.

The downstream effects of damming a river

When the gradient of a river suddenly flattens, so that the rate of flow of the water is sharply reduced, the river becomes wider and shallower. At times of high water, as at snow melt or during the rainy season, the river overflows the banks of its main channel and floods out across the flat country on one or both sides. The inundated area is called the floodplain. Within the floodplain area the river may change its course over the centuries and leave oxbow lakes to mark its previous line of flow. Although these and other lakes on the floodplain are isolated from the main river for most of the time, they will periodically receive flood water from the river. Those nearest to the present river channel will be flooded every year but those further away may be flooded only every few years. The ecology of these lakes is closely associated with the frequency of flooding and where people depend on them for water, fish and other resources their lifestyle revolves around the flood regime of the river.

A floodplain upstream of a dam will become inundated forever, but downstream the dam will reduce or eliminate the regular floodplain inundation unless those who now control the outflow from the reservoir make some attempt to simulate the natural flood. Frequently this is contrary to the reasons for building the dam in the first place, which was to store the water that would otherwise be 'wasted' during the flood. One of the prime reasons for controlling rivers is to prevent flooding of settled, agricultural land. Banks are raised and channels deepened for this purpose along rivers such as the Mississippi, the Danube and many great lowland rivers all over the developed world. Control of the Hwang Ho as it crosses the Great Plain of central China was the basis of state organisation in ancient China and, for centuries, defences against the power of the river have involved many thousands of people building dykes. Care of these dykes reflected the strengths and weaknesses of the ruling dynasty: with a weak regime and neglect of the dykes came fearful disasters and enormous loss of life.

Many people still depend on a river flood cycle for their livelihood and, to them, the flood is a blessing not a curse. Floods bring not only fresh water but also silt which, as the current slackens, is deposited on the floodplain and provides new soil when the flood recedes. This rich bounty can be cultivated or left to grow grass for grazing. However, floods, like fire, are valuable servants but disastrous masters. Balanced against the benefit is the fact that use of the floodplain risks enormous loss of property and life if the floods are unexpectedly severe. In a few dry years people set up permanent homes, then just days of monsoon rain may cause a million people to lose their lives or become homeless. A dam offers the prospect of stability and the control of nature's capricious ways.

For millenia the River Nile deposited alluvial soil over the fields of Egypt, creating, with the help of primitive irrigation from the river, the green strip that winds its way through an otherwise parched and brown desert landscape. Most of this deposition has ceased since the building of dams higher up the river. The first dam was built at Aswan in 1902 and its height was raised in 1912 and again in 1933; in Sudan, the Sennar Dam was completed in 1925 and the Jebel Aulia Dam in 1937. Both these dams were subsequently raised by 1 m and 10 cm, respectively, in order to increase their storage capacities. The primary objective of these and other dams built on tributaries of the Nile was to store water that would enable crops such as cotton to be irrigated all year round, even in years of very low river flow. Huge irrigation schemes, such as that in the Gezira region of Sudan, were developed in association with these dams.

The most recent, and probably most famous, of the Nile dams is the High Dam at Aswan, which was completed in 1966. Again, the primary consideration was to ensure water supply at times of low flow, but the other major object in building this dam was the generation of hydroelectric power. The installed capacity for hydro-power generation is only 18% of the potential capacity within the whole Nile Basin, and most of that is due to the High Dam. The lake which formed behind the High Dam is known as Lake Nasser in Egypt and as Lake Nubia in Sudan.

Lake Nasser/Nubia

The High Dam at Aswan on the River Nile is well below the confluence of the White Nile and the Blue Nile at Khartoum. The White Nile drains the high land of East Africa: the most southerly source is in Burundi and flows via the Kagera River into Lake Victoria, out via the Victoria Nile into Lake Mobutu Sese Seko (formerly Lake Albert), then northwards through the Sudd swamps in Southern Sudan and so to Egypt. Another source is the water flowing from the permanent snowfields of the Ruwenzori Mountains into Lake George, in western Uganda, and then to Lake Edward whose outflow, the Semliki also flows into Lake Mobutu. Despite this enormous catchment, the White Nile supplies only 16% of the water entering Egypt. Much has been lost through evapotranspiration in the Sudd swamps and the water is relatively clear because its silt load has also been deposited in the Sudd. Eighty-four percent of the water in the Nile below Khartoum is from the Blue Nile, which drains the highlands of Ethiopia. Despite two dams, the water is still very turbid and its highly seasonal flow generates the Nile flood during July to September. This massive amount of water used to flow directly to the Mediterranean and the silt was deposited seasonally along the banks of the lower Nile and on the delta, where the land increased in height each year. Several metres of recent alluvium lie over coarse sands and gravels. Since the closure of the High Dam, this deposition has more or less ceased and the delta suffers from increased erosion on the seaward side, as the depth of offshore water begins to increase. The river now divides into two main branches but formerly had seven distributaries. How long will it be before the farmers of Egypt find they need expensive fertilisers to maintain the productivity of their farms?

In an average year the Nile flood carries 134 million tonnes of mud, but there is great variation from year to year depending on the rainfall in Ethiopia and consequent flow of the river. Much of the sediment brought down by the Blue Nile now accumulates on the bottom of Lake Nasser/Nubia, gradually filling it up. When the High Dam was first closed, turbid water was carried right up to the Dam and all fractions of the sediment deposited there, but as the lake began to fill the flow rate was decreased further and further to the south, so that the turbid water only reached about half-way along the

The behaviour of silt carried by the River Nile and trapped in Lake Nasser/Nubia upstream of the High Dam at Aswan. (Based on information in Latif, 1984.)

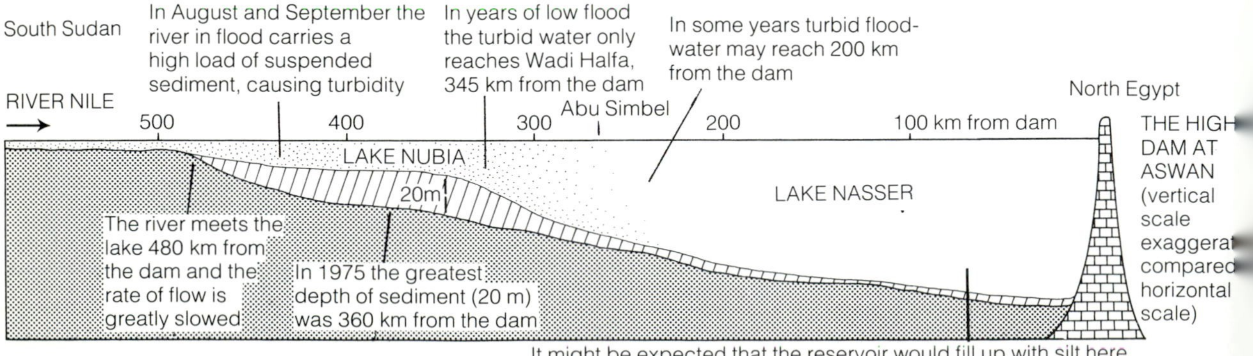

lake: in years of high flow it extends further than in drier years, such as 1972. Already the effect of the Dam on slowing river flow is perceptible 480 km upstream of the Dam itself. As flow rate decreases, the coarser, heavier components of the sediment are deposited first and the lighter parts carried furthest. This results in the early deposition of inorganic particles; the lighter, organic sediment accumulates in the northern parts of the reservoir. In the first ten years about one metre of sediment had accumulated on the bottom of the reservoir at Abu Simbel, but so huge is the volume of the reservoir that it is estimated to have a potential life of 500 years. Nevertheless, one day it will form a large fertile plain – but by then a new dam will be needed to generate electricity!

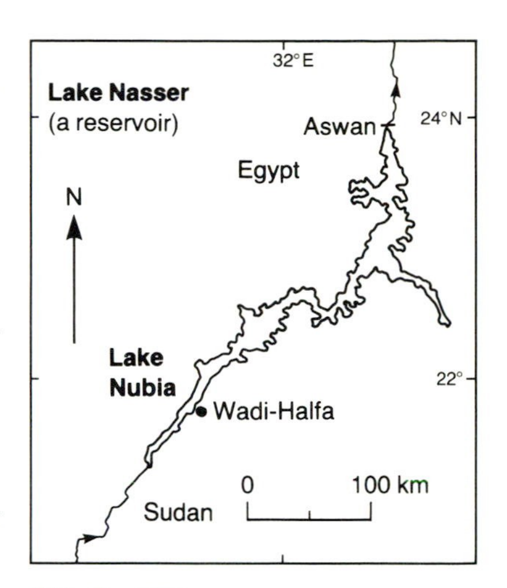

Altitude: 183 m
Catchment area: 2.88 million km^2
Principal inflow: River Nile
Lake area: 6216 km^2
Lake volume: 156.9 km^3
Maximum depth: 130 m
Mean depth: 25.2 m
Outflow: River Nile

Unlike many man-made lakes in the tropics, Lake Nasser/Nubia is in a region almost wholly devoid of vegetation, so the problems of deoxygenation after closure of the dam were minimal. Unlike the relatively shallow reservoirs of the Sennar and Jebel Aulia Dams, Lake Nasser/Nubia is deep enough to have an annual cycle of thermal stratification and mixing. The variation of oxygen with depth follows the thermal stratification and much of the hypolimnion becomes deoxygenated. Stratification of both oxygen and temperature is gradually broken down as the flood arrives in July, starting in the south of the lake.

As in all impoundments a plankton community is able to develop that is quite different, in its dominant components, from that carried by the river. Both the headwater lakes and the upstream reservoirs provide species which develop their populations within the reservoir. The Nile water has a higher conductivity (>200 μS/cm) than many African reservoirs, so it is hardly surprising that the productivity of Lake Nasser/Nubia is also higher. This has led to a gradual increase in catches by the fisheries developed on the lake. They have not experienced the initial boom in catches, followed by a marked decline, that has been recorded from other reservoirs. This may be because of the lack of drowned vegetation and the long filling time of the reservoir (>10 years). Fish such as *Distichodus, Citharinus* and *Bagrus*, which dominated the catches during the early years, are now much less abundant and species such as *Lates niloticus, Oreochromis niloticus* and *Alestes baremose* are now widely distributed throughout the lake. The species populations and distribution within the lake vary seasonally in relation to the flood. The increase in fish catches is an important bonus for both Egypt and Sudan.

The London reservoirs

Many British cities obtain their drinking water from reservoirs in the uplands of Wales or the Pennines. These reservoirs are formed by damming steep-sided river valleys and are small-scale equivalents of the large 'dam lakes' just described. They gather water from largely undeveloped catchments and contain unproductive water which needs little treatment before it can be supplied to the human population. However, London is too far from the upland to be supplied from reservoirs of this type, and instead it has been necessary to build

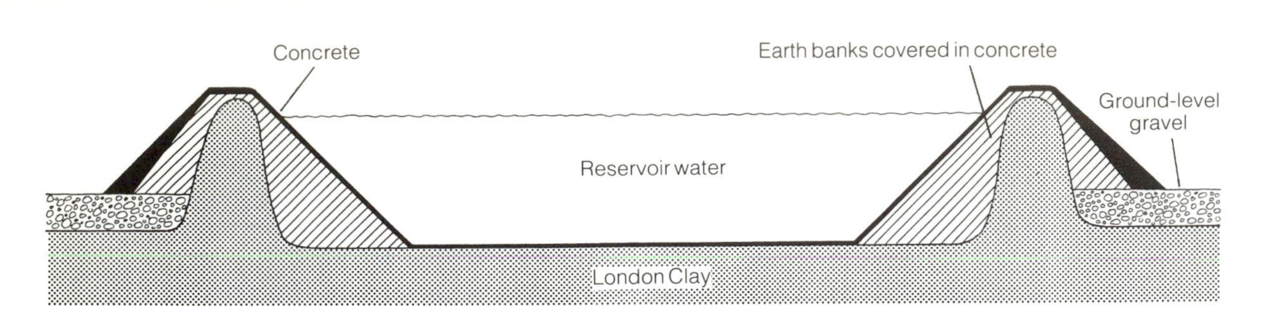

The construction of the Thames Valley reservoirs, in which water from the river is stored as the first stage in its purification for drinking.

artificial lakes on flat, low land to the west and north of London. These act as storage lagoons and are filled from the River Thames and its tributary the Lea. These London reservoirs are vital to the nation's capital, and also of biological interest, being the largest water bodies in South-East England. They are artificial lakes, of course, but also carefully managed to ensure continuous availability of clean water. The reservoir management is very much concerned with details of the chemistry and biology of the water they contain. The Thames is a lowland river and throughout its length runs through rich farmland and several major towns. Fertilisers run off the fields and the towns put their treated sewage into the river; the flow of the river is slowed by weirs and locks which were built to aid navigation but also allow planktonic algae time to take advantage of the nutrients supplied by farms and sewage works and build up their populations. This all results in a very productive water, containing a high load of suspended particles, both living and dead, which must be purified before it reaches the taps in London households.

The first stage in the process of purification is to pump the river water into a reservoir. The London reservoirs were made not by damming the river but as wholly artificial basins constructed on the floodplain close to the river. Water purification was developed by the Metropolitan Water Board (now part of the Thames Water Authority) before anywhere else in the world, and before London had expanded into the countryside, land could be bought cheaply for the construction of storage reservoirs. On each reservoir site the superficial gravel was excavated down to the level of the London Clay. Some of this was dug out and piled up to form a high bank which extended above the surface of the ground almost as far as it went down below it. The gravel was then put back on to these banks to create a slope of 45° from the top down into the basin and on the outside. The tops of the banks were capped with concrete to form clay-lined, water-tight basins, well above the level of the river and surrounding land. The oldest reservoirs, in the Lea Valley, are 10–12 m deep. Later reservoirs were built in the valley of the Thames itself, to the west of London, and the most recent are 17–20 m deep.

Reservoirs more than about 10 m deep become thermally stratified in summer, just like natural lakes, and, because the water is so productive, the hypolimnion becomes deoxygenated. Water can be drawn off from several levels. Because the upper layers are full of algae, which block the filters of the treatment works, the engineers would prefer to take off water from lower down, but they do not want it deoxygenated, as that too would harm the filters. Experimental mixing of the water during summer, by pumping from the hypolimnion to the surface, was tried in order to break down the stratification. The success of this scheme led to a major innovation in the design of the more recent reservoirs at the construction stage. The Queen Elizabeth II Reservoir, opened in 1963, was the first in which the river water was pumped into the basin through special jets, which caused continuous circulation of the water in the reservoir basin and prevented stratification becoming established. This also made it worthwhile for this and subsequent reservoirs to be deeper and thus able to store more water, an important consideration when land is now so expensive close to the city.

Water flows through the basin at such a speed that the average drop of water stays there for 40 days. During this time much of the particulate matter coming in from the river settles on the bottom of the reservoir and any harmful bacteria which have survived the sewage works upstream die. However, impounding the water in these reservoirs has just the same effect as damming a river, in that it slows the rate of flow and gives the plankton (both plants and animals) time to develop instead of being swept on towards the sea. Moreover, the simple structure of these basins results in very little variety of habitats for plants and animals. Almost all the water is the same depth, the substrate is the same all over the bottom, and the concrete sides are smooth and straight so there are no aquatic macrophytes and only a very limited community of benthic animals. The dominant habitat is the open water and, in those reservoirs where the temperature and oxygen are artificially kept the same throughout the water mass, the algae and zooplankton are also evenly spread out across the reservoir. This was shown by taking a large number of samples (49) with a net hauled from the bottom to the top of the water all over the reservoir, in July 1971, and weighing the mass (after drying) of animals sieved out of the water. The average was 130 mg dry weight per vertical net haul and all the results fell within 25% above or below this. A much greater variation would be expected in an unmixed lake because the algae and animals would tend to be clumped in certain parts of the lake (particularly downwind) and more spread out in other parts.

Table 9.3 *Changes in quantity and nature of particles suspended in the water when river water is impounded in a reservoir (from Duncan, 1975).*

	River	QE II Reservoir	
		1970	1972
Total suspended material (mg C/m^3)	*c.* 2500	570	520
Percentage composition by weight			
Non-living organic matter and bacteria	68	54	63
Phytoplankton	32	26	14
Zooplankton	1	20	21
Daphnia		18.2	18.9

Annual cycle of changes in the water of a London reservoir (based on information in Steel, 1975).

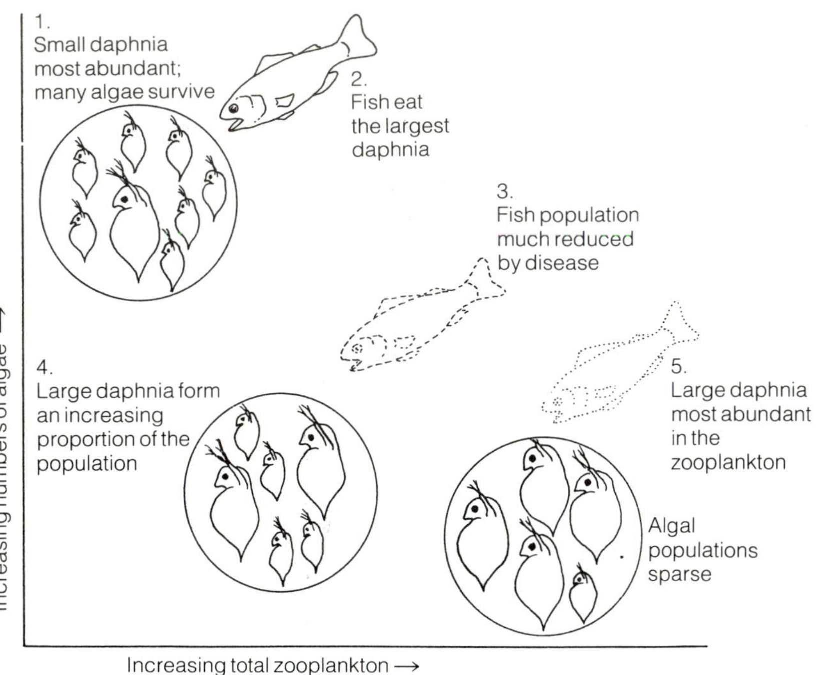

The zooplankton is usually dominated by large populations of very few species, particularly *Daphnia* and during the 1960s and 1970s some very intriguing changes were observed in the Queen Elizabeth II Reservoir. Initially there were three species: *Daphnia hyalina* (the smallest and most numerous), *D. pulex* and *D. magna* (the largest). Gradually the balance changed until *D. magna* was most abundant and, because of its large size, the total biomass of zooplankton greatly increased. In parallel with these changes in the zooplankton the abundance of algae had been declining and it seemed likely that the larger and now more abundant species *D. magna* was consuming far greater quantities of algae than the previous populations of *D. hyalina*. In fact it could be calculated that in July 1972 the population of *Daphnia* present in the reservoir was capable of filtering one third of the volume of the reservoir each day and, presumably, consuming most of the algae it contained. This was of

Changes in the quantity of algae (measured as Chlorophyll concentration) and in the populations of *Daphnia* spp. which fed on them in the Queen Elizabeth II Reservoir between 1968 and 1973. (Based on Steel, 1975.)

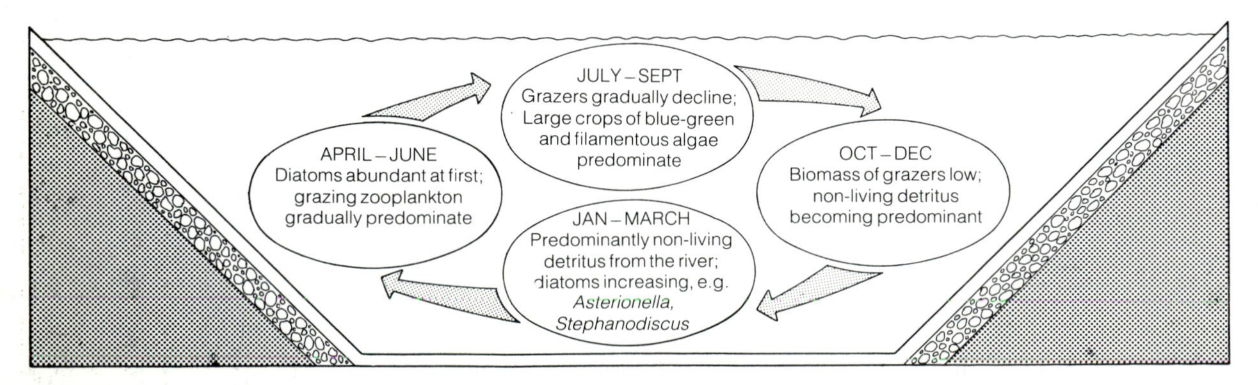

considerable significance to the water engineers: after it leaves the reservoir the water is filtered through two sets of sand filters and these rapidly become clogged when the water is full of algae. Cleaning the filters is costly, so any reduction in the quantity of algae in the reservoir saves money at the treatment works. The increased number of zooplankton would also tend to block the filters but not so quickly.

These changes were probably due to the fact that during the 1960s many fish were killed by disease in the south of England. Plankton-eating fish tend to pick out the largest and most obvious prey items first and thus it is probable that when fish are abundant in the reservoir the *D. magna* are preferentially preyed upon and not able to build up large populations. When the fish population was reduced, the larger *Daphnia* were released from predation pressure and were then able to out-compete their smaller relatives, their larger body size enabling them to filter out a wider size range of algal cells. As the fish populations gradually recovered, so the plankton began to revert to its former composition. There is another interesting practical problem here – should managers attempt to remove fish and drastically reduce their population to protect the plankton as a cheap and natural way of removing algae from the water? This would upset anglers, but if it was left to them to remove the fish, they would mainly catch the big ones, not the youngsters which are most numerous and mainly responsible for eating the plankton.

Gravel-pit lakes

We usually think of derelict industrial sites as blots on the landscape but in the case of gravel pits these 'wastelands' can often be of greater interest than the land all around or that which they replace. In time, the pits become colonised by plants and animals, many of them apparently finding the peculiar characteristics of gravel pits exactly what they need. Dry-worked pits and areas of open sand are attractive to primary colonisers of open, disturbed ground (itself a scarce and transient habitat). Some species will take advantage of the exposed gravelly substrate even while excavation is in progress: plants such as rosebay willowherb (*Epilobium*), ragwort (*Senecio jacobaea*) and poppy (*Papaver*) will spring up on bare gravel. The little ringed plover (*Charadrius dubius*) is a species which nests on bare sand and gravel. It bred for the first time in Britain in 1938, but steadily increased to over 250 pairs in 30 years, three quarters of them nesting in sand and gravel pits. Sand martins (known in America as bank swallows *Riparia riparia*) are also colonisers of freshly exposed sand, but they use the banks not the ground. Once the pits flood and develop a fish and invertebrate fauna, they support increasing numbers of animals – frogs, newts, water voles, mink and even otters.

Birds seem to benefit most, with a clear succession over time. Even a small pit of 8 ha may support more than twenty bird species. Open-water birds like ducks are the most obvious. There are different kinds of ducks, each with its own feeding specialisation and thus

The great crested grebe

The bird that seems to have gained most from gravel-pit lakes in Britain is the great crested grebe (*Podiceps cristata*). This was not a common British bird and persecution for highly prized feathers (for trimming coats and hats) reduced the total population to less than a hundred birds. Full legal protection was granted in 1870 but despite this the population had risen to only 2700 birds by 1930; similar numbers were present throughout the 1940's, implying that this was probably about the full number that Britain's natural water bodies could support. After the Second World War there was an 80% increase in gravel digging and a 70% increase in crested grebe numbers. The population increased to 4500 birds in 1965 and by 1975 had reached about 7000. The great crested grebe is now a common species. It seems that the grebes find the depth of gravel-pit lakes convenient for fishing; they seek small fish that are abundantly available and most pits are just about the right size to provide the 2–3 ha of open water that are needed for the territory of a single pair of grebes, without a lot of bother from territorial incursions by the neighbours.

suited to a particular part of the pit at a particular time. For example, dabbling ducks like the mallard like to feed at the surface or by up-ending to search the bottom; they do best in shallow water around the edges. Diving ducks (e.g. tufted duck and pochard) increase in numbers on the more mature pits where there has been time for a rich bottom fauna to develop: there is no point in diving to feed in newly flooded pits because there will be little to eat at the bottom. While mallard can feed in about 30 cm of water and swans can reach the bottom at greater depth, the diving species manage in the deeper parts.

The development of a dense fringe of reeds and similar emergent vegetation will provide additional cover for shy species like teal and many birds other than ducks, such as rails (*Rallus*), bitterns (*Botaurus*) and warblers (*Muscicapidae*). Reeds also form a complex habitat, diversifying the physical structure of the lakeshore. As reeds

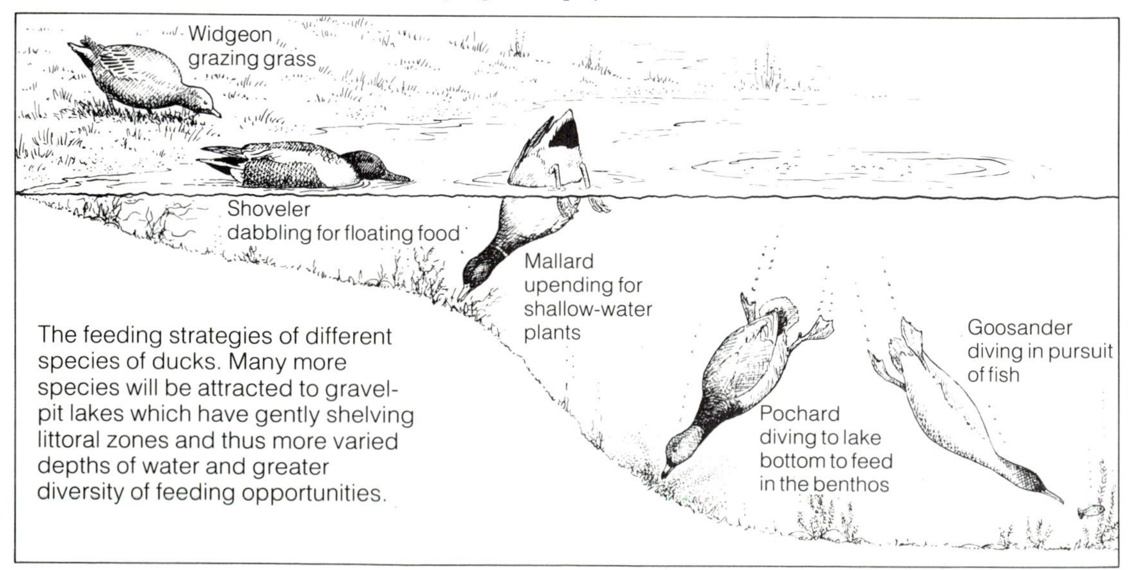

The feeding strategies of different species of ducks. Many more species will be attracted to gravel-pit lakes which have gently shelving littoral zones and thus more varied depths of water and greater diversity of feeding opportunities.

develop, so the number of bird species increases. On a mature pit about one third of the breeding birds will be reed-dwelling warblers and buntings. Reeds also provide shelter in winter for insects and birds, especially from the chilling effects of the wind. The value of the reeds cannot be overstated, and in time gravel pits may become one of the principal refuges of reed-habitat species as they are lost from natural water bodies elsewhere.

In contrast to reeds, which once established tend to spread, the habitat provided by open mud rarely persists for long, so waders such as snipe (*Gallinago*) tend to be abundant early in the life of a gravel pit, but disappear as the water deepens. Passing visitors, such as migrating black terns, are often seen over gravel pits and reservoirs in spring and autumn, swooping for insects on and over the surface. Gravel pits are not just attractive in summer; many waterfowl breeding in the Arctic and northern Soviet Union do not migrate south because they would then remain in areas of harsh, mid-continental winters. Instead they migrate south-west or west to the milder maritime climate of Britain where our gravel pits and reservoirs offer winter retreats. The Canada goose (*Branta canadensis*) and many ducks remain in summer, during which they moult their flight feathers. For a while they are completely flightless and need to remain hidden or safe out in the open water. Gravel-pit lakes provide for both strategies.

The suitability of the new habitat for wildlife can be enhanced by speeding up the natural development of the flora by appropriate planting and also by ensuring the greatest possible diversity of physical conditions. For example, if various ducks like to feed in different depths of water it is valuable to bulldoze waste earth into the pit to form shallows (where the macrophytes will also develop more quickly). If islands are made, these will be attractive and secure nest sites. Promontories increase the length of total shoreline; bays and curved islands form shelter away from the wind and waves and make good 'loafing spots' for ducks.

Further studies have shown that ducklings need a diet high in animal protein, difficult to find in newly established gravel pits. So waste straw, finely chopped, is added to the water to supply food and substrate for benthic invertebrates to grow and diversify: more food for the growing ducks. Rafts offer safe nest sites for terns and even perhaps shy species like grebes (*Podiceps*) and divers (*Gaviidae*). The advantage of rafts is that they do not need the valuable heap of gravel or earth that underlies an island, they also go up and down with the water level so that birds can nest at the water's edge but without the risk of floods rising and drowning the nests.

Much of the pioneer work on this sort of cosmetic treatment for old pits was developed by Redland Aggregates and by wildfowlers on gravel pits near Sevenoaks in Kent. Their success can be seen from the increase in numbers of 'waterfowl days' (number of birds × number of days sighted): 6000+ in 1956–57, rising to 73 000 in 1969. In the same time, 284 breeding pairs of forty bird species increased to 1069 pairs of fifty-seven species. The lessons learnt about ecological landscaping for waterfowl have now been put into practice at many other sites. Here they help improve the consequences of necessary gravel exploitation. Habitat restoration also improves the public image of gravel-digging companies and offers a way of responsible long-term use of sites left by transient mineral extraction. In some places, restored gravel pits have become so rich as wildlife habitats that they are now carefully protected as nature reserves.

The majority of gravel-pit lakes in South-East England are less than 30 years old. Frequently they are also rather small and shallow. Studies on eight such lakes situated in two groups of four, just 4 km apart, in the valley of the River Blackwater (a tributary of the Thames) show just how different they can be, even when so close together and similar in size and age. One of the most obvious variable features is the amount of aquatic plants found in each lake. The emergent plants occur around the shore wherever there is a shallow littoral zone, however restricted; common reed, reedmace, iris and many others form a narrow fringe around all or part of the shoreline. The extent of their development depends upon the shape of the lake basin, and hence the amount of shallow water. The variation in the submerged and floating plant species is more difficult to explain. Some lakes have none at all, some have extensive development of Canadian pondweed (*Elodea*), some have dense isolated patches of water lilies (the exotic *Nymphaea* in one case, the native *Nuphar* in another) and yet another has lilies scattered over almost the whole of its surface; by late summer two are almost filled to the surface with hornwort (*Ceratophyllum*) but only one of these develops dense mats of blanket weed on the surface. The lilies tend to persist but the submerged macrophytes show great variation from year to year, in both species composition and abundance.

The emergent and floating plants greatly enhance the visual attractiveness of the lakes and are important for increasing habitat diversity of the site. The submerged species play an important role in supporting plankton and young fish populations, both of crucial importance to anglers.

New gravel pits provide habitat for anglers, relieving pressure on natural waterside places and helping to satisfy the enormous demand for this activity. More people in Britain go fishing than participate in any other sport and the majority of them are fishing for what are known as 'coarse fish'. This includes species such as roach (*Rutilus rutilus*), perch (*Perca fluviatilis*), tench (*Tinca tinca*), bream (*Abramis brama*), carp (usually *Cyprinus carpio*) and pike (*Esox lucius*), all of which occur in lowland rivers and still waters rather than the cold, clear waters where salmon and trout ('sport fish') are found. A survey of thirty-nine gravel-pit lakes in the south-east of England in 1972 found that roach occurred in thirty-one, perch in twenty-nine, pike in twenty-four and tench in twenty-three of these lakes. Altogether, seventeen species were recorded, indicating the diversity of gravel-pit fish communities, but this included small species such as sticklebacks (*Gasterosteus aculeatus*) which are of little interest to anglers. It is probable that most of these fish colonised the lakes naturally but carp, extremely popular with anglers because they grow very large (the British record is 20 kg), are often deliberately

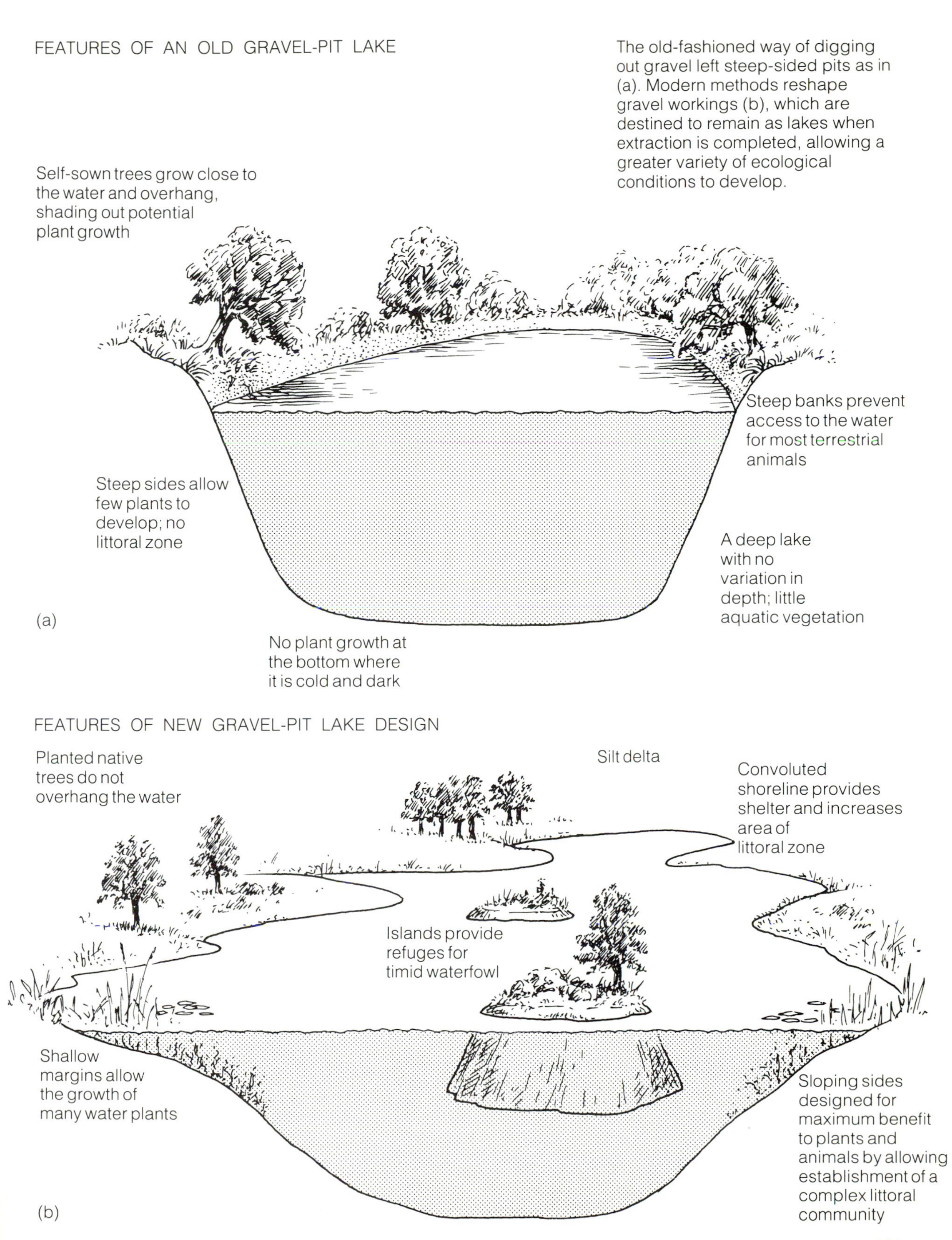
FEATURES OF AN OLD GRAVEL-PIT LAKE
The old-fashioned way of digging out gravel left steep-sided pits as in (a). Modern methods reshape gravel workings (b), which are destined to remain as lakes when extraction is completed, allowing a greater variety of ecological conditions to develop.
Self-sown trees grow close to the water and overhang, shading out potential plant growth
Steep banks prevent access to the water for most terrestrial animals
Steep sides allow few plants to develop; no littoral zone
A deep lake with no variation in depth; little aquatic vegetation
(a)
No plant growth at the bottom where it is cold and dark
FEATURES OF NEW GRAVEL-PIT LAKE DESIGN
Planted native trees do not overhang the water
Silt delta
Convoluted shoreline provides shelter and increases area of littoral zone
Islands provide refuges for timid waterfowl
Shallow margins allow the growth of many water plants
Sloping sides designed for maximum benefit to plants and animals by allowing establishment of a complex littoral community
(b)

stocked into gravel-pit lakes where fishing competitions are held. Some specimens are caught repeatedly throughout their long lives.

Adult carp and tench feed primarily on animals that live in the mud at the bottom of the lake, but most species of fish feed on zooplankton such as *Daphnia* during their first few weeks or months of life and many of them select the largest and most obvious individuals first. In the gravel-pit lakes, large species of Cladocera such as *Sida crystalina* and *Simocephalus vetulus* do not occur in the open water but are found only in the weed beds, where the fish are less likely to get them. Other species such as *Daphnia* are on average larger among the macrophytes than in the open water. Moreover, numbers of these larger forms in the weed beds remain high throughout the summer whereas in the open water they decline quite rapidly as the young fish eat them all. The macrophytes provide essential refuges to enable the Cladocera to survive intense predation pressure and to breed to supply more individuals to replace the stock of fish food in the open water. Without this essential refuge, the larger Cladocera would be wiped out completely. In turn this would mean that the lake would be less able to support young fish and maintain the species that anglers want. Many fish lay their eggs on the underwater parts of plants and the newly hatched fry rely on the dense protective macrophyte nursery to shelter them from bigger predatory fish such as perch and pike. So, even though water weeds are a confounded nuisance to fishermen, especially when they tangle lines and trap fish hooks, the plants are an essential part of a good fishing water.

Weed beds also harbour larger invertebrates, such as insect larvae and snails, which are important in the diet of some older fish. These are particularly significant in new lakes which have not yet developed a rich benthic fauna for the fish to feed on. But anglers do not like water weeds and try to drag them out to increase the areas of open water. Fishery managers therefore need to seek a compromise between clearing enough weed to satisfy the fishermen and leaving sufficient to ensure that there will be enough to support the fish they wish to catch. The quantification of this compromise is a problem yet to be solved.

There are other problems posed by anglers; one is that their lead weights are swallowed by ducks and swans, leading to poisoning and death. This is of serious concern to bird lovers and it seems likely that lead fishing weights will soon be phased out, hopefully along with certain bait dyes that cause cancer in fishermen and probably in fish too. Nylon fishing line is another wildlife hazard. It is not biodegradable and tangled masses of it in shallow water ensnare the feet of birds or strangle herons. There is also the paradox that anglers are among the quietest and least obtrusive of visitors yet they can potentially cause the greatest disturbance, especially to birds. This is because of their density. If there are anglers every 20 m around a lake, then birds nesting round the shore can be kept off their nest all day. This need only happen once a week to ensure failure of nesting, especially if parents are kept off partially incubated eggs or caused to leave the nest too often. So anglers are a threat even when they sit still doing nothing!

Anglers are only a part of a wider problem concerned with waterside recreation. Increased leisure time has meant increased demand for watersport facilities of all kinds, yet the availability of natural lakes and rivers is very limited. Moreover, many water-based activities are themselves mutually exclusive: the bird watcher and the water-skier cannot both be accommodated in the same place and yachtsmen get in the way of both. Gravel pits offer a solution to these problems: as *extra* aquatic habitats, they can take the pressure off natural water bodies. Also many pits are discrete entities, so each can be assigned to a particular use. One gravel-pit lake can be used for sailing or skiing, more or less wrecking its potential for wildlife, but allowing another site to be left alone for birds and other wildlife to live in peace.

Gravel-pit lakes have considerable potential as sites for wildlife conservation and this includes their communities of aquatic plants. As natural and semi-natural wetlands disappear or are altered by eutrophication, it is possible that some compensation may be gained from gravel-pit lakes acting as refuges for aquatic plants and invertebrates suffering from habitat loss elsewhere. As new habitats, the gravel pits offer lake conditions formerly common in natural water bodies but now altered by external factors. For example, the Blackwater Valley gravel pits contain very low concentrations of phosphate, presumably because they receive no sewage effluent and, being in a largely urban area, no agricultural run-off. This may be the reason why they contain relatively small populations of planktonic algae and sometimes large populations of macrophytes, often more than in natural lakes nowadays. Should their phosphate loading be increased, this balance might well be altered in favour of the algae because with high fish populations removing all the Cladocera, there would be little hope of the zooplankton reducing the algae and keeping the water clear.

Already the total area of gravel-pit lakes in Britain exceeds by five times that of the Lake District lakes and they are an important resource for water-based recreation of many kinds. They are also of great interest to freshwater biologists and the very differences between these apparently similar lakes provide opportunities for study of some intriguing problems which have relevance to the way in which we use artificial ecosystems.

10
The use and abuse of LAKES

Some of the earliest human communities were sited along the shores of East African lakes and it is likely that lakes have been exploited throughout the history of mankind. Water and food are the two principal resources a lake can provide and all around the world large numbers of people are still dependent on lakes as sources of both. It would be unreasonable to suggest, in the name of 'wildlife protection', that lakes should cease to be used. Water, fish, macrophytes and other lake products are renewable resources and, if wisely managed, could be of value indefinitely. This constitutes real conservation; all aspects of the life of lakes would be retained.

Nowadays, lakes have an ever-increasing value as the base for a wide variety of recreational interests. These include wildlife watching, which supports an enormous tourist industry and is of particular importance to many Third World countries. But affluent populations, while appreciating the aesthetic and recreational aspects of lakes, also dump their waste into them. Fortunately, many of these people have come to realise that the latter compromises the other uses of the lake. It is important that people of developed and developing nations do not jeopardise the future of lake ecosystems through thoughtless use at the present time. Lakes must be used wisely, or the life they support will be in danger. This book has attempted to show that a lake *needs* its living community, not just for sentimental or aesthetic reasons, but in order for its whole ecosystem to function properly. If the system fails, the consequences can often be unpleasant, expensive and irreversible.

Water is needed for drinking, agriculture and industry. The abstraction of water, either from the lake itself or from its inflowing rivers, will tend to have the same effects whatever the purpose for which the water is required. All these uses alter the nature of the water, which then often returns to the lake whence it came, either directly or via a river.

Drinking water must be pure – free of animals and plants, and free of chemicals as far as possible. Most important, it must be free of organisms that cause disease. If water is taken from a lake for domestic consumption then it will need less expensive treatment if the lake is already largely free of these things. For this reason, drinking-water reservoirs are best sited in upland areas where the hard rocks of the catchment area yield few salts to the water, which then nurtures few unwanted plants and animals. Productivity can be kept low if erosion from the catchment is minimised by forbidding public access and protecting the vegetation from damage. A blanket of thick grass or trees prevents soil and nutrients being washed off the land and into the lake. When Thirlmere in the English Lake District first became a reservoir for the city of Manchester in 1890, the level was raised by 15 m and the steep sides of the valley planted with conifers to stabilise the soil. Thirlmere water was so pure that it could be piped to the consumer without any treatment but, to be on the safe side, it was chlorinated. Even in this important tourist area the public was not allowed near the lakeshore for fear of water contamination and the consequent cost of more thorough treatment. In the 1960s the water company needed to take water from Windermere also, but the water of that lake was already productive enough for it to need extensive purification. The National Park Authority insisted that, since a treatment plant would have to be built to process the water from Windermere it should be made rather bigger and process the water from Thirlmere too; the public could then be allowed access to both lakes. This was made a condition of the consent to take water from Windermere.

Clear Lake, California: a classic example of the cumulative poisoning of an ecosystem. Insecticide was originally used to kill midges which were a nuisance to tourists. The diagram shows progressive accumulation of DDD (a chlorinated hydrocarbon similar to DDT) in the food chain, with figures indicating the increasing concentrations of pesticide found in animal tissues at each successive trophic level. The poisonous chemicals and their residues are stored in body fat where they reach high concentrations. At each stage in the food chain, the feeder accumulates poisons from its food so that levels in plankton may be insignificant, but the grebes (top carnivores) end up with dangerous amounts. Chlorinated hydrocarbons are released when fat is metabolised, such as during breeding or food shortage, thus killing the animal some time after insecticides were used. Before 1950 there were about 1000 pairs of grebes on Clear Lake but many died after the 1954 and 1957 applications of insecticide. Sub-lethal accumulations are also dangerous, causing sterility and fatal thinning of eggshells. The surviving grebes were unable to reproduce successfully until after 1969. This example provided the first, now classic, evidence that certain insecticides accumulate in an ecosystem and kill non-target species of carnivorous birds and mammals, even when used at very low concentrations in the water.

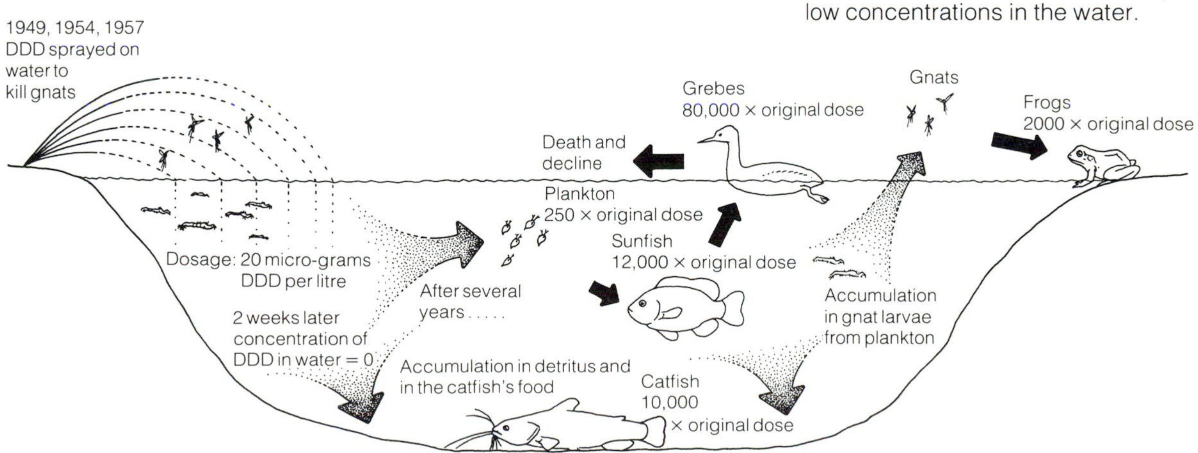

Concentrations calculated from multiplication effect: × original application dosage.

Using a lake as a reservoir frequently results in a barren zone of bare earth or mud around the lakeshore because water tends to be drawn off at times when inflow is low. This 'draw-down' kills the vegetation of the littoral zone and its associated animals and is also very unsightly. In order to avoid this problem at Windermere, a sill was installed across the outflow to ensure that the water in the lake does not fall below a certain level. Only surplus water can be taken from the river downstream.

Water does not have to be taken directly out of a lake to affect its contents. Pyramid Lake in western Nevada receives water from Lake Tahoe via the Truckee River but the river has been dammed and so much water has been diverted for irrigation that the level of the lake is now declining and its salinity increasing. The native fish are dwindling as the salinity of Pyramid Lake rises and the trout, which formerly provided income for the local Indians, are no longer able to ascend the river for spawning. Many fish-eating birds, notably the breeding colony of pelicans, now face local extinction.

Industrial water needs

Industry needs water, often in very large quantities. In addition to taking it away from lakes, industry also has significant effects when the water is put back as effluent. Users of water inevitably want to get rid of waste and the water they discharge is never in the same condition as when it started: something has always been added. Some of these additions (e.g. insecticides) are directly poisonous to plants and animals, but most have more insidious effects. Power stations and factories which have used water for cooling add no chemicals, but they discharge water that has been warmed by a few degrees. This encourages plant growth and speeds up the life cycles of insects and plankton. It might help a species which would die back in winter to flourish for longer in the year. It may allow a species introduced from warmer climates to grow and breed where it would not have done under the normal temperature regime. Although these changes do not sound damaging, they are radical alterations to the natural ecosystem. Moreover, since warm water carries less oxygen than cold, even small changes in temperature can have widespread effects. In temperate zones such thermal pollution does not actually kill species wholesale, because the temperatures reached do not approach the limits which animals and plants can tolerate, but species adapted to low temperatures, such as many of the salmonids, are easily eliminated because there is less oxygen in the warmer water. In tropical climates where natural temperatures approach the upper limit that animals can tolerate, a few extra degrees may be lethal to whole communities because the water may be heated beyond the limit of tolerance. With only a few specialised exceptions, animals cannot live above about 35–37°C and where water temperatures rise naturally above 30°C at the water surface it would not take much additional heat to raise the temperature to dangerous levels.

Many industries, such as logging and mining, wash huge

quantities of particulate matter into lakes and rivers. Although not directly poisonous, sawdust or silt has a devastating effect on the structure of the environment. It settles like a blanket over mud and stones on the bottom and few animals, even those which burrow in soft mud, can survive being constantly inundated with silt. On rocky shores and substrates, all the living spaces are filled in and the habitat is totally changed. Thus, although the animals are not killed by poisoning they are as surely eliminated by alteration of their habitat. Moreover, even moderate amounts of suspended matter increase turbidity and reduce photosynthesis. Fish cannot see to feed even on those animals which remain, and their gills are damaged by the particles as they breathe. The gills secrete more mucus to try and protect their tissues and eventually the animals die of suffocation. Suspended organic particles, such as sawdust or debris from a paper-pulp factory, undergo bacterial decomposition which uses up oxygen from the water.

Many wastes contain poisons such as heavy metals (e.g. lead, copper, mercury) and pesticides which, at low concentrations, do not kill directly but are taken in by animals with their food. They are retained in the body and in time can reach toxic levels. Meanwhile, if the animal is eaten by a carnivore, the latter receives a more concentrated dose than it would do from the water directly. Thus carnivores accumulate these poisons even more quickly than their prey and they succumb more rapidly. The selective death of carnivorous fish is an indication that cumulative poisons are present in the water. In parts of North America and other industrial countries fishermen are sometimes forbidden to take home their catch to eat because of these insidious poisons.

Sewage and Agricultural uses of water

Intensive agriculture uses large quantities of pesticides to control insects and fungi which may spoil or reduce the yield of the crops. Even when applied sparingly, and with due regard to the weather, these substances are washed off the fields by rain and eventually drain down into lakes from throughout the whole catchment area.

Modern agriculture also uses artificial fertilisers which wash off the fields into rivers and lakes. They contain both nitrogen and phosphorus to stimulate the growth of crops so it is hardly surprising that the run-off stimulates the growth of water plants. The other major source of these fertilizing nutrients is sewage. As the human population continues to increase, greater volumes of sewage have to be disposed of; where the population is concentrated into towns and cities, the waste cannot just be absorbed into the soil. Coastal towns dispose of their sewage into the sea but inland it must either be carried to the sea by a river or put into a lake. Whether the lake receives sewage directly or via its inflows, it will have the same effect. Sewage contains not only the nitrogen and phosphorus excreted by the human population but also extra phosphorus from detergents plus, in many places, heavy metals and other poisons washed from the streets by storm water. Nevertheless it is the artificial addition of nitrogen and, most particularly, phosphorus to the natural lake ecosystem, resulting in eutrophication, that is of greatest concern.

It is technically possible, though expensive, to remove phosphorus from sewage effluent before it is discharged ('phosphate stripping'). This is now done in many places, including Switzerland (but only after considerable changes – increased algal biomass and deoxygenation of the hypolimnion – had taken place in several Swiss lakes). Fortunately, it has been shown that in some cases (but by no means all) the removal of phosphate or the diversion of effluent away from the lake can result in a reversal of the eutrophication process. Much depends on the natural productivity of the basin as well as its physical characteristics and behaviour. One of the critical factors in the eutrophication of the Lake of Zurich, for example, was that the effluent discharged into the epilimnion, causing massive blooms of blue-green algae in summer. Phosphate stripping started in 1967 and by 1970 the situation had already improved so that the oxygen content of the hypolimnion was higher than it had been since 1896.

Lake Washington

Lake Washington occupies a glacially formed trough just inland from the north-west coast of the United States. It was a relatively unproductive lake in its natural state, with a maximum depth of 62.5 m and surface area of 88 km^2. The city of Seattle has spread to the western shore of the lake, and numerous smaller towns and villages were also built around the shore. The construction across the lake of two multi-lane roads on floating pontoon bridges encouraged commuter settlement on the far side. About 64% of the shoreline is occupied by residential development and in 1977 just over half a million people lived within its catchment area. The lake is no longer used as a source of drinking water but is used for water-based recreation – sailing, swimming, fishing, water-skiing, and so on.

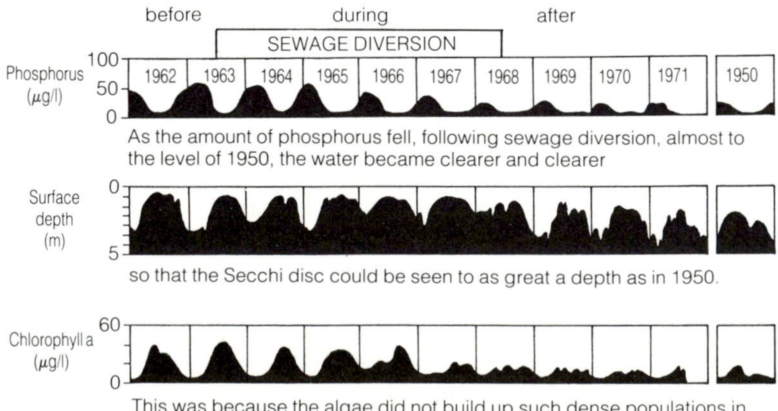

The phosphorus concentration, water transparency and the amount of algae (measured as chlorophyll concentration) in Lake Washington before, during and after the diversion of sewage. By 1971 the condition of the lake had almost returned to that in 1950 before the big expansion of urban development around the lake. (From Edmondson, 1972*b*.)

Although the proximity of the rapidly growing human population to the lake was the cause of its deterioration, it also ensured that the local people noticed the changes in its condition which were caused by their own sewage and were keen to do something about it.

With its proximity to the ocean, the lake never freezes and therefore has one mixing season per year, in winter. It stratifies in summer and although it never becomes deoxygenated the concentration of oxygen in the deepest water did fall to 5 mg/l as the amount of sewage put into the lake increased. There have been two phases in the eutrophication of the lake but the second is much better documented than the first. By 1926 there were thirty outfalls discharging the raw sewage from 50 000 people into the lake. Most of the city sewage went straight into Puget Sound, so the waste from the outlying communities was gathered together and diverted, also into Puget Sound. This first diversion of sewage was completed by 1936. The respite was short lived because, with continued expansion of the population around the lake, more sewage treatment works were built and from 1955–62 there was another gradual deterioration in the condition of the lake. The concentration of algae increased sufficiently to be noticed by the public during the summer. The appearance in 1955 of blue-green species such as *Oscillatoria rubescens* stimulated more detailed study of the lake and made people more aware of its deterioration. Fortunately, studies had been made of conditions in the lake during 1933 (when the first diversion was almost complete) and in 1950. These indicated that not only had the abundance of summer algae increased but the dominant species had changed. The transparency of the water had decreased and the concentration of phosphorus present in the lake during the winter had gone up, implying that there was more phosphorus available for algal growth in the spring than there had been earlier. At its maximum in 1962, 76 000 m^3 of treated sewage was entering the lake each day and the maximum annual mean concentration of total phosphorus in the lake was 65.7 μg/l, in contrast to the 8 μg/l recorded in 1933.

In order to organise a second diversion of sewage from the lake, it was necessary to pass a law in the State Legislature which would enable some of the smaller communities to form a municipality which could carry out and finance joint projects that none of them could undertake alone. A majority of the citizens had to support this law thus, in effect, voting in favour of the expensive sewage diversion scheme. The main campaign was led by a volunteer group concerned about what would happen if pollution of the lake was allowed to continue. Their main task was education of the politicians and local public so that voting would be on a basis of understanding why diversion of the sewage was necessary. Freshwater biologists, particularly Professor Edmondson of the University of Washington in Seattle, played a crucial role in this. The fact that the old sewage works was rapidly becoming obsolete and would soon have to be rebuilt was also an important consideration. A majority vote was obtained at the second attempt in 1958 and METRO, the municipal authority, was formed.

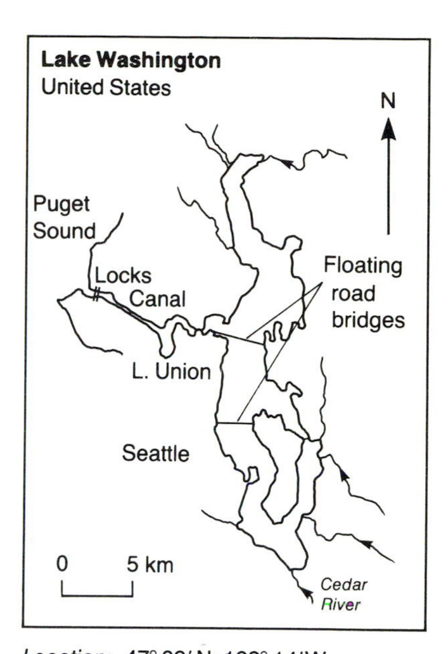

Location: 47° 38′ N; 122° 14′ W
Altitude: about 6 m (regulated)
Catchment area: 1588 km^2
Principal inflow: Cedar River
Lake area: 87.6 km^2
Lake volume: 2.88 km^3
Maximum depth: 62.5 m
Mean depth: 32.9 m
Outflow: via L. Union and the Ship Canal to Puget Sound

Simple monitoring of eutrophication

The most obvious consequence of nutrient enrichment is the increasing density of algae, particularly in summer. Algae reduce the transparency of the water and this can be measured easily using a Secchi disc. This is a white disc, about 20 cm in diameter, held flat and lowered into the water. When it has disappeared from view it is pulled gently back until it can just be seen. The depth of that point is measured on the string.

In Minnesota, which is estimated to have about 12 000 lakes, it is impossible for scientists to monitor the condition of every lake, so a scheme was devised to involve the public in recording the Secchi-disc depths in their local lakes throughout the summer. Each volunteer was supplied with a Secchi disc and a card on which to send measurements back to the laboratory. An average Secchi-disc depth could then be calculated for each lake for the summer of that year.

The relation between the amount of algae and the Secchi-disc depth is not a simple one and small differences in relatively sparse concentrations of algae make a larger difference to the Secchi-disc reading than small changes in algal density in already dense populations. In order to explain to the volunteers whether or not any changes they observed from year to year were significant, and how their lake compared with others, a 'Trophic State Index' was devised, which incorporated the relation between Secchi-disc depth and amount of algae. The 'Trophic State Index' was calculated for all the Minnesota lakes for which Secchi-disc depth records had been kept and values plotted to show how many lakes came into each category. They could also be compared with some other famous lakes. Notice that each division of the scale along the bottom indicates a doubling of the index.

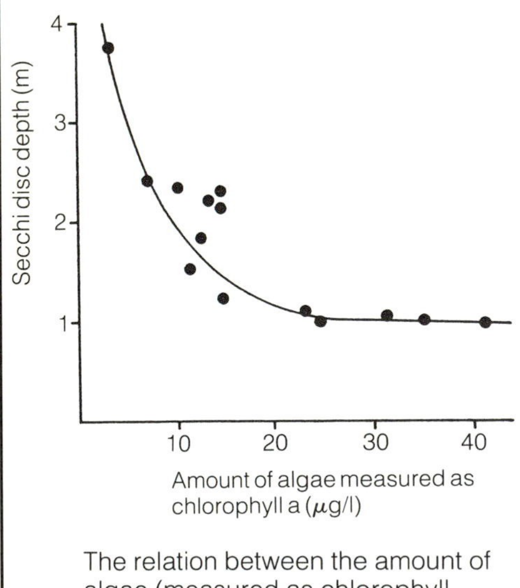

The relation between the amount of algae (measured as chlorophyll concentration) and the depth at which a Secchi disc was visible in Lake Washington during July and August (after Edmondson 1972*a*).

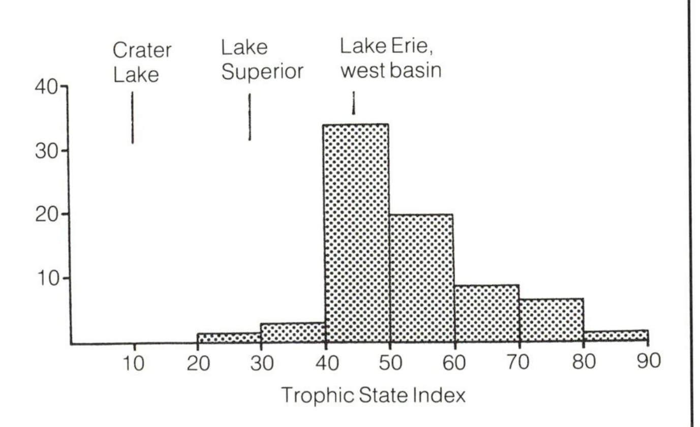

The number of lakes in Minnesota which fell into each class of the 'Trophic State Index', and the position of three other famous lakes on the index scale. Note that an increase of ten units on this scale is equivalent to halving the transparency of the water. (After Shapiro *et al.*, 1975.)

Diversion of the sewage took place in three stages and the whole project was completed in 1968. An additional environmental bonus was gained from building treatment works large enough to process also the raw sewage from the city of Seattle, which until that time was still entering Puget Sound untreated. Moreover, after extensive study of water movements in the Sound, the outfall was moved to deeper water offshore so that the effluent would be rapidly dispersed and well mixed into the ocean water of the Sound, which is already rich in nutrients from upwelling of water from the deep Pacific.

Throughout this period, conditions in the lake were monitored in detail by Professor Edmondson and his assistants. The completed diversion removed 50% of the incoming phosphorus; the rest comes from natural sources. The lake began to respond immediately after the first stage of the diversion and by 1970 had reached conditions that had not been recorded since 1950. The winter concentration of phosphorus had declined, the summer concentration of algae was reduced and, more obviously from the public's point of view, water clarity had improved. The mean summer transparency, measured by lowering a Secchi disc into the water, was 1 m in 1963; in 1969 it was 2.5 m and in 1971 3.5 m.

The success of this sewage diversion scheme showed that, in Lake Washington at least, eutrophication is reversible. It is essential to stress that before such schemes are undertaken the cause of the eutrophication must be clearly identified. In shallow lakes, or lakes with a much longer renewal time than Lake Washington (only 2–3 years) such a scheme might not work at all if there was sufficient phosphorus stored in the sediments to supply abundant populations of algae regardless of sewage. A good example of this is provided by Lake Trummen in Sweden. This lake is only 1 km^2 in area, with a maximum depth of 2.2 m. The lake was severely polluted and sewage diversion was undertaken, but even after 10 years the quality of the water had not improved. It became evident that the algae were being maintained by nutrients stored in the highly organic sediments. In 1970 and 1971 a suction dredge was used to remove about 40 cm of the surface layers of the sediment and restore the lake to the depth it was before pollution started. Since then the lake has shown a marked improvement: both phosphorus and nitrogen levels have decreased and much less is released from the mud when it does become deoxygenated. This has resulted in reduced algal growth to the extent that not all the silica is used by the diatoms in spring.

The Great Lakes of North America

The five huge stretches of water which comprise the Laurentian Great Lakes form the largest body of fresh water on earth. Unfortunately, these lakes are well known not only for their great size but also for the story of their deterioration as a result of abuse. Their enormous basins were formed by glacial action about 10 000 – 15 000 years ago (see Chapter 1) but they now receive drainage from a relatively small area so that water passes through them rather slowly. This is especially true of the three largest lakes and particularly Lake

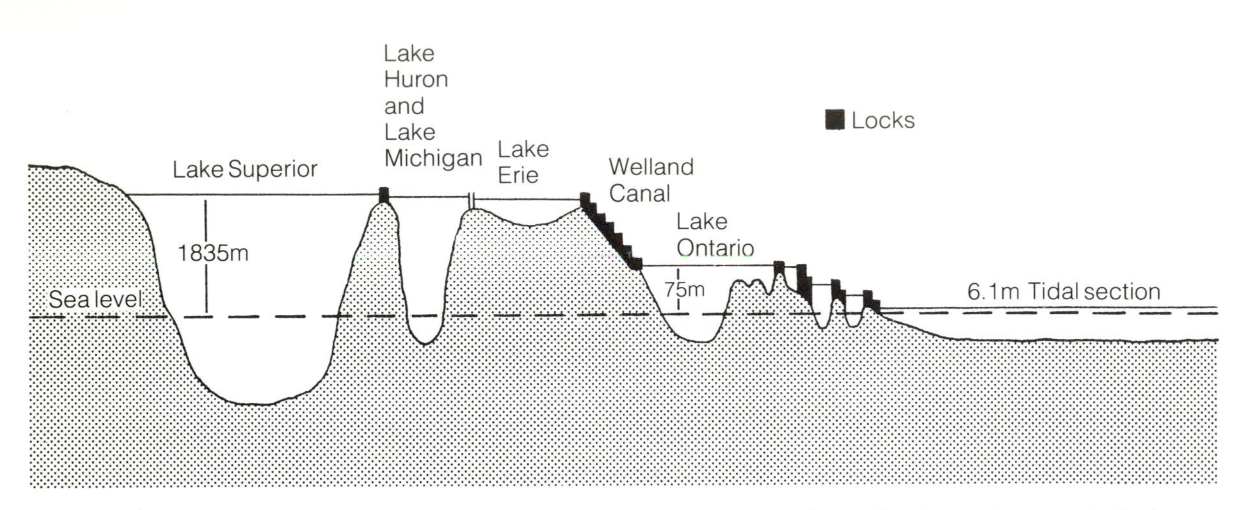

A diagrammatic profile of the Great Lakes of North America to show the differences in altitude which result in a flow of water from the highest, Lake Superior, into Lake Huron. The waters from Lake Huron and Lake Michigan flow into Lake Erie and thence, via the Niagara falls to Lake Ontario, the lowest in the chain whose outflow is the St Lawrence River.

Superior. This and Lake Huron have few large cities on their shores and their catchments are too cool and rocky for intensive agriculture, so their waters are very unproductive. The two lowest lakes, Ontario and Erie, are smaller and more responsive to environmental effects, whether natural or not. They and Lake Michigan have some of America's largest industrial cities on their shores which have discharged their waste into the lakes for many years.

All the lakes have suffered from being used for indiscriminate disposal of human and industrial waste. The effects are localised in Lakes Superior and Huron, which have enormous volume and largely undeveloped shorelines, but further downstream the problems are more serious. Lake Erie has suffered most because it is relatively shallow (mean depth only 19 m), and has industry and urbanisation concentrated along its shores, particularly at the western end where the water is shallowest. By 1970, nearly 12 million people were living within the catchment area of Lake Erie, many of them concentrated in cities such as Detroit and Cleveland. Their sewage and the effluent from more than 300 major industrial sites was discharged into the lake or its inflows. This resulted in dense algal blooms and severe deoxygenation of the hypolimnion, both

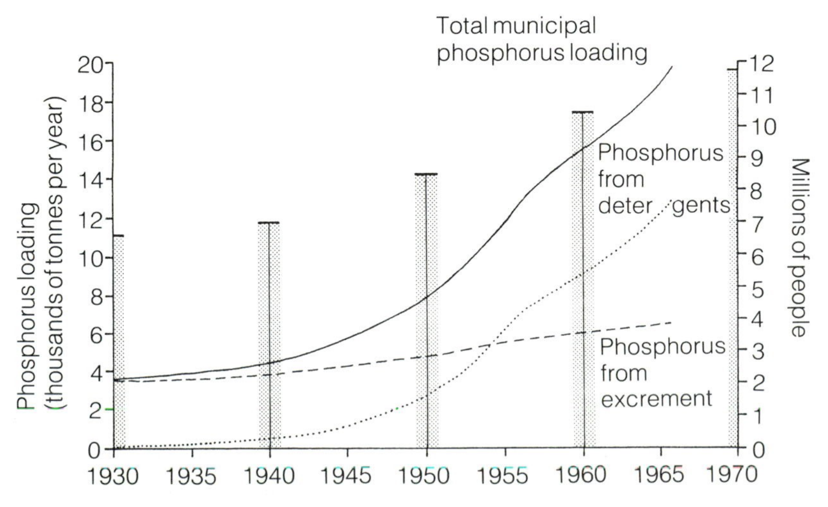

The phosphorus input to Lake Erie between 1930 and 1970, showing the proportion due to human excrement and that due to increasing use of detergents (from Golterman, 1975). The increasing population living in the catchment area of the lake is shown by the vertical bars (from Robertson and Scavia, 1984).

evidence of accelerating eutrophication. The worst effects were seen close to the cities and inspired such descriptions as 'foul smelling scum, mixed with bodies of waterfowl, fish and assorted jetsam'. Lake Erie was declared 'dead' and its epitaph written in the world's conservation literature. The main culprit was again phosphorus, particularly from detergents. Here, in the centre of a continent, sewage is not easily diverted to the sea, so the cost of dealing with it properly will be high. Fortunately, the shallowness and small size that made Erie so vulnerable also ensure a more rapid flow-through (short retention time), as the lake is flushed with relatively clean water from Lake Huron. This helps Erie to respond positively to any reduction in its pollution load. Lake Ontario, downstream, is smaller than Erie, but much deeper (maximum 244 m), so although it receives nutrient-enriched water from Erie and has its own polluted catchment, it suffers less severe eutrophication.

So large is their volume of water that the Great Lakes have a marked influence on the climate of the land around them. In summer the lake water is slow to warm and causes cool breezes; during winter, heat stored in the lakes is only gradually lost. None of the lakes freezes over completely, except Lake Erie in the coldest years, so they have one period of mixing and one period of summer stratification. Nevertheless, thick ice forms on the inshore waters and blocks the ports and canals.

The natural fish population of the Great Lakes contains about 103 species, including the paddlefish (*Polyodon*), lake sturgeon (*Acipenser fulvescens*) and bowfin (*Amia calva*). There are no flocks of endemic species, as there are in the great lakes of Africa, because the lakes are too young, but the community is rich in species and supports a successful fishery. The major commercial fish of the Great Lakes are all salmonids; the mainstay for over 100 years was the lake trout (*Salvelinus namaycush*) until its precipitous decline in all five lakes during the decade from 1945–1955.

The construction of the Welland Canal allowed ocean-going vessels to bypass Niagara Falls, and traverse the whole 'chain' of lakes as far as the western shore of Lake Superior, more than 3700 km from the Atlantic Ocean. The canal not only provides passage for ships but also allowed alien species of fish to pass from Lake Ontario to the rest of the system. Two of the invaders have become particularly notorious: the sea lamprey (*Petromyzon marinus*) and the alewife (*Alosa pseudoharengus*).

Lampreys are eel-like animals which live as parasites on fish. They have no jaws, only a round, sucker-like mouth with which they cling to the body of their prey. Using their horny teeth and tongue they rasp away the host's tissues. If the prey is large, healthy and not attacked too often, it will survive; smaller fish succumb quickly. If the population of larger fish is reduced so that the remaining smaller individuals are frequently attacked, mortality is much higher. Lampreys first penetrated into Lake Erie in 1930 and reached Lake Superior by 1946. Their population rapidly increased and they were blamed for the decline in lake trout stocks, despite the fact that lampreys have always occurred in Lake Ontario alongside the lake trout, and that the lake trout decline began in Lake Michigan before

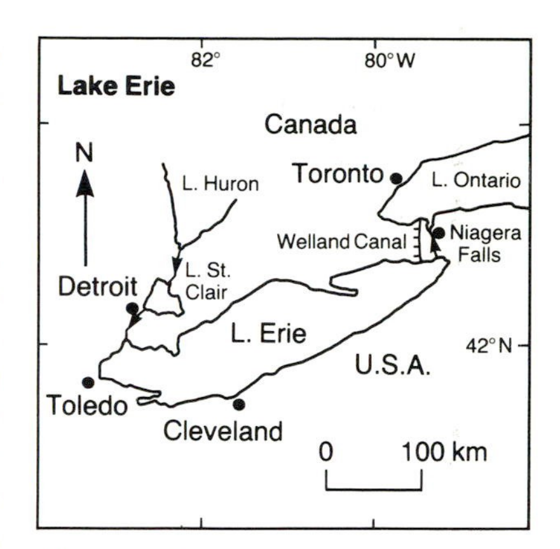

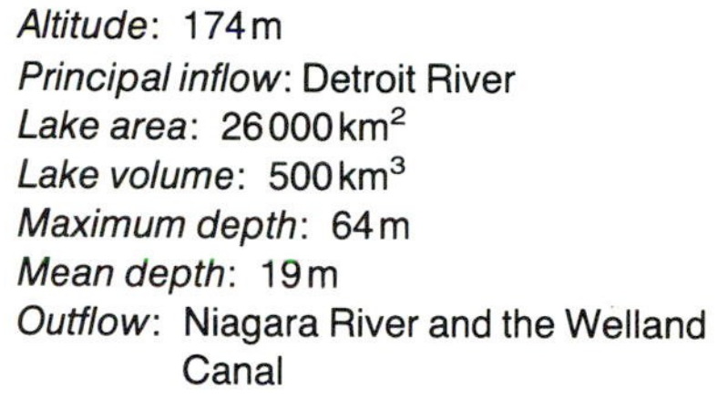
Altitude: 174 m
Principal inflow: Detroit River
Lake area: 26 000 km^2
Lake volume: 500 km^3
Maximum depth: 64 m
Mean depth: 19 m
Outflow: Niagara River and the Welland Canal

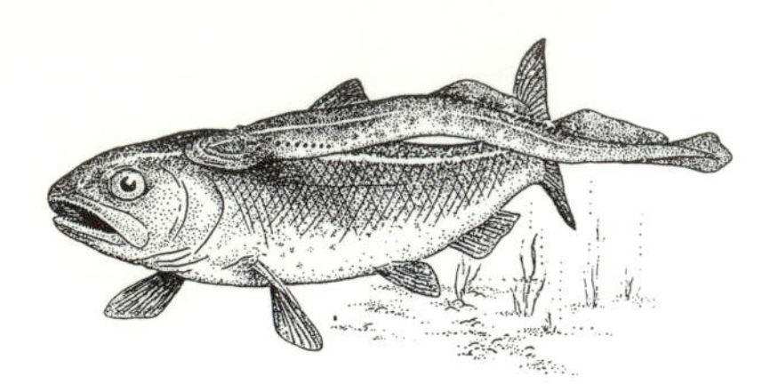

The marine lamprey, *Petromyzon marinus*, clings with its horny-toothed sucker to a lake trout, *Salvelinus namaycush*, and feeds on the living fish. There parasites reached the Great Lakes above Niagara Falls via the Welland Canal and almost certainly caused the final decline of the trout populations which were already suffering from heavy fishing mortality.

the lamprey was very common there. The most likely explanation is that the lake trout populations were already under stress from fishing and the arrival of the lamprey tipped the balance. The fishery had always concentrated on the larger fish and, since this species does not reproduce until it is 6–7 years old, the breeding stock was depleted. The lampreys would have preyed first on the larger fish, which could then no longer replace the losses due to the fishery. The collapse of the fishery was very sudden and the fish population showed no sign of recovery until fishing was banned and trout were artificially re-stocked into the lakes. At the same time efforts were made to reduce the populations of lampreys by electrocuting and poisoning them on their spawning grounds.

The other major salmonid fishery of the Great Lakes was for the whitefish, lake herring and cisco (all *Coregonus* species). These have all been successively eliminated from the commercial fishery, starting with the larger species (about 40 cm in length) such as *Coregonus nigripinnis* which disappeared long before the lamprey arrived, down to *C. hoyi*, which is only 20 cm in length. The decline is partly attributable to the lamprey but is also probably due to a combination of overfishing and competition for food from the alewife which arrived at about the same time as the lampreys. The alewife is a zooplankton feeder and as such directly competes with the coregonids. In Lake Ontario, where the lake herring and the alewife had previously coexisted, fishing probably tipped the balance. In Lake Michigan the alewife did not build up a very big population until 1950–60 and was probably kept in check by the lake trout, but as these declined the alewife population increased. Another alien, the rainbow smelt (*Osmerus mordax*), also a plankton feeder, was introduced to Lake Michigan in the 1920s and 1930s and rapidly built up large populations in the shallow waters of Lakes Michigan and Huron. The coregonids thus faced competition in both deep and shallow water from these two introduced species. The fishery for the smaller coregonids did not collapse as rapidly as that for the lake trout because, initially, the fishermen switched from catching high-value food fish to taking smaller types, to make low-value fish meal, for which more species (including alewife and carp) are acceptable. More recently, alewife populations have fallen prey to the Pacific salmon (*Onchorhynchus*) which was introduced as a control measure.

The Great Lakes are not dead, except perhaps in some of the worst polluted inshore areas, but they are different: their phytoplankton, zooplankton and fish communities are all greatly changed from their natural state as a result of pollution, heedless exploitation and the thoughtless destruction of natural barriers between species.

The Norfolk Broads

In the easternmost part of England there are areas of shallow open water, surrounded by marshland, known as the Norfolk Broads. They lie along the lower reaches of several rivers, including the Yare, Bure, Ant and Waveney. Some Broads are simply widenings of the river but others are separate basins linked to the river by channels.

These lakes result from peat digging, which ceased in the fourteenth and fifteenth centuries. The peat must have been laid down during a period of high water levels and was excavated during a relatively dry phase; the workings were flooded by a change in sea level which backed up the rivers. The whole area became a mosaic of reed swamp, open water and wet grassland, inundated to varying degrees, with damp woodland on the higher ground. With naturally productive water, rich in nutrients, an impressively diverse community of plants and animals developed.

The local people harvested reeds for thatching and exploited fish and wildfowl populations. Farmers drained the higher areas by digging networks of ditches so that their cattle could graze the grass and herbs which took over as the water table fell slightly. Until the end of the nineteenth century the local population was sparse and the Broads continued to be relatively remote. The ecological diversity remained, enhanced by the effects of small-scale exploitation. For example, harvesting reeds prevented the natural drying out and succession to woodland which would otherwise have taken place; grazing allowed many pioneer species to remain.

Three changes in human activity have had profound effects on the ecology of the Broadland area since the beginning of this century. The city of Norwich and the population of its hinterland have increased greatly in size and sewage treatment plants of ever-increasing capacity put their effluent into the rivers. Late last century the area was also 'discovered' by tourists: the holiday industry now caters for more than a quarter of a million visitors each year, most of whom own or hire boats to cruise through the complex of waterways. When the boats had sails, little harm was done; but the motor boats which have taken over stir up the mud and drive their wash against fragile banks. Although forbidden to empty their waste directly into the water, tourists also add to the effluent of the sewage works. The third major change has a less direct influence on the lakes themselves but has profoundly altered the marshland around them. When steam pumps took over from windmills, marshland drainage became much more efficient and over the years more than 80% of the area has been embanked and drained. Land thus reclaimed has latterly been ploughed and then re-seeded, with either cereals or highly productive grasses for fodder. Both these crops attract Government subsidies and both require the application of fertilizers which inevitably drain off the land into the ditches and rivers. Both also turn species-rich wet meadows, full of herbs and invertebrates that provide rich pickings for wintering birds, into either monocultures or fields dominated by a few cultivated species.

Despite these changes the Norfolk Broads still form the largest area of semi-natural wetland in Britain. They are home to some of our rarest wetland birds, such as the marsh harrier and bittern, as well as rare plants and insects; indeed the area includes three National Nature Reserves. The problem is to maintain what wildlife interest remains and manage the Broads in such a way as to support or increase the present species diversity.

The Broads themselves are very shallow lakes, seldom reaching

4 m in depth and usually much less deep. Originally they all contained clear water and were mostly fringed by reeds and water lilies, with a host of submerged plants growing across the bottom. The first notable change was a substantial increase in the submerged macrophytes. They grew to the surface and blocked the passage of boats, except in the channels kept open by increasingly active and expensive weed cutters. This phase lasted some years but eventually the weeds disappeared and were replaced by thick masses of green algae which prevented light reaching the bottom of the lakes. When they died they caused the sediment to build up more rapidly than before, decreasing the depth of the water. The original characteristics of the Broads only persisted in one or two of the forty-two lakes.

The transformation of the Broads was thus in two phases, marked first by the increase in macrophytes, then by their virtual disappearance. One hypothesis concerning the mechanism by which macrophytes are replaced by planktonic algae as the dominant primary producers in shallow lakes suggests that while nutrient levels are relatively low, the macrophytes take up most of what they need from the substrate and secrete substances which suppress the phytoplankton. The water thus stays clear and macrophytes continue to predominate. If nutrient levels are increased, the epiphytic algae which grow on the surface of macrophyte leaves and the filamentous algae such as blanket weed (*Cladophora*) increase. They shade the macrophytes which then grow less well and release decreased amounts of the substances which suppress the phytoplankton. Planktonic algae then proliferate; this increases the turbidity of the water, reducing light penetration and further inhibits macrophyte growth. Once the change-over has started it becomes self-perpetuating.

Nevertheless, the balance between macrophytes and algae is not simply dictated by the absolute level of nutrients in the water, but may depend on more subtle factors. For example, sewage and agricultural fertilisers both add nitrogen and phosphorus to the water. But those broads which still have abundant macrophytes receive only land drainage, no sewage. Since land drainage water contains more nitrogen than phosphorus, it seems likely that phosphorus is the critical nutrient causing the transition from dominance by macrophytes to dominance by phytoplankton. The phosphorus loading (quantity received per year per unit area of the lake : see p. 40), as well as the average concentration of phosphorus in the water of the broads, has greatly increased, particularly since the 1940s. There is evidence for this not only in the present situation but also in the phosphorus content of the sediments which record the history of the lakes. However, the situation is still not that simple. There are some lakes in which macrophytes predominate and others in which phytoplankton is dominant, though both have similar levels of phosphorus.

The interactions between grazing zooplankton and fish also affect the balance between the plant communities. When fish stocks are low, filter feeders such as *Daphnia* are able to flourish, as are other large Cladocera like *Sida* and *Simocephalus*, especially if macrophytes are present to shelter them. These consume sufficient phytoplankton to prevent the build-up of algal populations. If the Cladocera are eaten by fish, the phytoplankton will be released from grazing pressure and proliferate. This too is self-perpetuating: once the macrophytes have been ousted, there is nowhere for the Cladocera to hide from their predators.

The situation is not irreversible. One small broad was experimentally isolated from the river channel which fed it phosphorus-loaded water, and it gradually reverted to its former condition. Removal of phosphorus from sewage effluent by additional treatement at the sewage works was also done experimentally. The phosphorus loading of the river certainly decreased but the response of the downstream broads was not as marked as had been hoped and it appears that there are at least two further complicating factors – birds and boats.

Hickling Broad, on the River Thurne which does not receive any sewage effluent, has suffered loss of its macrophytes and has an increased concentration of phosphorus. Martham Broad, on the same river system, still has a rich macrophyte flora. Hickling is the largest of the Broads and is a National Nature Reserve. Observation suggested that the increased phosphorus in the water of Hickling Broad might be due to the black-headed gulls which roost on the open water each night, particularly during winter. Numbers have risen from about 25 000 in the 1950s up to a quarter of a million during the 1970s, part of a general twentieth-century increase in the numbers of gulls throughout Europe. The gulls' excrement sinks to the sediment and is released into the water to be taken up rapidly by the phytoplankton as temperatures rise in the spring. This source of phosphorus is much less easy to manipulate than the sewage works.

The increase in boat traffic on the Broads has had widespread ecological effects and, while it should be easier to control boats than gulls, politically it has proved difficult to do so. During the first phase of change, when macrophyte growth greatly increased, the weeds hindered boat passage but also prevented them from moving outside the regular channels. Since the weeds have disappeared there is less limitation on the movement of boats, and their propellors churn up the mud all over the lakes. This not only increases the turbidity of the water but also releases phosphorus from the mud into the water. While the increased turbidity cuts down the light available to algae, it also cuts off light from any macrophytes struggling to re-establish themselves. The phosphorus released from the mud will for many years offset the effects of reducing the supply from the sewage works unless something is done to restrict the quantity and speed of the boats. Speed limits are set but to control the number of boats depends on cooperation from the local firms who own most of them. The boat business is a mainstay of the local economy.

The loss of macrophytes which has allowed boats to cruise widely over The Broads has also left the banks exposed to their wash. The fringing reed beds have died back and it is not clear whether this is a direct consequence of boat traffic or whether it has happened from some other cause. Either way, the soft peat banks are easily eroded and this too increases the turbidity of the water. In many places banks have had to be strengthened by steel piling, whose cost runs to millions of pounds. The control of erosion and the re-establishment of fringing reed beds would not only be ecologically desirable, but might also save a lot of money in the long run.

Management of the Norfolk Broads was in the hands of numerous separate organisations. Water quality, fishing and the aquatic environment were the responsibility of Anglian Water Authority, but navigation was controlled by the Great Yarmouth Port and Haven Commissioners. General planning matters were the responsibility of the County Councils, though land drainage was organised by numerous Internal Drainage Boards, encouraged and grant-aided by the Ministry of Agriculture intent on farm modernisation. The Countryside Commission was concerned with recreational activities and landscape protection, but wildlife conservation was the concern of the Nature Conservancy Council and a host of voluntary bodies.

As the evidence of change in both the aquatic and semi-terrestrial environments mounted, and some of the causes became clearer during the early 1970s, it became evident to all that some sort of coordinating body was needed to oversee The Broads as a whole. The idea of a National Park was rejected (as it had been in 1949 when National Parks were first established in Britain) because many saw it as too restrictive, and in any case a National Park Board could not have control over navigation or the water supply. As a compromise, the Broads Authority was created in 1978 as an experiment to run for 5 years. It is to be made permanent if a Bill now before Parliament becomes Law. But the Port and Haven Commissioners are reluctant to surrender to a new Authority their 300-year-old right to control navigation. They are supported by the boat owners who fear restriction of their trade and question why, since water quality is the root problem, they should be subject to the new Authority when the Anglian Water Authority will not. Responsibility for water quality cannot legally be delegated to the Broads Authority; nevertheless, the Water Authority now has a record of cooperation with the Broads Authority which the boat owners have yet to establish.

Thus, once again, the conservation of lakes enters the arena of politics and it is evident that people will only bring pressure to bear on the politicians who ultimately make the decisions if they understand the causes of change and the need for particular remedies. It is perhaps even more crucial that the politicians should understand the complexities of the natural environment and the effects of human interactions with natural processes.

Lake fisheries

In some parts of the world, freshwater fish contribute significantly to the protein supply of local people. This is particularly true for landlocked countries. At the end of the 1970s the yield of fish from freshwaters contributed just under one quarter of the total global harvest of aquatic species (10.2 out of 43 million tonnes). This was *commercial* fishing for food and quite separate from subsistence fishing and fishing for sport. Strenuous efforts are made all over the

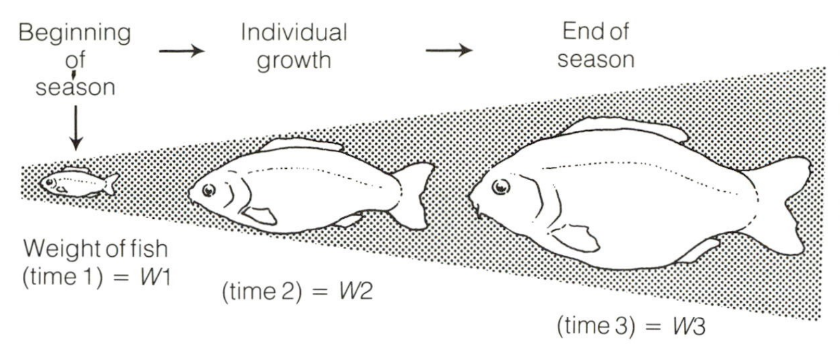

The meaning of 'production' and 'yield' as applied to an individual fish. As the fish grows, it gets bigger; its production is the extra amount of fish body grown per unit time. **Production** between Time 2 and Time 3 = $W3 - W2$/Time 3 − Time 2. However the **yield** (e.g. to a fisherman) at Time 3 is simply the weight at that time ($W3$).

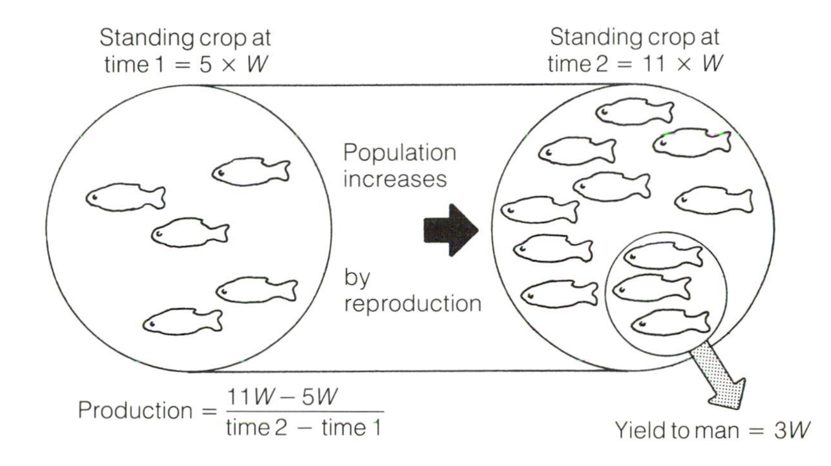

The difference between production and yield as applied to a population of fish. **Production** is the total weight of fish produced over a period of time, including all that die in the meantime, all that are eaten by predators and all that remain in the lake. **Yield** comprises only that part of the population harvested by man. The standing crop is the total weight of fish present at any one time.

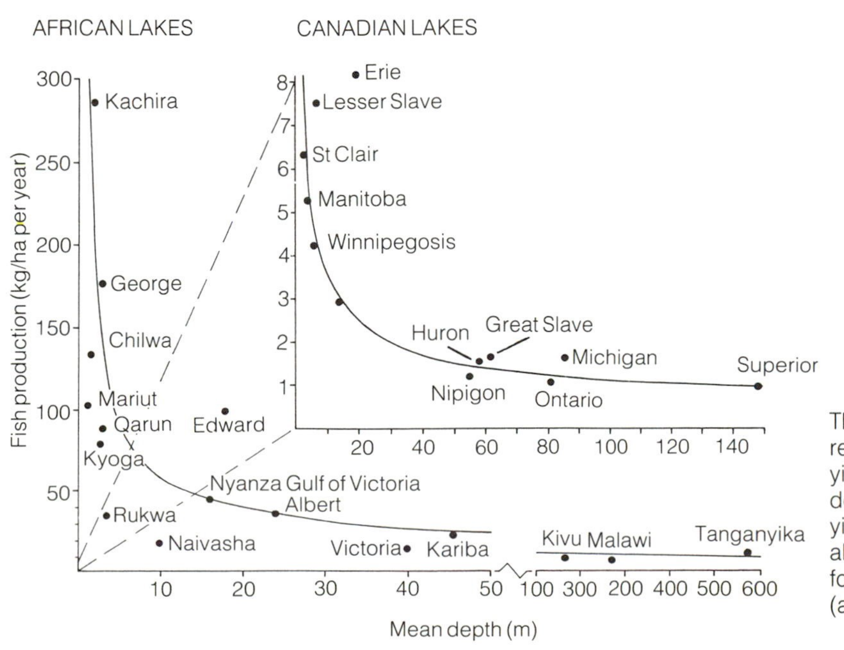

There seems to be a general relation between the total annual yield of fish per unit area and the depth of a lake, but notice that yields from Canadian lakes are about one fifth to one tenth of those for lakes of similar depth in Africa (after Fryer and Iles, 1972).

world to increase the size of the total harvest but more benefit is likely to come from intensively managed pond systems ('fish farming'), where much higher yields per unit area of water are possible than from natural lakes. Indeed in many lakes it is now probably too late to turn back the clock to earlier days of abundant harvests because pollution, overfishing or the establishment of introduced species (or all three in some cases) have altered the environment and community structure beyond the point of no return.

In Europe the traditional role of lakes as sources of cheap food has gradually declined over the centuries and has now been almost totally extinguished by pollution. There have also been severe declines in the lake fisheries of other industrialised areas such as Japan, the Soviet Union and the United States. Now, more than 50% of the total freshwater fish harvest is taken in Asia, mostly from highly managed systems of fish cultivation developed over more than 2000 years.

Traditional methods of catching fish include traps, baskets, hooks, spears and poisons, which vary in detail of manufacture and operation and are often unique to localised areas. In many places such methods are still employed but the most widely used method of catching fish in lakes is the gill net. This consists of a long strip of net whose mesh forms squares or diamonds with four sides of equal length. Along the top and bottom run ropes carrying floats or weights, which allow the nets to hang vertically in the water at a predetermined depth. Fish swim into the net, either passively or when frightened in by the fisherman beating the water, and are caught behind the bony operculum which covers the gills; hence the name of the net. Square holes are more suitable for catching round-bodied

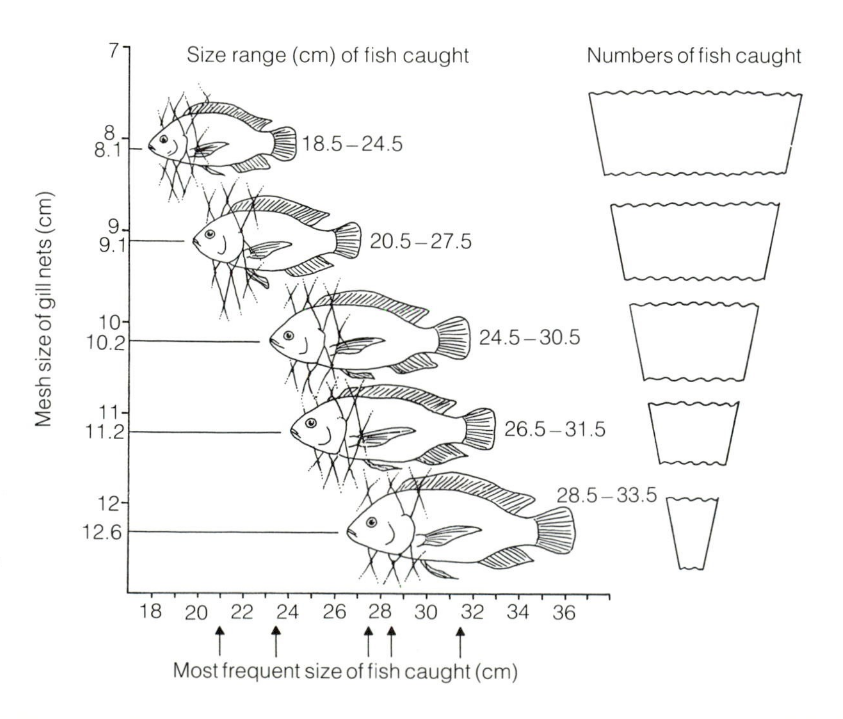

Gill nets are very widely used for catching fish in lakes. Depending on the size of the mesh, they catch a very restricted size range of fish. Larger mesh nets catch relatively few, large fish while smaller mesh nets catch many more but smaller fish. Care should be taken that the mesh size used is not such that it catches too many of the breeding fish, thus leaving none to reproduce and replace those caught. (Based on *Tilapia* catches in Lake Victoria; after Fryer and Iles, 1972.)

fish, while diamond-shaped mesh is more likely to catch deeper-bodied fish. Thus the nets can be used to fish selectively for desired species, depending on where they are set and on the shape of the mesh. The mesh size determines the size of fish caught. This varies for different species depending on the body shape: the same-sized holes will catch older specimens of a long slim form than of a short fat species. Juveniles of large fish will pass through some nets, while very large specimens may not be trapped at all because their heads will not enter the holes. On the other hand, nets set to catch large specimens of a small species may catch juveniles of a larger species.

Since the advent of nylon, light and durable gill nets have been available worldwide. They can be easily operated from the smallest boats, even by one person, so they are ideal for the diffuse, labour-intensive fisheries of many developing countries. Trawling for fish, in contrast, requires considerable capital investment in large boats and heavy gear. It is only possible on large lakes whereas gill nets are usable anywhere. Whatever methods are used, once a fishery starts to grow beyond subsistence level it becomes necessary to conserve the stocks by regulating the number and type of nets used and also the number of people involved. This is evident from the repeated failures of unregulated fisheries all over the world, but even now it is not clear if the lesson has been learned.

The Lake Victoria fishery

Lake Victoria, one of the world's largest lakes, contains a very diverse fish community which is dominated by cichlids, many of them endemic. At the beginning of the 1970s cichlids comprised nearly half the total fish tonnage taken from the lake and, of these, 38% belonged to the genus *Oreochromis* (which was formerly part of the genus *Tilapia*). These fish are very good to eat and are highly prized whether sold fresh or dried and smoked. Commercially, the most important species was *Oreochromis esculentus*.

After the arrival of imported gill nets the fishery began to boom. At the start, a standard length of net set for a particular length of time caught an average of about twenty-five fish, but this average soon began to decline. By the 1920s the average catch per ten was down to seven and caused sufficient concern that it became illegal to use nets with a mesh smaller than 5 inches (12.5 cm). With a burgeoning human population, increasing effort was put into the fishery but the catch per net declined further. By 1953 it was estimated that during the main fishing season a total of 2000 km of nets were set each night and that about 30 000 fishermen were involved in the fishery. By then the catch per net was so low that fishing was rapidly becoming unprofitable, a situation known as economic overfishing.

From 1956, illegal use of 4.5-inch nets increased and eventually the restrictions on mesh size were relaxed. Initially, catches rose but after only a few months they declined again to less than two fish per net. There was a small gain in total yield, but of smaller fish and with increased expenditure of effort: a sure sign that a stock is being fished too intensively.

In an unexploited fish population the mortality of eggs and young is extremely high because only a few need survive to replace the parents and keep the population constant. In an exploited stock the survival of eggs and young may increase because of the reduced competition in a depleted population. The extra survivors will compensate for those removed by fishermen. When a fishery takes so many fish that the remainder cannot produce sufficient young to replace them, the species is suffering from 'biological overfishing'. Fortunately, for many species, economic overfishing usually occurs before biological overfishing, so the fishermen give up and the remaining fish have a chance to build up numbers again. In some cases, as in Lake Victoria, fishing continues beyond the point of biological overfishing; then the fishery collapses totally. Or, as in the Great Lakes, a secondary biological disaster (the arrival of the marine lamprey) on top of intensive fishing combines to eliminate the stocks.

The length of the smallest mature tilapia varied between 19 and 26 cm in different parts of the lake: half of them were not mature until they reached 22–28 cm. Once the mesh size regulations on Lake Victoria had been relaxed, increasingly smaller nets were used in an effort to boost the catch and smaller and smaller fish were caught. The average length of fish retained by a net of 3.5-inch (9 cm) mesh is 22 cm. Thus the nets were now able to catch almost all the mature fish in the population, leaving none to breed. By 1959–60, the tilapia fishery in the Nyanza Gulf of Kenya had virtually collapsed, and a similar collapse occurred soon after in most other parts of the lake.

In less than half a century a most valuable resource was entirely used up through reluctance to recognise what was happening and to manage the situation in such a way that the tilapia population could sustain the fishery indefinitely. One difficulty of formulating such a management policy stems from the fact that one needs to know the size of the stock and the reproduction rate to work out how many fish it is reasonable to take. In the case of Lake Victoria tilapia, the research necessary to establish these facts was being carried out at the same time as the fishery was growing, and much of the information came from the fishery itself. Catch per unit effort (i.e. number of fish per net per night) can be taken as an indicator of relative stock density: a declining figure indicates that the stock is not being replaced. This evidence alone was sufficient to make the danger clear; the necessary remedies were obvious but required political decisions for their implementation. A decision to conserve stocks, whether of fish, metals or oil, rather than exploit them as rapidly as possible, ultimately rests with the politicians who are not renowned for giving precedence to long- rather than short-term benefits.

The Lake Victoria fish community now contains several species of tilapia, only one of which is native; the others have been introduced. *Oreochromis esculentus* was formerly thought to be two species, both of which fed on algae, but one of which was more frequently found in shallow sheltered bays, where it fed on dense phytoplankton suspended in the water and the other on exposed shores where it fed on sedimented algae. Both populations are now considered to be the same species. This simple situation has been complicated by the introduction of other closely related species to enhance the fishery. One of these, *Tilapia zillii*, feeds on water weeds and has become well established. With different feeding and quite different breeding habits, it was not expected to compete with the native species but it turns out that the juveniles of *T. zillii* feed and behave in a very similar way to those of some populations of the native *O. esculentus* and congregate in the same places. Only gradually do they become weed eaters; as juveniles they feed on sedimented algae and small benthic animals in exactly the same nursery areas as *O. esculentus*. The extent of these areas expands and contracts with variation in the water level, so at times of low water level there may well be competition between the native and the introduced species that was unsuspected by those who introduced *T. zillii* to what was apparently a vacant feeding niche in the lake.

The tilapia are only a few of the hundreds of cichlid species found in Lake Victoria; the majority of cichlids belong to the genus *Haplochromis*. With the increasing use of small mesh nets many of these species began to be exploited too and, in the search for other fish stocks, trawling began in the deeper waters of the lake. This revealed even more cichlid species and there was concern at the possible damage that would be done by trawling among the unique assemblage of species found in the lake. But the haplochromines were in even greater danger from another form of interference: the introduction of Nile perch (*Lates niloticus*). This occurs widely in African lakes, but was absent from Lake Victoria until about 1960. It was never quite clear how it arrived but it had already been introduced to Lake Kyoga, close-by. Advocates of the introduction of this large voracious carnivore argued that it would cause a change from a large population of 'tiddlers' into a smaller population of larger, more useful fish. They ignored the fact that some unique species of small fish might thus become extinct. The Nile perch is a top carnivore, so its population could never sustain as much fishing (in terms of weight taken per unit time) as smaller, more rapidly reproducing species lower down the food chain, because of the inevitable losses as energy passes from one link of the chain to the next.

Nile perch seem to be spreading rapidly down the eastern side of the lake, but it was not until the late 1970s that the population began to increase greatly and dominate the trawl catches in the Nyanza Gulf. But by 1984 a trawl which would formerly have caught thousands of haplochromines would catch only two. At the same time increased numbers of Nile perch began to appear in catches at Mwanza at the south end of the lake. Reports from the south-west corner indicate that the perch has arrived there too and that the *Haplochromis* have almost disappeared. Many species have probably gone for good, some of them perhaps unknown to Science. Even now, new species are still being discovered and the 'species flock' in Lake Victoria is of exceptional interest in the study of evolution. To save something for posterity, zoologists are trying to breed as many as possible in aquaria and to make collections of preserved samples so that at least we shall know something of what has been lost.

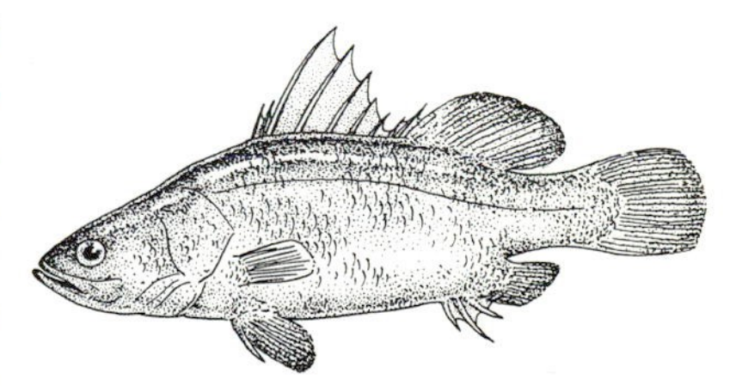

The nile perch *Lates niloticus* occurs naturally in many African lakes but was absent from Lake Victoria until its introduction in 1960. In just over 20 years this voracious carnivore has wrought havoc among the natural community of fish in this lake, the majority of whose species are endemic.

Lakes for recreation

In Europe and North America, demand for water-based recreation has increased enormously over the last decade. Sports such as water-skiing, sailing and windsurfing are booming and fishing (from the bank and from boats) is increasing in popularity. Interest in nature conservation and bird watching has also increased, as well as just walking through the countryside; lakes are frequently a focus for these activities too. There are potential conflicts between these uses and, even at a distance, many species of birds are easily disturbed. Teal, goldeneye and shoveler are much more susceptible than tufted duck, pochard and mallard. Goldeneye are put to flight by sailing boats up to 700 m away while even tufted duck and pochard often take flight at 300–400 m. Power boats are likely to be much worse and boats of any sort disturb fishermen. The presence of bankside anglers has a marked effect on the distribution of wildfowl, both within a lake and between adjacent water bodies. Nevertheless the demands of birds and all these people can be met by regulating who does what where. What is much more difficult to visualise (and costly to deal with) is the threat to the lake itself, resulting from the impact of increasing use by people, many of whom may not touch the water or even go near it.

Lake Tahoe
Lake Tahoe, on the border between California and Nevada, is one of the deepest in North America. With a maximum depth of 501 m, it contains uniquely clear water and is extremely unproductive. A

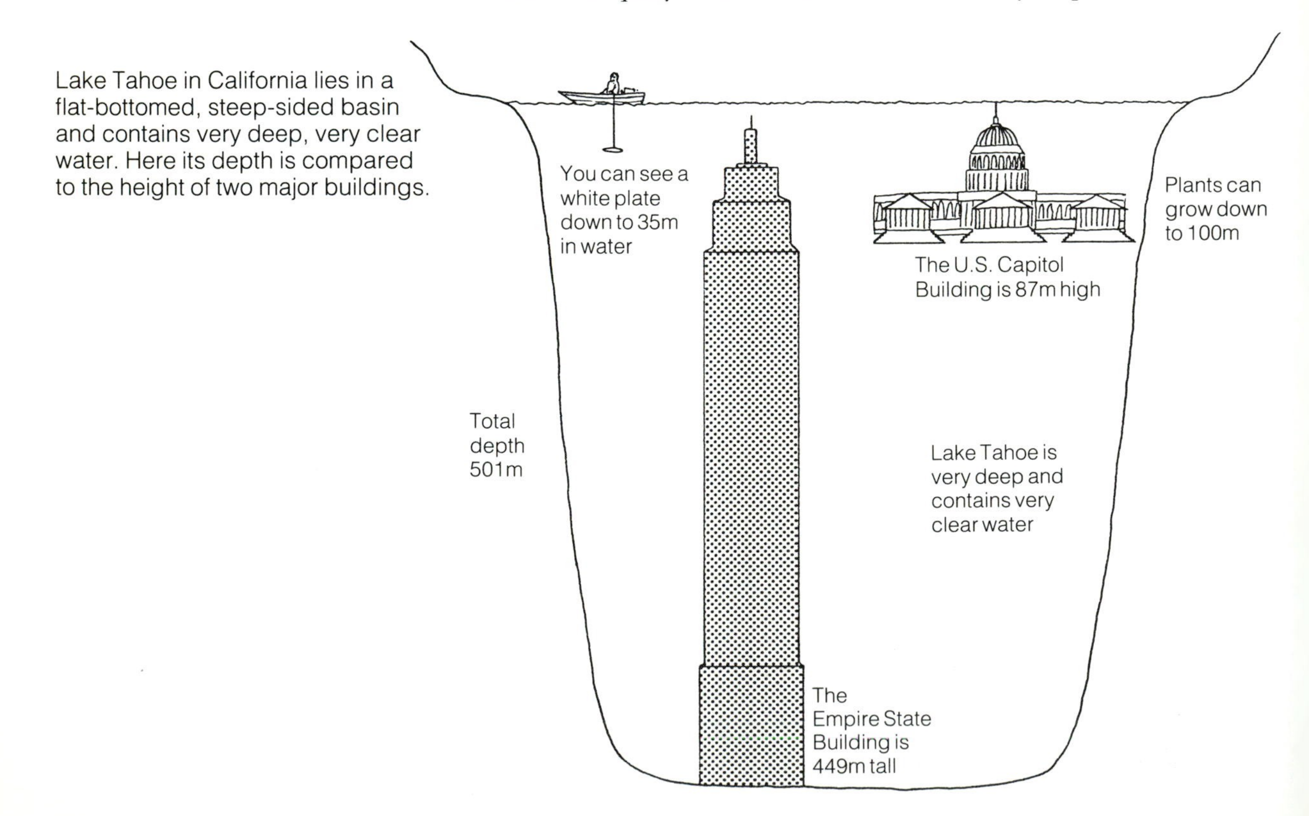

Lake Tahoe in California lies in a flat-bottomed, steep-sided basin and contains very deep, very clear water. Here its depth is compared to the height of two major buildings.

Secchi disc can be seen down to 40 m : photosynthesis can take place down to 100 m. At this depth, about 0.1% of the surface light is available and the algae are highly adapted to function at very low light intensities. This wonderful clarity and the spectacular location in the Sierra Nevada mountains gives Lake Tahoe its tourist appeal. It is the centrepiece of a huge recreational boom, a playground close to the Pacific coast cities inhabited by some of the world's richest people. But its glass-like clarity is matched by its ecological fragility.

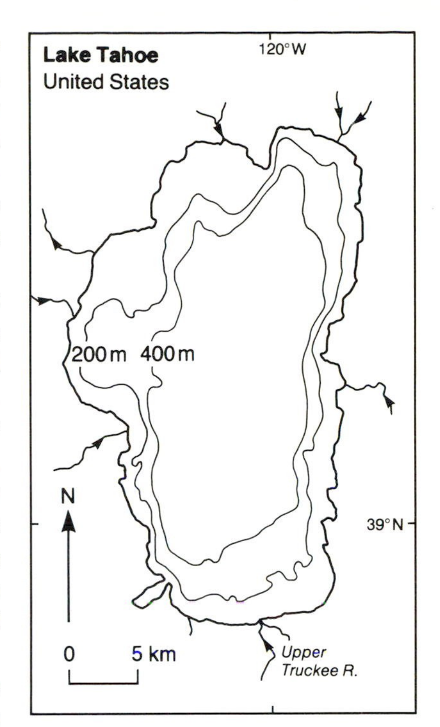

Altitude: 1898 m
Catchment area: 1242 km²
Lake area: 499 km²
Lake volume: 156 km³
Maximum depth: 501 m
Mean depth: 313 m
Outflow: Truckee River

Lake Tahoe was first seen by a white man in 1844 and the first settlers arrived in 1851. The route over the mountains to the Comstock silver mines opened up the south end of the lake, and completion of the railway in 1868 brought traffic to the north. Tourism started soon after and through the 1920s was led by wealthy Californians who built summer homes along the shores. Since then, chalets, hotels and casinos have proliferated. Recreational activity goes on right through the year, only the emphasis changing. The *Lake Tahoe Guide* lists cruises in glass-bottomed boats (of which at least one holds 350 passengers), rafting on the rivers, fishing from boat and shore, thirty beaches and picnic areas, and boating of all descriptions from fourteen marinas and three public launch ramps. There is also hiking through the wilderness scenery and along the lakeshore, cycling on more than 75 km of bicycle trails, and camping at twenty-two sites. All this is quite apart from the many other activities for which the lake is unecessary, such as golf, horse riding and winter sports, for which there are 163 ski lifts within the Tahoe catchment.

During the 1960s there was a development boom, bringing more people to Tahoe's shores. About 50 000 people now live there permanently and far more come seasonally. Even though they may not use the lake itself, they bring to its catchment a great many things that were not there before. This includes nutrients, whose impact on the lake, largely devoid of them, has been considerable. Although the level of primary production in the lake was extremely low, it became apparent that it was increasing in the phytoplankton and that green algae were beginning to grow on the rocks of the littoral zone. This prompted a campaign to 'Save Lake Tahoe'. The haphazard disposal of sewage was stopped and it is now pumped right out of the catchment altogether for treatment and disposal. By 1971, despite the diversion of sewage, the total annual primary production was some 50% higher than in 1959–60. No natural change had occurred in the weather, light or water chemistry which could account for the increase; presumably the nutrient supply to the lake was still being enhanced due to erosion resulting from continued building within the catchment area. In 1969 The Tahoe Regional Planning Agency was created to control development around the lake and it is now difficult to obtain permission for any further building. All properties must be connected to a sewage collection system for which the 1983 charge was US $4000.

Another form of interference with the lake is also related to recreational demand, this time from anglers. From 1949 to 1960, over 11 million fry of kokanee salmon (*Oncorhynchus nerka*) were introduced to the lake to improve the sport fishing. Regular stocking continued until 1969. In 1963–65, 333 000 opossum shrimp (*Mysis relicta*) were also added to the lake to provide food for fish and improve the catch. Not until 1970 was it noticed that the *Daphnia*, previously abundant in the lake zooplankton, had almost disappeared. Both lake trout (*Salvelinus namaycush*) and kokanee switched from eating *Daphnia* to *Mysis*. It seems likely that predation by the shrimps, in addition to that by the trout and salmon, had tipped the balance against the *Daphnia* population. *Mysis* has been introduced into other lakes where it has also been implicated in causing changes to the resident zooplankton. The great depth of Lake Tahoe probably allows the shrimps to take refuge in dark water during the day and thereby escape from salmon and trout, which feed by sight. At night they rise towards the surface where they feed on cladocerans. Although the fish eat *Mysis*, they do not consume sufficient to control the effect of their predation on *Daphnia*. The drastic reduction of *Daphnia* in the open water of the lake might partially account for the increasing populations of algae, now that these are relieved of cladoceran grazing.

While this is a speculative explanation for the continued increase in the phytoplankton of Lake Tahoe, it does again highlight the dangers and complexities of introducing species to lakes where they do not occur naturally. The consequences are never as simple as predicted and are often quite unforseen; frequently they are disastrous for the native fauna. Nevertheless people who should know better continue to suggest further introductions into even the most precious and fragile ecosystems.

11

The conservation of LAKES some concluding comments

Despite the contemporary importance of conservation issues, we have deliberately chosen to say little directly about such matters. This in no way implies that we consider conservation unimportant, but it is not the purpose of this book to tell people what they should or should not do with their lakes. Instead it has been our objective to explain how lakes work and how they respond to what is done to, and around, them. If policy decisions regarding the use of lakes could be based on better understanding, perhaps there would be less need for a discussion of lake conservation. The basis of that understanding is what we have tried to provide.

Hopefully, the conservation message is implicit throughout this book. Lakes are complex, dynamic systems; doing things to them is bound to have effects and they need protection from abuse. Many useful lessons may be learned from past mistakes and positive encouragement can also be gained from successful applications of ecological information to rectify past environmental damage (as in the case of Lake Washington, see p. 188).

We have tried to show that the life of a lake extends beyond its shores. Lakes are not isolated and are critically dependent upon the health of their catchments. In some cases, such as the East African Rift Valley lakes, conservation must take into account the relationship of one lake to the whole group. Even the protection of individual species, like flamingos or pelicans, cannot be considered in isolation. It must involve not just the ecology of one lake but also the key role played by other lakes nearby. The same principle applies to the lakes and wetlands that constitute the 'flyways' for migratory wildfowl in the United States. The loss of a few lakes in the chain may not seem significant, but removal of too many stopover points could be disastrous to some birds even if they only used those places for a few critical days a year. In the same way, today's longhaul airliners would become more vulnerable on trans-Pacific flights if Honolulu International airport was closed; it is only one airport among the world's hundreds, but is important beyond the immediate context of Hawaii.

A central theme of our book must be that you do not conserve a lake by erecting a fence around its shore and forbidding people to use it. Such a policy is unpopular, unscientific and probably unsuccessful. Sometimes such action is also unnecessary, because some lakes are less sensitive than others to certain things. Again, the understanding of a few relatively simple ecological principles offers the prospect of predicting what might or might not be harmful, and thereby avoiding expensive errors. For example, putting warmed water back into a large cold lake in Wales may do less harm than a similar operation in, say, Brazil where the lake is already warmer and its oxygen levels therefore lower, and an increase of only a couple of degrees may be sufficient to push a major proportion of the fauna beyond its physiological limits.

The placid appearance of a lake surface and perhaps our unconscious assumption that water is an unreactive substance may mislead people into imagining that the lake itself is inert. But in practice a lake behaves almost like a living creature, constantly sensitive to what goes on around it. The lake responds directly to what is done to it and its inseparable partner, the catchment area. Everyone understands this principle of 'response to stimulus' perfectly well in the treatment of, say, a pet dog; it would help a lot if lakes were thought of in the same vein.

The problem with lakes is that what goes on inside is largely out of sight and so long as nothing disturbs the surface no problem is perceived. Looking at a lake is the same as looking at a factory standing on its shore. There is an awful lot of activity going on inside, even though there may be no externally visible signs. One small spanner could bring the whole works to a halt, sometimes causing visible consternation, but often the effects remain hidden from the observer outside. One such invisible spanner in the works of a lake is sewage. This can be cleaned up to the extent of being clear and even drinkable, as a proud Chief Water Engineer might demonstrate. But if that effluent is heavily loaded with invisible dissolved phosphates, the receiving lake system is in dire danger even if the effects do not become noticeable for a decade or more.

As so many of the lake's processes are normally invisible and unfamiliar, perhaps a greater use of analogies might aid understanding and avoid conservation disasters. For example, we all know not to pour petrol on a fire because we understand what would happen if we did. The addition of sewage to an unproductive lake is an analogous act. Similarly, almost without debate, carnivorous trout have been widely planted in remote alpine lakes to amuse fishermen;

but who would agree to liberating tigers to liven up a city park? A gardener would not introduce bullocks to his rose beds, despite the promise of free hamburgers for years to come, yet the voracious Nile perch has been introduced to Lake Victoria as a 'Good Thing'. In fact predatory fish do not ruin the lake immediately, and their effects may never actually be visible above the surface. This is like a farmer keeping pigs in a small copse. From the air the trees still look the same, though the undergrowth has been devastated and birds no longer inhabit the shrub layer. The long-term damage is subtle and unseen: the pigs prevent growth of regenerating trees, so that when the tall trees finally die there is no replacement for them. Only then will the full scale of the disaster be visible from the air above.

In fact the introduction of new things to lakes is just like the taking of alien plants and animals to remote oceanic islands. We know that this has often resulted in the extinction of many unique island species, and indeed the study of extinction rates on islands has led to the formulation of a series of ideas known collectively as 'Island Biogeography Theory'. These theories offer predictions about the number of species that can survive on islands of given size and remoteness. Perhaps the same ideas might be usefully applied to lakes as similarly separate ecosystems, and give us some indications as to their vulnerability and what kind of conservation measures would be most appropriate for them.

To conserve lake ecosystems it is often not necessary to prevent all forms of use. Some lakes can be used for a variety of purposes simultaneously without long-term threat. The aim should be to make wise use of a lake's resources, on a renewable and self-sustaining basis, not to squander them all at once. Limnologists and other specialist scientists can advise on how to accomplish this. Fish stocks, for example can be conserved by establishing close seasons, or by imposing catch quotas or other limitations; but the implementation of such measures is a matter for the local community and its leaders. Lake management should prevent excessive exploitation of one kind that compromises other uses and the health of the lake itself. For example on some of Austria's alpine lakes petrol- and diesel-driven boats are not permitted, protecting peace and quiet as well as the lake water. Again, scientific advice is available to assist conservation based upon multiple use, but unfortunately the principle of planning for diverse utility is rejected by some conservationists who seek, unrealistically, to preserve habitats for one use only (e.g. birds). This is as narrow-minded (and short-sighted, in a political sense) as the industrialists, power-boat enthusiasts or other selfish interests who want only to use the lake for their own needs, regardless of others. Conservationists must accept that people need what lakes can provide, especially in the developing world. Often this can be accommodated, along with wildlife conservation, provided that each camp can control its own activities to allow the other's to be maintained in perpetuity. That way everyone benefits and the lake continues to be of use to all.

Perhaps a final analogy here would be most apposite. Back in 1968 (in *Science* vol. **162**) Garrett Hardin described what he called 'The

Tragedy of the Commons'. He envisaged a group of herdsmen, each grazing one cow on the common land. Each man could almost double the benefits gained if he kept a second cow; yet if they all did so, the land would soon become so overgrazed that they would all end up worse off. The tragedy would not just be the destruction of the commons, but the dreadful inevitability of the process. If a herdsman gains from a second cow he naturally buys another, and another . . . This modern-day parable serves as an analogue of countless different aspects of human activity where the use of common property (e.g. by dumping waste in the water or dispelling it in the air) benefits some people just so long as too many others don't try to do the same. In this context a lake is a part of the 'commons'. If the ferryman is successful transporting passengers, their fares soon allow him to put aside his harmless oars and buy a noisy polluting outboard engine, which then means he can carry more people, save more money and get a diesel-powered boat. The fisherman spends his profit on more nets to enlarge his catch at the expense of the pelicans. If the conservationist succeeds in protecting the birds, they swiftly increase in numbers (as he wishes) at the expense of the fisherman's livelihood.

The lake is a dynamic and delicately poised system. It is easily perturbed if the self-fuelling components of Hardin's parable are not all kept in check. The lake is resilient, up to a point, and can accommodate many users. But if one faction gets too greedy and pushes the ecosystem beyond its capability to cope, all lose out – and so does the lake itself.

Further reading about lakes

There are many technical books about lakes, some more readable than others, and a few are included below, but there are not many for the general reader. All those listed here include material concerned with either some of the organisms or some of the ecosystems dealt with in this book.

Angel, H. and Wolseley, P. (1982) *The Family Water Naturalist*. London: Michael Joseph

Beadle, L.C. (1981) *The Inland Waters of Tropical Africa*. London: Longman

Belcher, H. and Swale, E. (1976) *A Beginner's Guide to Freshwater Algae*. London: HMSO

Brown, L. (1971) *East African Mountains and Lakes*. Nairobi: East African Publishing House

Clegg, J. (1965) *The Freshwater Life of the British Isles*. 3rd edn. London: Warne

Fryer, G. and Iles, T.D. (1972) *The Cichlid Fishes of the Great Lakes of Africa*. Edinburgh: Oliver and Boyd

Goldman, C.R. and Horne, A.J. (1983) *Limnology*. Japan: McGraw Hill

Hancock, J. (1984) *Birds of the Wetlands*. Beckenham, Kent: Croom Helm

Haslam, S.M., Sinker, C.A. and Wolseley, P.A. (1981) *British Water Plants*. (Reprinted from *Field Studies,* **4**: 243-351.) Preston Montford, Shropshire: Field Studies Council

Hutchinson, G.E. *A Treatise on Limnology*. Vol. 1 (1957) *Geography, Physics and Chemistry*. Vol. 2 (1967) *Introduction to Lake Biology and the Limnoplankton*. Vol. 3 (1975) *Limnological Botany*. New York: John Wiley & Sons

Löffler, H. (1974) *Der Neusiedlersee*. Vienna: Fritz Molden

Macan, T.T. (1959-66) *Guide to Freshwater Invertebrate Animals*. London: Longman

Macan, T.T. (1970) *Biological Studies of the English Lakes*. London: Longman

Macan, T.T. and Worthington, (1951; paperback 1972) *Life in Lakes and Rivers*. London: Collins New Naturalist Series

Maitland, P.S. (1977) *The Hamlyn Guide to Freshwater Fishes of Britain and Europe*. London: Hamlyn

Moss, B. (1980) *Ecology of Freshwaters*. Oxford and Boston: Blackwell Scientific Publications

Muus, B.J. and Dahlstrom, P. (1971) *Collins Guide to the Freshwater Fishes of Britain and Europe*. London: Collins

North, R. (1983) *Wild Britain: The Century Book of Marshes, Fens and Broads*. London: Century Publishing

Readers Digest Nature Lovers' Library (1984): *Field Guide to the Water Life of Britain*.

Wetzel, R.G. (1975) *Limnology*. Philadelphia: W.B. Saunders

Williams, W.D. (1983) *Life in Inland Waters*. Melbourne: Blackwell (written for Australian readers)

Sources of information used in tables and figures

Beadle, L.C. 1981. *The Inland Waters of Tropical Africa*, 2nd edn. Harlow: Longman.

Biro, P. 1984. Lake Balaton: a shallow Pannonian water in the Carpathian Basin. In *Lakes and Reservoirs*, Ecosystems of the World, vol. 23, ed. F. Taub, pp. 231–45. Amsterdam: Elsevier.

Boughey, A.S. 1975. *Man and the Environment*, 2nd edn. London: Macmillan.

Brittain, J.E. 1978. The Ephemeroptera of Øvre Heimdalsvatn. *Holarctic Ecology*, **1**, 239–54.

Brooks, J.L. 1950. Speciation in ancient lakes. *Quarterly Review of Biology*, **25**, 30–60, 131–76.

Duncan, A. 1975. Production and biomass of three species of *Daphnia* co-existing in London reservoirs. *Verhandlungen Internationale Verein Limnologie*, **19**, 2858–67.

Edmondson, W.T. 1972*a*. Nutrients and phytoplankton in Lake Washington. *Nutrients and Eutrophication Special Symposia*, no. 1, 172–93. Washington, D.C.: American Society of Oceanography and Limnology Inc.

Edmondson, W.T. 1972*b*. The present condition of Lake Washington. *Verhandlungen Internationale Verein Limnologie*, **18**, 284–91.

Franks, J.W. & Pennington, W. (Mrs. T.G. Tutin) 1961. The Late-Glacial and Post-Glacial deposits of the Esthwaite Basin, North Lancashire. *The New Phytologist*, **60**, 27–42.

Fryer, G. & Iles, T.D. 1972. *The cichlid fishes of the Great Lakes of Africa*. Edinburgh: Oliver and Boyd.

Gardarsson, A. 1979. Waterfowl populations of Lake Mývatn and recent changes in numbers and food habits. *Oikos*, **32**, 250–70.

Gaudet, J.J. & Melack, J.M. 1981. Major ion chemistry in a tropical African lake basin. *Freshwater Biology*, **11**, 309–33.

Golterman, H.L. 1975. *Physiological Limnology*. Amsterdam: Elsevier.

Gudmundsson, F. 1979. The past status and exploitation of the Mývatn waterfowl populations. *Oikos*, **32**, 232–49.

Haslam, S.M. 1973. The management of British wetlands, II Conservation. *Journal of Environmental Management*, **1**, 345–61.

Kalff, J. & Welch, H.E. 1974. Phytoplankton production in Char Lake, a natural Polar lake, and in Maretta Lake, a polluted Polar lake, Cornwallis Island, Northwest Territories. *Journal of the Fisheries Research Board of Canada*, **31**, 621–36.

Kozhov, M. 1963. *Lake Baikal and its life*. Monographiae Biologicae, vol. 11. The Hague: Dr W. Junk.

Latif, A.F.A. 1984. Lake Nasser – the new man-made lake in Egypt (with reference to Lake Nubia). In *Lakes and Reservoirs*, Ecosystems of the World, vol. 23, ed. F. Taub, pp. 385–410. Amsterdam: Elsevier.

Liu, J.K. 1984. Lakes of the middle and lower basins of the Chang Jiang (China). In *Lakes and Reservoirs*, Ecosystems of the World, vol. 23, ed. F. Taub, pp. 331–55. Amsterdam: Elsevier.

Lowe-McConnell, R.H. 1975. *Fish communities in tropical freshwaters*. London: Longman.

Macan, T.T. 1970. *Biological Studies of the English Lakes*. London: Longman.
Maitland, P.S. 1977. *The Hamlyn Guide to Freshwater Fishes of Britain and Europe*. London: Hamlyn.
Maitland, P.S. 1979. *Synoptic Limnology: the analysis of British freshwater ecosystems*. Cambridge: Institute of Terrestrial Ecology.
Maitland, P.S. ed. 1981. *The Ecology of Scotland's Largest Lochs*. Monographiae Biologicae, vol. 44. The Hague: Dr W. Junk.
Marzolf, G.R. 1984. Reservoirs in the Great Plains of North America. In *Lakes and Reservoirs*, Ecosystems of the World, vol. 23, ed. F. Taub, pp. 291–302. Amsterdam: Elsevier.
Moriarty, D.J.W. 1973. The physiology of digestion of blue-green algae in the cichlid fish *Tilapia nilotica*. *Journal of Zoology*, **171**, 25–39.
Moss, B. 1980. *Ecology of Freshwaters*. Oxford: Blackwell Scientific Publications.
Pieczynska, E. ed. 1976. *Selected problems of lake littoral ecology*. Warsaw: Department of Hydrobiology, Institute of Zoology, University of Warsaw.
Ramsbottom, A.E. 1976. Depth charts of the Cumbrian lakes. *Scientific Publications of the Freshwater Biological Association*, **33**, 1–39.
Reynolds, C.S. 1984. *The ecology of freshwater phytoplankton*. Cambridge: Cambridge University Press.
Robertson, A. & Scavia, D. 1984. North American Great Lakes. In *Lakes and Reservoirs*, Ecosystems of the World, vol. 23, ed. F. Taub, pp. 135–76. Amsterdam: Elsevier.
Ruttner, F. 1963. *Fundamentals of Limnology*. Translated by D.G. Frey and F.E.J. Fry. Toronto: University of Toronto Press.
Shapiro, J., Lundquist, J.B. & Carlson, R.E. 1975. Involving the public in Limnology – an approach to communication. *Verhandlungen Internationale Verein Limnologie*, **19**, 866–74.
Smith, I. & Lyle, A. 1979. *Distribution of freshwaters in Great Britain*. Cambridge: Institute of Terrestrial Ecology.
Steel, J.A. 1975. The management of Thames Valley reservoirs. In *The Effects of Storage on Water Quality*, Water Research Centre, Medmenham, Henley, England.
Welch, H.E. & Kalff, J. 1974. Benthic photosynthesis and respiration in Char Lake. *Journal of the Fisheries Research Board of Canada*, **31**, 609–20.
Wetzel, R.G., Rich, P.H., Miller, M.C. & Allen, H.L. 1972. Metabolism of dissolved and particulate detrital carbon in a temperate hard-water lake. *Mem. Ist. Ital. Idrobiol.*, **29** Supplement, 185–243.

Index

Diagrams

Tables

Maps